Edited by Olaf Deutschmann

**Modeling and Simulation of
Heterogeneous Catalytic
Reactions: From the Molecular
Process to the Technical System**

Related Titles

Behr, A., Neubert, P.

Applied Homogeneous Catalysis

2012
Hardcover
978-3-527-32641-9

Beller, M., Renken, A., van Santen, R. A. (eds.)

Catalysis

From Principles to Applications

2012
Hardcover
978-3-527-32349-4

Drauz, K.-H., May, O., Gröger, H. (eds.)

Enzyme Catalysis in Organic Synthesis

3 Volumes

2012
Hardcover
978-3-527-32547-4

Astruc, D. (ed.)

Nanoparticles and Catalysis

2008
Hardcover
ISBN: 978-3-527-31572-7

Ertl, G., Knözinger, H., Schüth, F., Weitkamp, J. (eds.)

Handbook of Heterogeneous Catalysis

8 Volumes

2008
Hardcover
ISBN: 978-3-527-31241-2

Astruc, D. (ed.)

Nanoparticles and Catalysis

2008
Hardcover
ISBN: 978-3-527-31572-7

Edited by Olaf Deutschmann

Modeling and Simulation of Heterogeneous Catalytic Reactions

From the Molecular Process to the Technical System

WILEY-VCH Verlag GmbH & Co. KGaA

The Editors

Prof. Dr. Olaf Deutschmann
Karlsruhe Institute of Technology (KIT)
Institute for Chemical Technology and Polymer
Chemistry
Engesserstr. 20
76131 Karlsruhe
Germany

All books published by **Wiley-VCH** are carefully produced. Nevertheless, authors, editors, and publisher do not warrant the information contained in these books, including this book, to be free of errors. Readers are advised to keep in mind that statements, data, illustrations, procedural details or other items may inadvertently be inaccurate.

Library of Congress Card No.: applied for

British Library Cataloguing-in-Publication Data
A catalogue record for this book is available from the British Library.

Bibliographic information published by the Deutsche Nationalbibliothek
The Deutsche Nationalbibliothek lists this publication in the Deutsche Nationalbibliografie; detailed bibliographic data are available on the Internet at http://dnb.d-nb.de.

© 2012 Wiley-VCH Verlag & Co. KGaA, Boschstr. 12, 69469 Weinheim, Germany

All rights reserved (including those of translation into other languages). No part of this book may be reproduced in any form – by photoprinting, microfilm, or any other means – nor transmitted or translated into a machine language without written permission from the publishers. Registered names, trademarks, etc. used in this book, even when not specifically marked as such, are not to be considered unprotected by law.

Typesetting   Thomson Digital, Noida, India
Printing and Binding   Fabulous Printers Pte Ltd, Singapore
Cover Design   Adam-Design, Weinheim

Printed in Singapore
Printed on acid-free paper

Print ISBN: 978-3-527-32120-9
ePDF ISBN: 978-3-527-63989-2
oBook ISBN: 978-3-527-63987-8
ePub ISBN: 978-3-527-63988-5
Mobi ISBN: 978-3-527-63990-8

Contents

Preface *XI*
List of Contributors *XV*

1	**Modeling Catalytic Reactions on Surfaces with Density Functional Theory** *1*	
	John A. Keith, Josef Anton, Payam Kaghazchi, and Timo Jacob	
1.1	Introduction *1*	
1.2	Theoretical Background *2*	
1.2.1	The Many-Body Problem *2*	
1.2.2	Born–Oppenheimer Approximation *3*	
1.2.3	Wave Function-Based Methods *4*	
1.2.3.1	Hartree–Fock Approximation *4*	
1.2.3.2	Post Hartree–Fock Methods *5*	
1.2.4	Density-Based Methods *6*	
1.2.4.1	The Thomas–Fermi Model *7*	
1.2.4.2	The Hohenberg–Kohn Theorems *7*	
1.2.4.3	The Kohn–Sham Equations *9*	
1.2.4.4	Exchange–Correlation Functionals *10*	
1.2.5	Technical Aspects of Modeling Catalytic Reactions *13*	
1.2.5.1	Geometry Optimizations *13*	
1.2.5.2	Transition-State Optimizations *14*	
1.2.5.3	Vibrational Frequencies *14*	
1.2.5.4	Thermodynamic Treatments of Molecules *16*	
1.2.5.5	Considering Solvation *17*	
1.2.6	Model Representation *19*	
1.2.6.1	Slab/Supercell Approach *19*	
1.2.6.2	Cluster Approach *21*	
1.3	The Electrocatalytic Oxygen Reduction Reaction on Pt(111) *22*	
1.3.1	Water Formation from Gaseous O_2 and H_2 *24*	
1.3.1.1	O_2 Dissociation *25*	
1.3.1.2	OOH Formation *27*	
1.3.1.3	HOOH Formation *28*	
1.3.2	Simulations Including Water Solvation *28*	

1.3.2.1	Langmuir–Hinshelwood Mechanisms 30
1.3.2.2	Eley–Rideal Reactions 31
1.3.3	Including Thermodynamical Quantities 32
1.3.3.1	Langmuir–Hinshelwood and Eley–Rideal Mechanisms 33
1.3.4	Including an Electrode Potential 35
1.4	Conclusions 36
	References 37

2	**Dynamics of Reactions at Surfaces** 39
	Axel Groß
2.1	Introduction 39
2.2	Theoretical and Computational Foundations of Dynamical Simulations 41
2.3	Interpolation of Potential Energy Surfaces 43
2.4	Quantum Dynamics of Reactions at Surfaces 45
2.5	Nondissociative Molecular Adsorption Dynamics 49
2.6	Adsorption Dynamics on Precovered Surfaces 55
2.7	Relaxation Dynamics of Dissociated H_2 Molecules 59
2.8	Electronically Nonadiabatic Reaction Dynamics 62
2.9	Conclusions 66
	References 67

3	**First-Principles Kinetic Monte Carlo Simulations for Heterogeneous Catalysis: Concepts, Status, and Frontiers** 71
	Karsten Reuter
3.1	Introduction 71
3.2	Concepts and Methodology 73
3.2.1	The Problem of a Rare Event Dynamics 73
3.2.2	State-to-State Dynamics and kMC Trajectories 75
3.2.3	kMC Algorithms: from Basics to Efficiency 77
3.2.4	Transition State Theory 80
3.2.5	First-Principles Rate Constants and the Lattice Approximation 84
3.3	A Showcase 88
3.3.1	Setting up the Model: Lattice, Energetics, and Rate Constant Catalog 88
3.3.2	Steady-State Surface Structure and Composition 90
3.3.3	Parameter-Free Turnover Frequencies 95
3.3.4	Temperature-Programmed Reaction Spectroscopy 99
3.4	Frontiers 102
3.5	Conclusions 107
	References 108

4	**Modeling the Rate of Heterogeneous Reactions** 113
	Lothar Kunz, Lubow Maier, Steffen Tischer, and Olaf Deutschmann
4.1	Introduction 113
4.2	Modeling the Rates of Chemical Reactions in the Gas Phase 115
4.3	Computation of Surface Reaction Rates on a Molecular Basis 116

4.3.1	Kinetic Monte Carlo Simulations	116
4.3.2	Extension of MC Simulations to Nanoparticles	120
4.3.3	Reaction Rates Derived from MC Simulations	124
4.3.4	Particle–Support Interaction and Spillover	125
4.3.5	Potentials and Limitations of MC Simulations for Derivation of Overall Reaction Rates 125	
4.4	Models Applicable for Numerical Simulation of Technical Catalytic Reactors 128	
4.4.1	Mean Field Approximation and Reaction Kinetics	129
4.4.2	Thermodynamic Consistency	131
4.4.3	Practicable Method for Development of Multistep Surface Reaction Mechanisms 134	
4.4.4	Potentials and Limitations of the Mean Field Approximation	139
4.5	Simplifying Complex Kinetic Schemes	141
4.6	Summary and Outlook	142
	References	143
5	**Modeling Reactions in Porous Media**	149
	Frerich J. Keil	
5.1	Introduction	149
5.2	Modeling Porous Structures and Surface Roughness	152
5.3	Diffusion	158
5.4	Diffusion and Reaction	163
5.5	Pore Structure Optimization: Synthesis	173
5.6	Conclusion	175
	References	175
6	**Modeling Porous Media Transport, Heterogeneous Thermal Chemistry, and Electrochemical Charge Transfer** 187	
	Robert J. Kee and Huayang Zhu	
6.1	Introduction	187
6.2	Qualitative Illustration	189
6.3	Gas-Phase Conservation Equations	190
6.3.1	Gas-Phase Transport	191
6.3.2	Chemical Reaction Rates	191
6.3.3	Boundary Conditions	192
6.4	Ion and Electron Transport	192
6.5	Charge Conservation	194
6.5.1	Effective Properties	195
6.5.2	Boundary Conditions	195
6.5.3	Current Density and Cell Potential	196
6.6	Thermal Energy	196
6.7	Chemical Kinetics	196
6.7.1	Thermal Heterogeneous Kinetics	197
6.7.2	Charge Transfer Kinetics	198

6.7.3	Butler–Volmer Formulation	204
6.7.4	Elementary and Butler–Volmer Formulations	206
6.8	Computational Algorithm	207
6.9	Button Cell Example	207
6.9.1	Polarization Characteristics	208
6.9.2	Electric Potentials and Charged Species Fluxes	208
6.9.3	Anode Gas-Phase Profiles	212
6.9.4	Anode Surface Species Profiles	213
6.9.5	Applicability and Extensibility	214
6.10	Summary and Conclusions	214
6.10.1	Greek Letters	217
	References	218

7 Evaluation of Models for Heterogeneous Catalysis 221
John Mantzaras

7.1	Introduction	221
7.2	Surface and Gas-Phase Diagnostic Methods	222
7.2.1	Surface Science Diagnostics	222
7.2.2	*In Situ* Gas-Phase Diagnostics	223
7.3	Evaluation of Hetero/Homogeneous Chemical Reaction Schemes	225
7.3.1	Fuel-Lean Combustion of Methane/Air on Platinum	225
7.3.1.1	Heterogeneous Kinetics	225
7.3.1.2	Gas-Phase Kinetics	228
7.3.2	Fuel-Lean Combustion of Propane/Air on Platinum	231
7.3.3	Fuel-Lean Combustion of Hydrogen/Air on Platinum	234
7.3.4	Fuel-Rich Combustion of Methane/Air on Rhodium	238
7.3.5	Application of Kinetic Schemes in Models for Technical Systems	240
7.4	Evaluation of Transport	242
7.4.1	Turbulent Transport in Catalytic Systems	243
7.4.2	Modeling Directions in Intraphase Transport	245
7.5	Conclusions	246
	References	248

8 Computational Fluid Dynamics of Catalytic Reactors 251
Vinod M. Janardhanan and Olaf Deutschmann

8.1	Introduction	251
8.2	Modeling of Reactive Flows	253
8.2.1	Governing Equations of Multicomponent Flows	253
8.2.2	Turbulent Flows	256
8.2.3	Three-Phase Flow	256
8.2.4	Momentum and Energy Equations for Porous Media	257
8.3	Coupling of the Flow Field with Heterogeneous Chemical Reactions	258
8.3.1	Given Spatial Resolution of Catalyst Structure	258
8.3.2	Simple Approach for Modeling the Catalyst Structure	259

8.3.3	Reaction Diffusion Equations 260
8.3.4	Dusty Gas Model 261
8.4	Numerical Methods and Computational Tools 262
8.4.1	Numerical Methods for the Solution of the Governing Equations 263
8.4.2	CFD Software 264
8.4.3	Solvers for Stiff ODE and DAE Systems 264
8.5	Reactor Simulations 264
8.5.1	Flow through Channels 265
8.5.2	Monolithic Reactors 268
8.5.3	Fixed Bed Reactors 271
8.5.4	Wire Gauzes 273
8.5.5	Catalytic Reactors with Multiphase Fluids 273
8.5.6	Material Synthesis 275
8.5.7	Electrocatalytic Devices 277
8.6	Summary and Outlook 278
	References 279

9	**Perspective of Industry on Modeling Catalysis** 283
	Jens R. Rostrup-Nielsen
9.1	The Industrial Challenge 283
9.2	The Dual Approach 285
9.3	The Role of Modeling 287
9.3.1	Reactor Models 287
9.3.2	Surface Science and Breakdown of the Simplified Approach 288
9.3.3	Theoretical Methods 290
9.4	Examples of Modeling and Scale-Up of Industrial Processes 291
9.4.1	Ammonia Synthesis 291
9.4.2	Syngas Manufacture 294
9.4.2.1	Steam Reforming 294
9.4.2.2	Autothermal Reforming 297
9.5	Conclusions 298
	References 300

10	**Perspectives of the Automotive Industry on the Modeling of Exhaust Gas Aftertreatment Catalysts** 303
	Daniel Chatterjee, Volker Schmeißer, Marcus Frey, and Michel Weibel
10.1	Introduction 303
10.2	Emission Legislation 304
10.3	Exhaust Gas Aftertreatment Technologies 306
10.4	Modeling of Catalytic Monoliths 308
10.5	Modeling of Diesel Particulate Filters 313
10.6	Selective Catalytic Reduction by NH_3 (Urea-SCR) Modeling 315
10.6.1	Kinetic Analysis and Chemical Reaction Modeling 316
10.6.1.1	NH_3 Adsorption, Desorption, and Oxidation 316
10.6.1.2	NO-SCR Reaction 316

10.6.1.3	NH$_3$–NO–NO$_2$ Reactions *317*
10.6.2	Influence of Washcoat Diffusion *319*
10.7	Diesel Oxidation Catalyst, Three-Way Catalyst, and NO$_x$ Storage and Reduction Catalyst Modeling *319*
10.7.1	Diesel Oxidation Catalyst *320*
10.7.2	Three-Way Catalyst *321*
10.7.3	NO$_x$ Storage and Reduction Catalyst *321*
10.7.3.1	Species Transport Effects Related to NSCR: Shrinking Core Model *326*
10.7.3.2	NH$_3$ Formation During Rich Operation within a NSRC *327*
10.8	Modeling Catalytic Effects in Diesel Particulate Filters *328*
10.9	Determination of Global Kinetic Parameters *329*
10.10	Challenges for Global Kinetic Models *330*
10.11	System Modeling of Combined Exhaust Aftertreatment Systems *331*
10.12	Conclusion *335*
	References *339*

Index *345*

Preface

The Nobel Prize in Chemistry 2007 awarded to Gerhard Ertl for his groundbreaking studies in surface chemistry highlighted the importance of heterogeneous catalysis not only for modern chemical industry but also for environmental protection. Today, heterogeneous catalysis is also expected to be the key technology to solve the challenges associated with the increasing diversification of raw materials and energy sources. Heterogeneous catalysis is the decisive step in most chemical industry processes, it is the major way to reduce pollutant emissions from mobile sources, and it is present in fuel cells to produce electricity and in many systems for the use of solar energy (photocatalysis).

With the increasing power of computers over the past decades and the development of numerical algorithms to solve highly coupled, nonlinear, and stiff equation systems, modeling and numerical simulation also have developed into valuable tools in heterogeneous catalysis. These tools were applied to study the molecular processes in very detail by quantum mechanical computations, density functional theory (DFT), molecular dynamics, and Monte Carlo simulations, but often neglecting the engineering aspects of catalytic reactors such as the interaction of chemistry and mass and heat transfer on one side. On the other side, mixing, flow structures, and heat transport in technical reactors and processes have been analyzed by computational fluid dynamics (CFD) in very detail, neglecting, however, the details of the microkinetics.

One objective of this book is to span bridges over the still existing gaps between both communities regarding modeling of heterogeneous catalytic reactions. In the past years, quite frequently, research proposal and programs on catalysis claim to work on bridging those gaps between surface science and industrial catalysis and indeed some progress has been made. Surface science studies, in experiment, theory, and simulation, more and more include technically relevant conditions. Reaction engineering of technical processes now often tries to understand the underlying molecular processes and even include quantum mechanical simulations in the search and development of new catalysts and catalytic reactors. However, convergence here is a slow process. One major reason is the gap between the high complexity of catalysts used in practice and the many approximations still to be made in molecular simulations. Furthermore, using kinetic data derived from numerical simulations in scale-up of technical systems might indeed become risky if the engineer does not take into account the simplifying assumptions and the computational uncertainties of the numerical simulations. Actually, this warning holds for all simulations relevant for heterogeneous catalysis, from DFT to CFD. In many chapters of this book, the authors will lead the reader not only to the potentials but also to the limitations of the modeling approaches and show where the use of the models presented are beneficial and where they are rather risky.

In this book, the state of the art in modeling and simulations of heterogeneous catalytic reactions will be discussed on a molecular level and from an engineering perspective. Special attention is given to the potentials and – even more important – to the limitations of the approaches used. The reader will become familiar not only with the principal ideas of modeling in heterogeneous catalysis but also with the benefits, challenges, and still open issues. The book is organized as follows: from chapter to chapter, time- and length scales as well as complexity increase on the expense of details of the molecular processes bridging all the way from the surface science to the industrial view on modeling heterogeneous catalytic reactions (Figure 1).

The book starts with a chapter on density functional theory presenting the concept of theoretical calculations of surface reactions. The electrocatalytic oxygen reduction is used as an example, showing the potential of DFT to study different mechanistic aspects, also including environmental effects. On the basis of the energy diagram derived and the ambient conditions, the likelihood of the realization of a specific reaction pathway can be estimated. Chapter 2 focuses on the computation of the dynamics of reactions at surfaces from first principles.

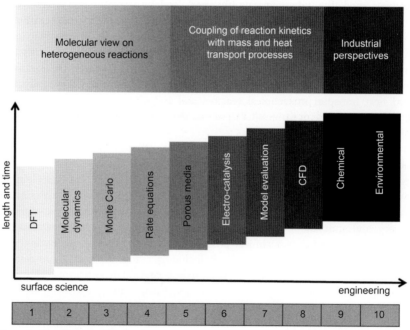

Figure 1 Organization of the book.

These dynamical simulations reveal the actual trajectory of the movement of a molecule on a catalytic surface but are in general restricted to the computation of this trajectory within few nanoseconds. If the interest is rather in the probability whether or not a certain process (adsorption, diffusion, or reaction) takes place, then Monte Carlos (MC) simulations are the model of choice, which are discussed in Chapter 3. Using information derived from DFT, first-principles kinetic Monte Carlo simulations lead directly to the surface reaction rate as function of ambient conditions and the state of the catalyst. They combine elementary kinetics with statistics to properly evaluate the kinetics of the catalytic particle. The diverse morphology of catalytic particles in technical systems calls for a simpler approach because MC simulations are too expensive to be directly included in simulation of catalytic reactors including mass and heat transfer. Chapter 4 therefore moves from MC simulations to rate equations to estimate surface reaction rates and homogeneous gas-phase reaction rates that also play a role in many catalytic processes. In this so-called mean-field approximation, the details of the molecular process such as diffusion of adsorbates and the dependence of the reaction rate on the crystal phases and defects are not taken directly into account. We now move from the microscopic to the mesoscopic processes. Chapter 5 is then the first part of the book in which the interaction of surface reactions and molecular transport of the reactants and products in the ambience of the catalytic particle is considered. This chapter covers modeling of processes in catalytic porous media and the interaction of diffusion and reactions in those pore structures. The last chapter on fundamentals (Chapter 6) also includes electrochemical charge transfer and couples it with the heterogeneous (thermo-) catalytic reactions and transport in porous media using processes in solid-oxide fuel cells as an illustrating example. Chapter 7 discusses the applicability of molecular-based models, in particular of rate equations, in reactive flow systems and their coupling to the surrounding mass transport processes. The comparison of spatially and temporally resolved species profiles in catalytic laboratory reactors using sophisticated experimental techniques with the profiles computed by coupling chemistry and mass transport models can be used for the evaluation of kinetic schemes developed. The coupling of chemistry and heat and mass transport in catalytic reactors is discussed in Chapter 8, using several multiphase flow systems as examples. Now the book has reached the macroscopic view. The last two Chapters 9 and 10 critically discuss the use, benefits, and limitations of modeling tools in chemical and automobile industry today.

Karlsruhe, August 2011 *Olaf Deutschmann*

List of Contributors

Josef Anton
Universität Ulm
Institut für Elektrochemie
Albert-Einstein-Allee 47
D-89069 Ulm
Germany

Daniel Chatterjee
MTU Friedrichshafen GmbH
Maybachplatz 1
88045 Friedrichshafen
Germany

Olaf Deutschmann
Karlsruhe Institute of Technology (KIT)
Institute of Chemical Technology and
Polymer Chemistry
Engesserstr. 20
76131 Karlsruhe
Germany

Marcus Frey
Daimler AG
Department GR/APE
HPC 010-G206
70546 Stuttgart
Germany

Axel Groß
Universität Ulm
Institut für Theoretische Chemie
Albert-Einstein-Allee 11
89069 Ulm
Germany

Timo Jacob
Universität Ulm
Institut für Elektrochemie
Albert-Einstein-Allee 47
D-89069 Ulm
Germany

Vinod M. Janardhanan
Indian Institute of Technology
Ordnance Factory Estate
Yeddumailaram
502205 Hyderabad
Andhra Pradesh
India

Payam Kaghazchi
Universität Ulm
Institut für Elektrochemie
Albert-Einstein-Allee 47
D-89069 Ulm
Germany

Robert J. Kee
Colorado School of Mines
Engineering Division
Office BB-306
Golden, CO 80401
USA

Frerich J. Keil
Hamburg University of Technology
Institute of Chemical Reaction
Engineering
AB VT 4, FSP 6-05
Eissendorfer Str. 38
21073 Hamburg
Germany

John A. Keith
Mechanical & Aerospace Engineering
Department
Princeton University, D320 Engineering
Quadrangle
Princeton, NJ 08544
USA

Lothar Kunz
Karlsruhe Institute of Technology (KIT)
Institute of Chemical Technology and
Polymer Chemistry
Engesserstr. 20
76131 Karlsruhe
Germany

Lubow Maier
Karlsruhe Institute of Technology (KIT)
Institute of Catalysis Research and
Technology (IKFT)
76128 Karlsruhe
Germany

John Mantzaras
Paul Scherrer Institute
Combustion Research
5232 Villigen PSI
Switzerland

Karsten Reuter
Fritz-Haber-Institut der Max-Planck-
Gesellschaft
Faradayweg 4–6
14195 Berlin
Germany

and

Technische Universität München
Lehrstuhl für Theoretische Chemie
Lichtenbergstr. 4–6
85747 Garching
Germany

Jens R. Rostrup-Nielsen
Haldor Topsoe A/S
R&D Division
Nymoellevej 55
2800 Lyngby
Denmark

Volker Schmeißer
Daimler AG
Department GR/APE
HPC 010-G206
70546 Stuttgart
Germany

Steffen Tischer
Karlsruhe Institute of Technology (KIT)
Institute of Catalysis Research and
Technology (IKFT)
76128 Karlsruhe
Germany

Michel Weibel
Daimler AG
Department GR/APE
HPC 010-G206
70546 Stuttgart
Germany

Huayang Zhu
Colorado School of Mines
Engineering Division
Office BB-306
Golden, CO 80401
USA

1
Modeling Catalytic Reactions on Surfaces with Density Functional Theory

John A. Keith, Josef Anton, Payam Kaghazchi, and Timo Jacob

1.1
Introduction

Predicting the reactivity of catalytic systems is a nontrivial process that usually requires knowledge about its geometric and electronic structure, properties determined by quantum mechanics (QM). Solving the Schrödinger equation[1] is a nontrivial task even for small systems, and it becomes especially arduous when the system involves multiple phases as is the case in a surface reaction. Theoretical calculations nevertheless provide useful and important perspectives on chemical reactions that are not accessible through experimental observations alone. Figure 1.1 schematically shows the hierarchy of multiscale modeling, starting from the subatomic regime, over the electronic and atomistic regimes, to the meso- and finally the macroscale. Different theoretical methods have been established to address questions related to each regime (or timescale and length scale); however, realistic processes usually involve effects from all scales. In this chapter, we will focus on the electronic and atomistic regimes, which not only provide the basis for climbing up the hierarchy of multiscale modeling but also provide important mechanistic information on catalytic reactions.

Quantum chemistry is the application of QM to better understand chemical systems. In its purest form, QM calculations solve the Schrödinger equation, which provides the energy of a given configuration of nuclei and their electrons. There are two general approaches to do this. One way is to solve the energy given by the Schrödinger equation approximately using a nonclassical wavefunction. Another popular approach, density functional theory (DFT), uses the electronic density to evaluate the energy of a system via an approximate functional. Both approaches have their merits and provide a nonthermodynamical representation of the energy of a system of electrons.

This chapter gives an introductory overview of the essential concepts behind theoretical calculations of surface reactions, which we then apply to better understand one of the key features in energy conversion and fuel cell technology:

---

1) For the relativistic case, one has to consider the Dirac equation.

Modeling and Simulation of Heterogeneous Catalytic Reactions: From the Molecular Process to the Technical System,
First Edition. Edited by Olaf Deutschmann.
© 2012 Wiley-VCH Verlag GmbH & Co. KGaA. Published 2012 by Wiley-VCH Verlag GmbH & Co. KGaA.

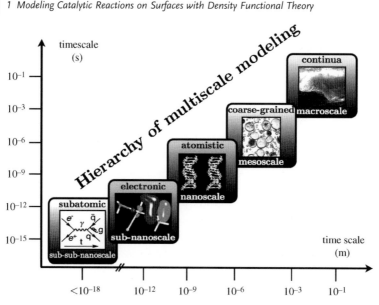

Figure 1.1 Hierarchy of multiscale modeling of different time and length regimes.

the electrocatalytic oxygen reduction reaction (ORR). After describing the multibody problem in quantum mechanics, we will discuss the fundamentals behind wave function- and density-based methods used to solve this problem. We will then focus on density functional theory and its advantages and disadvantages in applications to catalysis. Section 1.2, the first part of this chapter, ends with brief technical details one has to grapple with when modeling surface reactions. This includes thermodynamic approximations and other considerations that extend calculated values from theoretical studies to be more comparable to experiment. Section 1.3, the second part of this chapter, shows an example how quantum mechanical DFT calculations can resolve mechanistic details of a rather complex surface reaction.

1.2
Theoretical Background

1.2.1
The Many-Body Problem

Many material properties of interest to physicists and chemists can be obtained by solving the many-body Schrödinger equation. In stationary, nonrelativistic terms it can be written as

$$\hat{H}\Psi(\mathbf{r}_i\sigma_i, \mathbf{R}_\nu) = E\Psi(\mathbf{r}_i\sigma_i, \mathbf{R}_\nu), \tag{1.1}$$

where $\hat{H}$ is the Hamilton operator, $\Psi(\mathbf{r}_i\sigma_i, \mathbf{R}_\nu)$ is the many-body wave function, E is the total energy of the system, σ_i is the spin coordinate of electron i, and finally $\mathbf{r}_i$

and $\mathbf{R}_\nu$ are the spatial coordinates of electron i and nucleus ν, respectively. The Hamiltonian for a system consisting of a set of nuclei and electrons can be written as[2]

$$\hat{H} = \sum_{\nu=1}^{M} \hat{T}_\nu + \underbrace{\sum_{i=1}^{N} \hat{t}_i - \sum_{i=1}^{M}\sum_{\nu=1}^{N} \frac{Z_\nu}{|\mathbf{r}_i - \mathbf{R}_\nu|} + \frac{1}{2}\sum_{\substack{i,j=1 \\ i\neq j}} \frac{1}{|\mathbf{r}_i - \mathbf{r}_j|}}_{\hat{H}_e} + \frac{1}{2}\sum_{\substack{\nu,\mu=1 \\ \nu\neq\mu}}^{N} \frac{Z_\nu Z_\mu}{|\mathbf{R}_\nu - \mathbf{R}_\mu|},$$

(1.2)

where $\hat{T}_\nu$ and $\hat{t}_i$ are the respective kinetic energy operators for nuclei and electrons, and Z_ν is the charge on a nucleus. The two species (electrons and nuclei) interact with each other and themselves. It is difficult to solve such a coupled system since the motion of any particle is influenced by all other particles. Except for simple systems such as a hydrogen atom, solving Eq. (1.1) with the corresponding Hamiltonian, Eq. (1.2), is nontrivial for most materials that consist of several electrons and nuclei. Therefore, simulations of realistic systems require different applications of approximations.

1.2.2
Born–Oppenheimer Approximation

The Born–Oppenheimer (BO) approximation (introduced by Max Born and J. Robert Oppenheimer in 1927 [1]) treats the electronic and nuclear degrees of freedom as decoupled from each other. Nuclei are much more massive than electrons, so electrons are assumed to instantaneously follow the motion of the nuclei. Consequently, on the timescale of the motion of the electrons, the nuclei appear almost stationary. The total wave function of Eq. (1.1) can then be written as[3]

$$\Psi(\mathbf{r}_i\sigma_i, \mathbf{R}_\nu) = \psi_e(\mathbf{r}_i\sigma_i, \{\mathbf{R}_\nu\})\psi_n(\mathbf{R}_\nu) \quad (1.3)$$

where $\psi_e(\mathbf{r}_i\sigma_i, \{\mathbf{R}_\nu\})$ and $\psi_n(\mathbf{R}_\nu)$ are the electronic and nuclei wave functions, and $\{\mathbf{R}_\nu\}$ denotes that nuclear spatial coordinates are parameters and not variables. We then can divide the electronic and nuclear parts into

$$\hat{H}_e \psi_e(\mathbf{r}_i\sigma_i, \{\mathbf{R}_\nu\}) = \left(-\sum_{i=1}^{M} \frac{1}{2}\nabla_i^2 + \frac{1}{2}\sum_{\substack{i,j=1 \\ i\neq j}}^{M} \frac{1}{|\mathbf{r}_i - \mathbf{r}_j|} - \sum_{i=1}^{M}\sum_{\nu=1}^{N} \frac{Z_\nu}{|\mathbf{r}_i - \mathbf{R}_\nu|} \right) \psi_e(\mathbf{r}_i\sigma_i, \{\mathbf{R}_\nu\})$$

$$= E_e(\mathbf{R}_\nu)\psi_e(\mathbf{r}_i\sigma_i, \{\mathbf{R}_\nu\}) \quad (1.4)$$

---

2) Throughout this chapter, atomic units are assumed.
3) This is an expansion of $\Psi(\mathbf{r}_i\sigma_i, \mathbf{R}_\nu)$ in a series of eigenfunctions of the electronic Hamilton operator $\hat{H}_e$. Since in chemistry electronic excitations usually do not play a substantial role, we restrict our discussion to the eigenfunctions with the lowest energy of the electronic system (ground state) $\psi_e(\mathbf{r}_i\sigma_i, \{\mathbf{R}_\nu\})$.

and

$$\left(-\sum_{\nu=1}^{N}\frac{1}{2m_\nu}\nabla_\nu^2 + \frac{1}{2}\sum_{\substack{\nu,\mu=1\\\nu\neq\mu}}^{N}\frac{Z_\nu Z_\mu}{|\mathbf{R}_\nu - \mathbf{R}_\mu|} + E_e(\mathbf{R}_\nu)\right)\psi_n(\mathbf{R}_\nu) = E\psi_n(\mathbf{R}_\nu). \quad (1.5)$$

In Eq. (1.4), $\hat{H}_e$ and ψ_e depend only on the positions of the nuclei. In the case of negligible nonadiabatic effects, this approximation introduces a very small error into the energies, and this inaccuracy becomes even smaller for heavier elements [2]. Applying this approximation, we can restrict ourselves to the electronic part[4] (Eq. (1.4)), which can be solved exactly for one-electron systems only. Thus, we need further approximations for systems with many electrons. Many techniques solve the Schrödinger equation approximately from first principles (*ab initio*). Two common types of *ab initio* methods are the wave function-based and the density-based approaches [3]. Both have been applied extensively in material science and catalysis.

1.2.3
Wave Function-Based Methods

The Born–Oppenheimer approximation reduces the many-body problem to the electronic part only (with the positions of the nuclei as parameters), but the electronic wave function is still a function of the spatial coordinates of the electrons and their spin variables. We now describe the Hartree–Fock (HF) approximation, which can be viewed as the basis for all practical *ab initio* developments. This approach does not include any correlation effects, and we will later describe different approaches to account for these.

1.2.3.1 Hartree–Fock Approximation
In 1927, Hartree [4] considered the electron motions as independent (uncorrelated). Each electron could then be treated as moving in an averaged field originating from all other electrons.[5] Three years later, Fock [5] followed Hartree's idea to express the overall electronic wave function as simple product of single-particle wave functions, but he introduced the fermionic character of the electrons by using an antisymmetric sum product of single-particle wave functions. The most simple *ansatz* for such a representation is given by a single Slater determinant [6]:

$$\psi_e = \frac{1}{\sqrt{M!}}\begin{vmatrix} \varphi_1(\mathbf{r}_1\sigma_1) & \varphi_1(\mathbf{r}_2\sigma_2) & \cdots & \varphi_1(\mathbf{r}_M\sigma_M) \\ \varphi_2(\mathbf{r}_1\sigma_1) & \varphi_2(\mathbf{r}_2\sigma_2) & \cdots & \varphi_2(\mathbf{r}_M\sigma_M) \\ \vdots & \vdots & & \vdots \\ \varphi_M(\mathbf{r}_1\sigma_1) & \varphi_M(\mathbf{r}_2\sigma_2) & \cdots & \varphi_M(\mathbf{r}_M\sigma_M) \end{vmatrix}, \quad (1.6)$$

where $\varphi_i(\mathbf{r}_j\sigma_j)$ describes electron i at the position of electron j. According to Ritz [7], the lowest-energy system state corresponds to the ground state, which is obtainable

---

[4] Although Eq. (1.4) is the main electronic contribution, in practice we usually also consider the second term of Eq. (1.5).
[5] This is usually called the model of independent electrons or the effective one-particle model.

by variation under the constraint of an orthonormal[6] set of single-particle wave functions $\varphi_i(\mathbf{r}_j \sigma_j)$:

$$\delta \left\{ E_e - \sum_{i,j=1}^{M} \varepsilon_{ij} \langle \varphi_i | \varphi_j \rangle \right\} = 0. \tag{1.7}$$

This finally leads to the well-known Hartree–Fock equations:

$$\hat{F}(\mathbf{r}_i)|\varphi_i\rangle = \sum_{j=1}^{M} \varepsilon_{ij} |\varphi_j\rangle, \quad i = 1, 2, \ldots, M. \tag{1.8}$$

Here, $\hat{F}$ is the Fock operator, which can be written as[7]

$$\hat{F}(\mathbf{r}_i) = \hat{t}_i - \sum_{\nu=1}^{N} \frac{Z_\nu}{|\mathbf{r}_i - \mathbf{R}_\nu|} + \sum_{j=1}^{M} \int \varphi_j^*(\mathbf{r}') \frac{1}{|\mathbf{r}_i - \mathbf{r}'|} \varphi_j(\mathbf{r}') \, d\mathbf{r}' + V_x(\mathbf{r}_i). \tag{1.9}$$

The last term originates from the Pauli principle and describes the nonlocal exchange, for which there is no classical analogue:

$$V_x(\mathbf{r}_i) = -\sum_{j=1}^{M} \int \frac{1}{|\mathbf{r}_i - \mathbf{r}'|} \frac{\varphi_j^*(\mathbf{r}') \varphi_j(\mathbf{r}_i) \varphi_i^*(\mathbf{r}_i) \varphi_i(\mathbf{r}')}{\varphi_i^*(\mathbf{r}_i) \varphi_i(\mathbf{r}_i)} \, d\mathbf{r}'. \tag{1.10}$$

Due to the aforementioned nonlocal character of the exchange, practical calculation of this term is extremely demanding. Dirac [8] and Bloch [9] independently showed that the exchange integral for a free electron gas can be expressed as function of the electronic density. In order to generalize this idea, Slater [10, 11] added a scaling factor X_α to the expression for the free electron gas

$$V_x(\mathbf{r}_i) = -3X_\alpha \left(\frac{3}{8\pi} \varrho(\mathbf{r}_i) \right)^{1/3}. \tag{1.11}$$

For the free electron gas, the scaling factor is 2/3. However, a value of $X_\alpha = 0.7$ led to an improved accuracy for atoms [12, 13].

Insertion of the Slater approximation into the exchange term of the HF equations finally leads to the so-called Hartree–Fock–Slater equation (HFS).

1.2.3.2 Post Hartree–Fock Methods

A critical shortcoming of HF theory is its lack of *electronic correlation*, that is, the treatment of systems of electrons interacting with each other. This correlation is often separated into two parts. *Dynamical* correlation relates to responsiveness of electrons interacting with each other, while *nondynamical* electronic correlation relates to how a real system's energy is due to contributions from several accessible electronic states. HF theory treats neither and is not accurate enough to make reliable chemical

---

6) In the variation, the constraints are usually included as Lagrange multipliers ε_{ij}.
7) Spin variables have not been expressed explicitly.

determinations, but additional corrections have been developed in order to account for these shortcomings such as full CI (configurational interaction), Møller–Plesset perturbation theory (MPn), complete active space (CAS), or coupled cluster (CC) methods (see Refs. [14–16] for more details). The majority of these methods are under the umbrella of fully first-principles methods, that is, *ab initio* wave function methods that incorporate no empirical data into their calculation. For problems in catalysis, most of these methods are too expensive in practice; however, the MP2 method is occasionally useful in calculating dispersion forces and van der Waals interactions.

General perturbation theory presumes that the magnitude of a perturbation in a calculation is small compared to the unperturbed value of the calculation itself. This is valid for the electron correlation energy, which is small compared to the HF energy. According to perturbation theory in quantum mechanics, the total Hamiltonian is divided into the unperturbed reference Hamiltonian ($\hat{H}_0$) plus the Hamiltonian corresponding to its correction ($\hat{H}'$) times a scaling factor λ, which determines the strength of the perturbation:

$$\hat{H} = \hat{H}_0 + \lambda \hat{H}'. \tag{1.12}$$

Based on this separation, perturbation theory yields the different terms of the Taylor expansion of the electronic energy. Consequently, the first two orders of correction to the energy of an electronic state n become

$$E_{e,n}^{(1)} = \left\langle \psi_{e,n}^{(0)} | \hat{H}' | \psi_{e,n}^{(0)} \right\rangle$$
$$E_{e,n}^{(2)} = \sum_{j \neq n} \frac{\left| \left\langle \psi_{e,n}^{(0)} | \hat{H}' | \psi_{e,j}^{(0)} \right\rangle \right|^2}{E_{e,n}^{(0)} - E_{e,j}^{(0)}} \tag{1.13}$$

In Møller–Plesset perturbation theory, $\hat{H}_0$ is taken as the sum of one-electron Fock operators. The sum of $E_{e,n}^{(0)} + E_{e,n}^{(1)}$ is the electronic HF energy,[8] and additional $E_{e,n}^{(i>1)}$ corrects the HF energy for electronic correlation. Calculations including the first additional correction $E_{e,n}^{(2)}$ are called MP2 methods, while MP3, MP4, and so on also treat higher perturbation orders.

Although Møller–Plesset perturbation theory in principle allows for a full inclusion of electronic correlations, evaluating even the first correction terms becomes quite expensive as the number of electrons in the system increases. Even nowadays, MP2 is often considered too expensive for simulating surface reactions.

1.2.4
Density-Based Methods

Instead of employing many-body wave functions, density-based methods use the electron density as the basic variable to evaluate the total energy and other properties. A well-known density-based approach is the density functional theory, which was introduced by Hohenberg and Kohn in 1964 [17] and further developed by Kohn and

---

[8] In HF approximation, $E_{e,n}^{(1)} = 0$.

Sham in 1965 [18]. This method dates back to the works of Thomas and Fermi (TF). Before describing the DFT approach itself, we briefly review the TF model. Although this section focuses on the electronic part of the many-body problem, for simplicity, the index e is not given explicitly.

1.2.4.1 The Thomas–Fermi Model

The electronic part of the many-body wave function containing M electrons, $\psi(\mathbf{r}_1\sigma_1,\ldots,\mathbf{r}_M\sigma_M)$, is not easy to calculate since it depends on $4M$ coordinates ($3M$ coordinates if spin is not considered). Thomas and Fermi proposed a simpler approach, the TF model, using the electron density of the system $\varrho(\mathbf{r})$ as the basic variable [19, 20]. A key advantage of this approach is that the electron density of M electrons in a volume element $d\mathbf{r}$ depends only on three independent coordinates:

$$\varrho(\mathbf{r}) = \sum_{i=1}^{M} \int \psi(\mathbf{r}_1\sigma_1,\ldots,\mathbf{r}_M\sigma_M)^* \delta(\mathbf{r}_i - \mathbf{r}) \psi(\mathbf{r}_1\sigma_1,\ldots,\mathbf{r}_M\sigma_M) d(\mathbf{r}_1\sigma_1) \ldots d(\mathbf{r}_M\sigma_M). \quad (1.14)$$

Here, the ground-state total energy and other properties can be expressed as functionals of the electron density. By assuming that the kinetic energy density is locally equal to that of a homogeneous electron gas, Thomas and Fermi formulated the total energy functional $E^{TF}[\varrho(\mathbf{r})]$ as

$$E^{TF}[\varrho(\mathbf{r})] = \frac{3(3\pi^2)^{2/3}}{10} \int \varrho(\mathbf{r})^{5/3} d\mathbf{r} - \sum_{\nu=1}^{N} \int \frac{Z_\nu \varrho(\mathbf{r})}{|\mathbf{r} - \mathbf{R}_\nu|} d\mathbf{r} + \frac{1}{2} \int \int \frac{\varrho(\mathbf{r})\varrho(\mathbf{r}')}{|\mathbf{r} - \mathbf{r}'|} d\mathbf{r} d\mathbf{r}'. \quad (1.15)$$

The first term denotes the kinetic energy of noninteracting electrons, and the second and third terms are the classical electrostatic electron–nucleus attraction and electron–electron repulsion, respectively. Later on, Dirac included the exchange energy, which has no classical analogue, as an additional term to the TF model [8]:

$$E_x^{Dirac} = -\frac{3}{4}\left(\frac{3}{\pi}\right)^{1/3} \int \varrho(\mathbf{r})^{4/3} d\mathbf{r}. \quad (1.16)$$

TF calculated energies are usually too high. The Thomas–Fermi–Dirac (TFD) model accounts for this by adding an appropriate exchange term that gives a negative energy contribution. Despite this correction, calculated total energies are still not accurate for chemical predictions since the kinetic energy is poorly described. Although the TF model was not very successful in quantum chemistry or solid-state physics, it is the starting point for DFT in the sense of using the electron density for solving multielectron systems.

1.2.4.2 The Hohenberg–Kohn Theorems

The theoretical basis of DFT are two fundamental theorems, which were formulated and mathematically proven by Hohenberg and Kohn [17] for nondegenerate ground states. According to the first theorem, the electron density uniquely determines the external potential (the potential of the ions or nuclei) to within a constant. Therefore, the total electronic energy of a system E can be expressed as a functional of the electron density ϱ,

$$E[\varrho] = T[\varrho] + \int \varrho(\mathbf{r}) V_{\text{ext}}(\mathbf{r}) d\mathbf{r} + E_{\text{ee}}[\varrho], \tag{1.17}$$

where $T[\varrho]$ is the kinetic energy functional and $E_{\text{ee}}[\varrho]$ is the electron–electron interaction energy and also a functional. By defining the Hohenberg–Kohn functional

$$F_{\text{HK}}[\varrho] = T[\varrho] + E_{\text{ee}}[\varrho], \tag{1.18}$$

we obtain

$$E[\varrho] = \int \varrho(\mathbf{r}) V_{\text{ext}}(\mathbf{r}) d\mathbf{r} + F_{\text{HK}}[\varrho]. \tag{1.19}$$

The exact solution to the Schrödinger equation could be obtained with an explicit expression for the universal functional $F_{\text{HK}}[\varrho]$. Unfortunately, this expression is still unknown, but we will later show different approximations for this term. The remaining electron–electron interaction $E_{\text{ee}}[\varrho]$ can be written as

$$E_{\text{ee}}[\varrho] = J[\varrho] + E^{\text{noncl}}[\varrho], \tag{1.20}$$

where the first term is simply the classical Coulomb repulsion and the second term is the nonclassical part that contains self-interaction correction, exchange, and Coulomb correlation energy.

The second Hohenberg–Kohn theorem provides the energy variational principle for the exact functional. The ground-state density $\varrho_0(\mathbf{r})$ is the density that minimizes $E[\varrho]$

$$E_0 = E[\varrho_0(\mathbf{r})] \leq E[\varrho(\mathbf{r})] \tag{1.21}$$

when

$$\varrho(\mathbf{r}) \geq 0 \quad \text{and} \quad \int \varrho(\mathbf{r}) d\mathbf{r} - M = 0. \tag{1.22}$$

Energy minimization of the energy functional fulfills the Euler–Lagrange equation under the constraint of a constant number of electrons M (Eq. (1.22)), which can be written as

$$\mu = \frac{\delta E[\varrho(\mathbf{r})]}{\delta \varrho(\mathbf{r})} = V_{\text{ext}}(\mathbf{r}) + \frac{\delta F_{\text{HK}}[\varrho(\mathbf{r})]}{\delta \varrho(\mathbf{r})}, \tag{1.23}$$

where the Lagrange multiplier μ is the chemical potential of the electrons. In principle, this formulation provides all ground-state properties, but the Hohenberg–Kohn theorems do not tell us how to find the universal functional $F_{\text{HK}}[\varrho]$. Later on, Kohn and Sham [18] found rather accurate approximations for $F_{\text{HK}}[\varrho]$, and others still continue this development.

1.2.4.3 The Kohn–Sham Equations

Kohn and Sham decomposed the exact kinetic energy functional $T[\varrho]$ into two parts to approximate the universal functional $F_{HK}[\varrho]$ in terms of Eq. (1.18). The first term is the kinetic energy of a system of noninteracting electrons T_s given by

$$T_s = -\frac{1}{2}\sum_{i=1}^{N} <\phi_i|\nabla^2|\phi_i>, \qquad (1.24)$$

where ϕ_i are the so-called Kohn–Sham orbitals. The second term contains all remaining (and neglected) interactions in the noninteracting system $(T-T_s)$.

The second contribution is a small correction and it is included along with the nonclassical part of the electron–electron interaction E^{noncl} into a term defined as exchange–correlation (xc) energy:

$$E_{xc}[\varrho] = (T[\varrho]-T_s[\varrho]) + (E_{ee}[\varrho]-J[\varrho]). \qquad (1.25)$$

The energy functional Eq. (1.17) is now written as[9]

$$E[\varrho] = T_s[\varrho] + \int \varrho(\mathbf{r}) V_{ext}(\mathbf{r})d\mathbf{r} + J[\varrho] + E_{xc}[\varrho]. \qquad (1.26)$$

Finding suitable expressions for the $E_{xc}[\varrho]$ term is the main challenge in DFT development since it consists of all contributions that are not yet known exactly. The Euler–Lagrange equation now has the form

$$\varepsilon_i \frac{\delta\varrho(\mathbf{r})}{\delta\phi_i} = \frac{\delta E[\varrho(\mathbf{r})]}{\delta\varrho(\mathbf{r})}\frac{\delta\varrho(\mathbf{r})}{\delta\phi_i} = \frac{\delta T_s[\varrho(\mathbf{r})]}{\delta\phi_i} + V_{eff}(\mathbf{r})\frac{\delta\varrho(\mathbf{r})}{\delta\phi_i} \qquad (1.27)$$

with

$$V_{eff}(\mathbf{r}) = V_{ext}(\mathbf{r}) + \int \frac{\varrho(\mathbf{r'})}{|\mathbf{r}-\mathbf{r'}|}d\mathbf{r'} + V_{xc}(\mathbf{r}), \qquad (1.28)$$

where the exchange–correlation potential $V_{xc}(\mathbf{r})$ is

$$V_{xc}(\mathbf{r}) = \frac{\delta E_{xc}[\varrho(\mathbf{r})]}{\delta\varrho(\mathbf{r})}. \qquad (1.29)$$

The solution of Eq. (1.27) is obtained by solving the following set of one-particle equations

$$\left[-\frac{1}{2}\nabla^2 + V_{eff}(\mathbf{r})\right]\phi_i = \varepsilon_i\phi_i, \qquad (1.30)$$

---

[9] In principle, the kinetic energy term T_s is a functional of the Kohn–Sham orbitals ϕ_i.

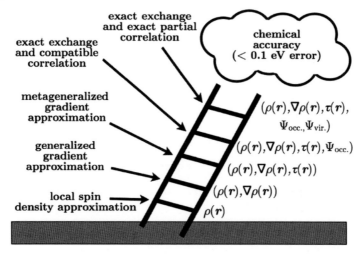

Figure 1.2 Jacob's ladder illustrating the hierarchy of exchange–correlation functionals.

where the electron density of the real system is constructed from the Kohn–Sham orbitals

$$\varrho(r) = \sum_{i=1}^{occ} |\phi_i(r)|^2. \tag{1.31}$$

Here, the summation runs over all occupied orbitals (occ). Equations (1.28), (1.30) and (1.31) are known as Kohn–Sham equations. The procedure to solve a particular problem is to start with a guessed $\varrho(\mathbf{r})$, determine V_{eff} from Eq. (1.28), and then obtain a new $\varrho(\mathbf{r})$ from Eqs. (1.30) and (1.31). This procedure is then repeated until $\varrho(\mathbf{r})$ is converged, and the total energy is obtained.[10]

1.2.4.4 Exchange–Correlation Functionals

Density-based formulations reduce the complexity of the many-body problem, but the resulting Kohn–Sham equation (1.30) still requires an xc functional. Different approximations of this term are the key distinction of all DFT implementations.

Perdew [21, 22] has illustratively formulated the hierarchy of different approximations of the xc functional as a "Jacob's ladder," rising from the earth of Hartree to the heaven of chemical accuracy (see Figure 1.2).

Starting from the earth of Hartree, the subsequent rungs of the ladder are defined as follows:

- **Local (spin) density approximation (L(S)DA):** Here the exchange–correlation energy of an inhomogeneous system is obtained by assuming that its density

---

10) This iterative process is known as the self-consistent field (SCF) approach.

can locally be treated as a uniform electron gas. The xc energy can then be written as

$$E_{xc}^{LDA}[\varrho] = \int \varrho(\mathbf{r})\varepsilon_{xc}(\varrho(\mathbf{r}))d\mathbf{r}, \quad (1.32)$$

where $\varepsilon_{xc}(\varrho(\mathbf{r}))$ is the xc energy per particle of the homogeneous electron gas, which can be split into exchange and correlation terms

$$\varepsilon_{xc}(\varrho(\mathbf{r})) = \varepsilon_{x}(\varrho(\mathbf{r})) + \varepsilon_{c}(\varrho(\mathbf{r})). \quad (1.33)$$

The exchange part is given by the Dirac expression (Eq. (1.16))

$$\varepsilon_{x}(\varrho(\mathbf{r})) = -\frac{3}{4}\left(\frac{3}{\pi}\right)^{1/3}\varrho(\mathbf{r})^{1/3}. \quad (1.34)$$

The correlation component ε_c has been determined by Monte Carlo (MC) calculations for a uniform electron gas considering a number of different densities [2]. Although one might expect that the LDA functional is valid only for a slowly varying density that might resemble a uniform electron gas, experience shows that this approximation is surprisingly valuable in a wide range of problems in solid-state physics and material science. LDA is noted to calculate molecular geometries and vibrational frequencies reasonably, but bond energies are strongly overestimated.

- **Generalized gradient approximation (GGA):** Inadequacies in LDA brought about a modified xc functional that, in addition to the density, contains terms for the density gradient:

$$E_{xc}^{GGA}[\varrho] = \int \varrho(\mathbf{r})\varepsilon_{xc}(\varrho(\mathbf{r}), \nabla\varrho(\mathbf{r}))d\mathbf{r}. \quad (1.35)$$

One form of GGA introduced by Perdew, Burke, and Ernzerhof (PBE) [23] is widely used in surface physics. In this approximation, the correlation energy is expressed as

$$E_{c}^{GGA}[\varrho] = \int \varrho\left[\varepsilon_{c}^{unif}(\varrho) + K(\varrho, t)\right]d\mathbf{r}, \quad (1.36)$$

with

$$K(\varrho, t) = \gamma \ln\left[1 + \frac{\beta t^2}{\gamma}\left(\frac{1 + At^2}{1 + At^2 + A^2 t^4}\right)\right] \quad (1.37)$$

and

$$\gamma \simeq 0.031091; \quad \beta \simeq 0.066725; \quad A = \frac{\beta}{\gamma}\frac{1}{e^{-\varepsilon_c^{unif}/\gamma} - 1}; \quad t = \frac{|\nabla\varrho|}{2\left(\sqrt{4(3\pi^2\varrho)^{1/3}/\pi}\right)\varrho}. \quad (1.38)$$

Here, t is a dimensionless density gradient. The exchange energy in terms of an enhancement factor F_x is written as

$$E_x^{GGA}[\varrho] = \int \varrho \varepsilon_x^{unif}(\varrho) F_x(s) d\mathbf{r}, \quad (1.39)$$

where $\varepsilon_x^{unif}(\varrho)$ and $\varepsilon_c^{unif}(\varrho)$ are, respectively, the exchange and correlation energies per particle of a homogeneous electron gas at point $\mathbf{r}$ (see Eq. (1.34)) and s is another dimensionless density gradient

$$s = \frac{|\nabla \varrho|}{2(3\pi^2 \varrho)^{1/3} \varrho}. \quad (1.40)$$

Finally, the function $F_x(s)$ is

$$F_x(s) = 1 + \kappa - \frac{\kappa}{1 + \mu s^2/\kappa}, \quad (1.41)$$

where $\kappa = 0.804$ and $\mu \simeq 0.21951$.

- **Meta-generalized gradient approximation (MGGA):** The kinetic energy density, $\tau(\mathbf{r})$, is an additional Kohn–Sham contribution that can be calculated. In general, MGGA functionals have the following form:

$$E_{xc}^{MGGA}[\varrho] = \int \varrho(\mathbf{r}) \varepsilon_{xc}(\varrho(\mathbf{r}), \nabla \varrho(\mathbf{r}), \tau(\mathbf{r})) d\mathbf{r}. \quad (1.42)$$

The main advantage of including kinetic energy densities is that it mostly eliminates self-interaction errors, causing inaccuracies with LDA and GGA functionals at low-density and strong interaction limits. In intermediate regions, however, MGGA functionals usually do not provide substantial improvement to corresponding GGAs.

- **Hybrid DFT with exact exchange:** An entirely different approach to improve deficiencies in GGA functionals is to incorporate the so-called exact exchange energy (EXX) contributions. The exact exchange energy E_x^{exact} is a derivative from the Hartree–Fock approximation (see Section 2.3.1), and is obtained by solving only the exchange part of the exchange–correlation functional exactly. The result is an energy value that when scaled according to

$$E_{xc}^{hybrid} = E_{xc}^{GGA} + a\left(E_x^{exact} - E_x^{GGA}\right) \quad (1.43)$$

provides a convenient cancellation of errors, making hybrid DFT methods surprisingly accurate. Based on this idea, the highly popular hybrid DFT method B3LYP combines exact HF exchange with the Slater [24] local exchange functional. In addition, it uses the Becke gradient correction [25], the local Vosko–Wilk–Nusair exchange functional [26], and the Lee–Yang–Parr local gradient-corrected functional [27]. Inclusion of the nonlocal exchange, however, limits these methods to be applicable only for finite systems and not in a periodic representation.

1.2.5
Technical Aspects of Modeling Catalytic Reactions

1.2.5.1 Geometry Optimizations

The following discussion of atomic properties in molecular configurations is simplified by considering a Born–Oppenheimer potential energy surface (PES). The PES is a hypersurface that is defined by the electronic energy as a function of the nuclei positions R_ν, which are $3N$ dimensional.[11] We now concentrate on the electronic part of the total energy that depends on the atom positions as parameters plus the nuclei–nuclei interaction ($E_e + E_{n-n}$). A PES can be expressed in the neighborhood of any spatial position R_ν^0 as

$$E(\mathbf{R}_\nu) = E(\mathbf{R}_\nu^0) + g(\mathbf{R}_\nu - \mathbf{R}_\nu^0) + \frac{1}{2}(\mathbf{R}_\nu - \mathbf{R}_\nu^0)^T \mathbf{H}(\mathbf{R}_\nu - \mathbf{R}_\nu^0) + \ldots, \quad (1.44)$$

where g and $\mathbf{H}$ are the gradients and Hesse matrix (or Hessian), whose components are defined as

$$g_\nu = \frac{\partial E}{\partial \mathbf{R}_\nu}\bigg|_{\mathbf{R}_\nu = \mathbf{R}_\nu^0} \quad \text{and} \quad H_{\nu\mu} = \frac{\partial^2 E}{\partial \mathbf{R}_\nu \partial \mathbf{R}_\mu}\bigg|_{\mathbf{R}_\nu = \mathbf{R}_\nu^0}. \quad (1.45)$$

While gradient calculations are typically manageable for large-scale calculations, Hessian calculations can be enormously time consuming.

After an electronic energy is calculated, a gradient calculation can be performed analytically or numerically. With this gradient, a cycle of procedures using any of the multitude of optimization schemes can change the atomic positions of the molecule R_ν until the gradient has reached a small value (below the desired convergence threshold), thereby indicating a stationary point on the PES. This process of calculating a gradient (and then adjusting the atomic positions accordingly) from an energy is usually referred to as geometry optimization step. Common procedures for geometry optimizations include the *steepest descent, conjugate gradient, Newton–Raphson*, or *Broyden–Fletcher–Goldfarb–Shanno(BFGS)* methods (see Ref. [28] for more details), some of which require only the first derivative of the energy or even use the Hesse matrix (see Eqs. (1.44) and (1.45)). Although each method has certain benefits, all are simply algorithms to reach the desired low-energy state faster in lieu of troublesome regions of the topology of the PES.

Stationary points define stable intermediates when located at a local minimum of the PES. In practice, geometry optimizations are accelerated by force constants associated with normal modes of the species at different points of the PES, and these force constants can be obtained from the Hessian. To avoid the expense of calculating a Hessian at every geometry iteration, methods are available to construct approximate Hessians from gradient calculations. The integrity of these Hessians is not always ideal, often still requiring a full (numerical) calculation of the Hessian at the end of the geometry optimization.

---

11) Note that this is not describing an *internal* coordinate system where one omits six coordinates related to system translations and rotations to result in $3N-6$ coordinates.

1.2.5.2 Transition-State Optimizations

Geometry optimizations can also be extended to seek transition states (TS), first-order saddle points where all but one internal coordinate are at a minimum. Unfortunately, these approaches are substantially more difficult since the local topology of a PES is usually required to find a first-order saddle point where the TS resides. Different automated routines have been developed to efficiently locate transition states; however, for many methods, there is no guarantee that a TS would ever be found.

A highly effective but cumbersome approach is to explicitly calculate a PES by manually varying specific constrained coordinates. In this approach, unconstrained coordinates are allowed to freely relax to minimum energy positions in a standard geometry optimization, and the result is a single data point on a PES surface. After multiple points have been sampled, the PES can be interpolated with minimum curvature or radial basis function algorithms.

The most ubiquitous TS searches use the modified Newton–Raphson routine and start from an atomic configuration close to that of the TS. The optimization procedure successively moves along the normal mode of the Hessian with the lowest magnitude in an attempt to find the point of maximum negative curvature. These methods are rather efficient if the starting geometry is near the real TS and the utilized Hessian contains only one negative mode.[12] When neither of these criteria is applicable, this procedure may fail spectacularly.

Other QM routines incorporate more information either by requiring coordinates from the initial reactant and product configurations or sometimes by including an initial guess for the TS as well. Such routines are quite efficient for locating transition states involving one or two bonds breaking in molecules, but for very large systems such as solids where atom movement influences a larger environment, alternative methods are needed.

For instance, nudge elastic band (NEB) methods are a notable and efficient technique to find these difficult TS without the use of a Hessian. NEB procedures linearly interpolate a series of atomic configurations (the so-called images) aligned in a row between the known initial and final states and then minimize their energies (see Figure 1.3). This method incorporates a limited description of the topology of the PES and thereby reveals the TS as the maximum energy image along the minimum energy reaction path. NEB is not the only procedure that uses a multitude of geometries coupled to each other. However, it is by far one of the most popular means to locate TSs for catalytic reactions happening on surfaces or in bulk environments.

1.2.5.3 Vibrational Frequencies

Vibrational frequencies corresponding to experimental IR spectra are often evaluated only around equilibrium structures (local minima on the PES) using the harmonic oscillator approximation for the potential. Using generalized coordinates[13] for the atom positions Q, Eq. (1.4) becomes

---

[12] Diagonalizing the Hessian would lead to a single negative eigenvalue.
[13] For example, Cartesian or internal coordinates.

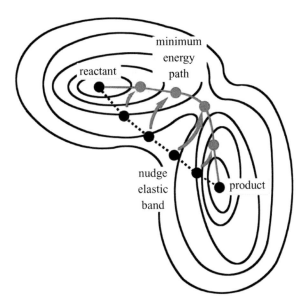

Figure 1.3 Schematic of the NEB method used to find a transition state between reactant and product states.

$$\left[-\sum_{\nu}^{N}\left(\nabla_{Q_\nu}^2 - \frac{1}{2}\omega_\nu^2 Q_\nu^2\right)\right]\psi_n(Q) = (E - E_e(Q^0))\psi_n(Q), \tag{1.46}$$

where $E_e(Q^0)$ is the electronic part of the energy for the system in its equilibrium structure. The quantum mechanical harmonic oscillator problem gives the eigenvalues of previous equations:

$$(E - E_e(Q^0))_\nu^\kappa = \left(\kappa + \frac{1}{2}\right)\hbar\omega_\nu \tag{1.47}$$

where κ is the vibrational quantum number[14] and ω_ν the vibrational frequency along direction ν:

$$\omega_\nu = \sqrt{\frac{k_\nu}{\mu_\nu}}, \tag{1.48}$$

with the corresponding force constant k_ν and effective mass μ_ν. The sign of the force constant shows the curvature of the PES along the particular direction (or mode). Calculated vibrational frequencies are, therefore, often analyzed in order to ensure that a structure optimized to either a stable intermediate or a TS. Indeed, vibrational frequencies should be calculated at the end of every geometry optimization for this very purpose. However, calculated frequencies are not expected to perfectly match

---

14) Usually, the index n is used for the quantum number, but this index had already been used to indicate nuclei-related expressions.

with experimental IR frequencies since the previously described approach does not account for anharmonicities of vibrational modes, regardless of the method used.

Equation (1.47) indicates that even in the ground-state $\kappa = 0$ there is a nonzero vibrational contribution to the energy of a system. Energy from this first thermodynamic correction is the zero-point vibrational energy (E_{ZPE}), defined as half of the sum of all vibrational normal modes ω_v:

$$E_{ZPE} = \frac{1}{2} \sum_v \hbar \omega_v. \tag{1.49}$$

1.2.5.4 Thermodynamic Treatments of Molecules

Macroscopic energies based on microscopic contributions can be obtained through traditional statistical thermodynamics using the ideal gas assumption. The rudimentary function that describes macroscopic properties is the partition function, which is separable into electronic, translational, rotational, and vibrational components:

$$E = E_e + E_{trans} + E_{rot} + E_{vib}. \tag{1.50}$$

Each of these energy contributions in turn has internal energy components, U, and entropy, S, which are also separable into components:

$$\begin{aligned} U &= U_e + U_{trans} + U_{rot} + U_{vib}, \\ S &= S_e + S_{trans} + S_{rot} + S_{vib}. \end{aligned} \tag{1.51}$$

We now discuss the different energy and entropy contribution terms as follows:[15]

- **Electronic contributions:** The electronic part of the internal energy U_e can be obtained by solving the electronic part of the many-body problem (Eq. (1.5)). In situations involving electronic degeneracies (i.e., doublet, triplet, and so on states) electronic entropy contributions come into play

$$S_e = k_B \ln(2S + 1), \tag{1.52}$$

where S is the total spin of the molecular state.
- **Translational contributions:** The temperature-dependent internal energy associated with molecular translations is $(1/2)k_B T$ for every translational degree of freedom of the molecule. In Cartesian space, this leads to $U_{trans} = (3/2)k_B T$. The entropy of translation is

$$S_{trans}^\circ = k_B \left\{ \ln \left[\left(\frac{m_v k_B T}{2\pi \hbar^2} \right)^{3/2} \frac{V^\circ}{N_A} \right] + \frac{5}{2} \right\} \tag{1.53}$$

where V° is the molar volume of a gas at its standard state: 24.5 ℓ.
- **Rotational contributions:** As for the translational contributions, the temperature-dependent internal energy associated with molecular rotations is also $(1/2)k_B T$

---

15) In this section, all energies are in $k_B T$ (per particle).

for each rotational degree of freedom. While single atoms receive no rotational energy, linear molecules receive $k_B T$, and nonlinear molecules receive $(3/2)k_B T$. The entropy of rotation is

$$S_{\text{rot}} = k_B \left\{ \ln\left[\frac{\sqrt{\pi I_A I_B I_C}}{\sigma}\left(\frac{2k_B T}{\hbar^2}\right)^{3/2}\right] + \frac{3}{2} \right\}, \tag{1.54}$$

where σ is the rotational symmetric number for the molecule's point group, and I_A, I_B, and I_C are the molecule's three principal moments of inertia.

- **Vibrational contributions:** Thermodynamic energy contributions from molecular vibrations are calculated per vibrational frequency, ω_v. Thus, they can be calculated only once vibrational frequencies are obtained from a Hessian calculation. Vibrational energies must be summed over all $3N-6$ molecular vibrations:[16]

$$U_{\text{vib}} = \sum_v \frac{\hbar\omega_v}{e^{\hbar\omega_v/k_B T} - 1} \tag{1.55}$$

and

$$S_{\text{vib}} = k_B \sum_v \left[\frac{\hbar\omega_v}{k_B T(e^{\hbar\omega_v/k_B T} - 1)} - \ln(1 - e^{-\hbar\omega_v/k_B T})\right]. \tag{1.56}$$

Using these energy contributions in addition to the calculated E_e from *ab initio* or DFT methods, the following thermocorrected energies can be obtained:

$$H_{0K} = E_e + E_{\text{ZPE}}, \tag{1.57}$$

$$H_T = H_{0K} + U_T \quad (+k_B T \quad \text{if in gas phase}), \tag{1.58}$$

$$G_T = H_T - T S_T. \tag{1.59}$$

1.2.5.5 Considering Solvation

Most chemical processes and catalytic reactions occur in the condensed phase rather than in gas phase. However, including the electronic structure of surrounding molecules implies more complex systems.

The methods used to model solvation can be basically categorized into two main approaches: the explicit and the implicit treatments (see Figure 1.4). In the explicit solvent model (see Refs. [29, 30] for further information), the solvent is described by individual molecules. Consequently, this approach provides detailed information about the structure of the solute/solvent interface for systems near equilibrium and along reaction pathways (e.g., Car–Parrinello *ab initio* molecular dynamics (CPMD) [31]). However, the demands of a purely quantum mechanical treatment requires a system size limited to a relatively small number of solvent molecules.

---

16) For linear molecules, there are $3N-5$ vibrational modes.

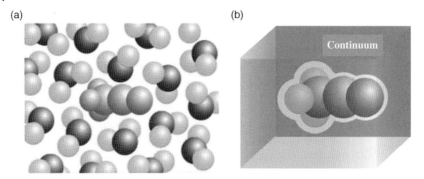

Figure 1.4 Illustrations of solvation methods. (a) Explicit solvation, where solvent molecules (here water) are treated as the molecule of interest. (b) Implicit solvation, where only the molecule of interest is surrounded by a continuum representing the solvent.

An alternative approach is to treat the entire or at least parts of the solvent with a lower level of theory than the solute, for example, with semiempirical force fields [29, 30, 32–34]. In these methods, usually known as QM/MM simulations, the solvent–solvent, and solute–solvent interactions are mostly given by a sum of electrostatic and van der Waals contributions. Further, the coupling between the quantum-mechanically treated solute and the molecular-mechanically treated solvent needs particular attention. Possible solutions are provided by a multipolar expansion of the QM wave functions [35, 36] or a solvent-averaged potential.

Much less demanding implicit solvent approaches [37, 38] can provide a useful correction accounting for solvation. These methods use a continuous electric field to represent a statistical average of the dielectric effects of the solvent over all the solvent degrees of freedom. The solute is then placed in a cavity, whose size and shape have to be chosen properly. Different ways exist to define these cavities, for example, using atom-centered interlocked spheres of certain radii [37, 39], the solvent-excluded surface (SES) [40], or the solvent-accessible surface (SAS) [41] method. Due to electrostatic interactions, the solute polarizes the solvent, which then forms the so-called reaction field that couples back to the solute. The corresponding electrostatic potential, which is required to evaluate the interaction energy, is obtained by solving the Poisson–Boltzmann equation (assuming a Boltzmann-type distribution of ions within the solvent):

$$\nabla[\varepsilon(\mathbf{r})\nabla V_{es}(\mathbf{r})] = -4\pi\left[\varrho^f(\mathbf{r}) + \sum_i c_i^\infty z_i \lambda(\mathbf{r}) e^{-z_i V_{es}(\mathbf{r})/k_B T}\right], \quad (1.60)$$

where $\varepsilon(\mathbf{r})$ is the position-dependent dielectric, V_{es} is the electrostatic potential, ϱ^f is the charge density of the solute, c_i^∞ is the bulk concentration of ion-type i with the net charge z_i, and finally $\lambda(\mathbf{r})$ is a factor for the position-dependent accessibility of the solution to position $\mathbf{r}$. This leads to a contribution to the overall free energy of a system:

$$\Delta G^{solv} = \Delta G^{chg} + \Delta G^{cav}, \quad (1.61)$$

where the two terms account for the free energy to generate the cavity and for charging the solvent near the solute.

In quantum mechanical calculations of a solute electronic structure, the solvent reaction field is incorporated into the iterative cycle, resulting in the so-called self-consistent reaction field (SCRF) approach. Instead of solving the Poisson–Boltzmann equation directly, different approximations exist to evaluate the solvation free energy by analytic expressions, for instance, the generalized Born model [42, 43], the Bell model, or the Onsager model [37, 38].

1.2.6
Model Representation

Studying surface-specific problems such as structures, adsorptions, and catalytic reactions is a formidable task and requires one of the two surface model approaches: the periodic slab/supercell approach and the cluster approximation. Both models have advantages and disadvantages. The choice depends on the particular physical or chemical question to be answered.

1.2.6.1 Slab/Supercell Approach

Although the many-body problem has been simplified by the formulation of DFT, calculating the electronic structure of extended systems (e.g., bulk, surfaces, and chains) with infinite number of electrons (e.g., in a solid) is of course impossible. However, by assuming periodic boundary conditions and applying Bloch's theorem [44], the calculation of extended systems becomes possible.

The Bloch theorem states that each electronic wave function[17] in a periodic solid can be written as the product of a periodic function, $u_{n,\mathbf{k}}(\mathbf{r})$, and a plane wave, leading to

$$\phi_{n,\mathbf{k}} = u_{n,\mathbf{k}}(\mathbf{r})e^{i\mathbf{k}\cdot\mathbf{r}}, \quad (1.62)$$

where $\mathbf{k}$ is the wave vector that lies inside the first Brillouin zone (BZ). The index n, the band index, labels the wave functions for a given $\mathbf{k}$. The function $u_{n,\mathbf{k}}(\mathbf{r})$ has the periodicity of the supercell and can be expanded using a set of plane waves

$$u_{n,\mathbf{k}}(\mathbf{r}) = \sum_{G} c_{n,\mathbf{k}}(G)e^{i\mathbf{G}\cdot\mathbf{r}}, \quad (1.63)$$

where the wave vectors $\mathbf{G}$ are reciprocal lattice vectors fulfilling the boundary condition of the unit cell, $\mathbf{G}\cdot\mathbf{L} = 2\pi\nu$.

The physical quantities of a system, such as the electron density and total energy, are obtained by performing an integration into reciprocal space. Numerically, the integral over the BZ can be transformed into a sum over only a finite number of **k**-points, called the **k**-point mesh:

---

17) Here, we show derivation for the density functional theory, thus expanding Kohn–Sham orbitals. However, the Bloch theorem is not restricted to DFT.

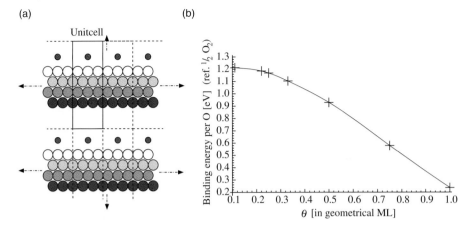

Figure 1.5 (a) Illustration of the supercell approach for surface calculations. The periodically repeated unit cell (2D slab) has to be large enough to minimize interactions with neighboring unit cells. (b) Binding energy of an oxygen atom as a function of surface coverage (with gas-phase 1/2 O_2 as reference).

$$\int_{BZ} \frac{1}{\Omega_{BZ}} d\mathbf{k} \rightarrow \sum_{\mathbf{k}} \omega_{\mathbf{k}}. \tag{1.64}$$

The error introduced by this approximation can be minimized if a sufficiently dense set of **k**-points is used, but the computational effort grows quickly with the number of **k**-points. Therefore, it is crucial to test the convergence of the results with respect to the number of **k**-points and choose an appropriate mesh size.[18]

In principle, an exact representation of the electronic wave functions should require an infinite number of plane waves. In practice, however, a finite number of plane waves up to a certain cutoff, E_{cutoff}, already provides sufficiently accurate results:[19]

$$\frac{1}{2}|\mathbf{k} + \mathbf{G}|^2 \leq E_{cutoff}. \tag{1.65}$$

Periodic systems are modeled by concentrating on the smallest possible unit cell and periodically repeating it over all space. For bulk materials, this box of atoms is mostly set up by using the primitive unit cell of the crystal. For a surface, however, the periodicity in direction perpendicular to the surface is broken. Surfaces are modeled by a periodic 3D structure containing crystal slabs separated by vacuum regions (see Figure 1.5) in order to maintain the periodicity in this direction. This so-called supercell approach represents the surfaces periodically, and they must be separated

---

[18] Different schemes have been proposed to distribute the **k**-points within the unit cell. The commonly employed scheme of a homogeneous grid of **k**-points was introduced by Monkhorst and Pack [45].

[19] In practical calculation, the cutoff energy is obtained by carefully checking the convergence of system properties with respect to the cutoff value.

by a vacuum wide enough to avoid unwanted interactions between slabs and their virtual copies above and below. Furthermore, the slab must be thick enough so that its center reproduces bulk-like behavior to avoid unphysical interactions between the surfaces on the top and the bottom of the slab. Depending on the system, this requires between 5 and sometimes even more than 20 atomic layers. Semiconductors and certain transition metals exhibit major surface properties, such as surface states or adsorbate binding energies, with only five-layer slabs. Some simple metals (e.g., Al) require more layers.

If the aim is to study catalytic reactions on surfaces, modifying the lateral extension of the unit cell also allows investigation of coverage effects. The zero-coverage limit requires rather extended unit cells to ensure that adsorbed species interact neither directly with their images in adjacent cells nor indirectly through changes in the electronic structure of the surface. Figure 1.5a shows the DFT-calculated binding energy of a single oxygen atom (with respect to half an O_2 molecule) on a face-centered cubic (fcc) site of a Pt(111) slab as a function of unit cell size (or coverage). Here, negligible adsorbate–adsorbate interactions are received with a 3×3 unit cell.

1.2.6.2 Cluster Approach

When simulating finite systems such as molecules or clusters, the cluster approach is more suitable. In particular cases, even surfaces can be modeled as cluster where a finite segment of the surface models the extended system (see Figure 1.6). Despite the lack of periodicity in this approach, the finite system avoids most issues connected with direct and indirect interactions of nearby slab images.

When modeling metallic surfaces with small clusters, most (if not all) atoms may be surface atoms. This causes unwanted border effects and an unpredictable behavior of the cluster that is why such clusters have to be built with special care. The graph in Figure 1.6b shows calculated binding energies for atomic oxygen (again

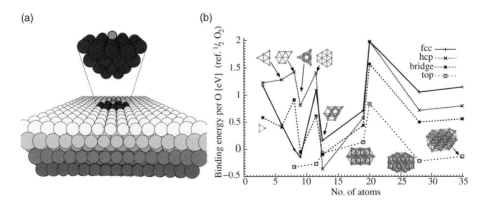

Figure 1.6 (a) A well-defined, finite cluster is used to model the surface in the cluster approach. (b) Binding energies (with respect to gas-phase 1/2 O_2) of atomic oxygen on differently sized and shaped clusters (Pt_3, Pt_6, Pt_8, Pt_{12}, $Pt_{6.3}$, $Pt_{12.8}$, $Pt_{5.10.5}$, $Pt_{9.10.9}$, $Pt_{14.13.8}$). For each cluster, the binding energies at fcc, hcp, bridge, and on-top sites are given.

with half an O_2 molecule as reference) on the (111) plane of differently sized and shaped Pt clusters [46].

On clusters of fewer than 20 atoms, the atomic oxygen binding energy and preferred adsorption site strongly depend on the system. Extending the cluster to more than 20 atoms results in a concordance of preferred adsorption sites and similar relative stabilities. However, convergence in binding energy and adsorption structure requires clusters of at least 3 atomic layers and 28 atoms. When such large clusters are used, calculated binding energies agree quite well with periodic calculations at low adsorbate coverages. Other metal surfaces sometimes require even more extended systems (e.g., 56 atoms for Cu(100) [47] or >100 atoms for Al(100) [48]). Although finite systems can mimic surfaces, systematic studies on cluster size convergence are necessary.

1.3
The Electrocatalytic Oxygen Reduction Reaction on Pt(111)

The ORR is a canonical chemical reaction due to its ubiquitous presence in corrosion, combustion, energy conversion, and storage processes. After describing the different aspects of the theoretical modeling of catalytic reactions, we now show how to apply these concepts to understand one of the most fundamental reactions in electrocatalysis. Besides its importance in basic electrochemistry, the oxygen reduction reaction is also relevant to energy conversion in polymer electrolyte membrane fuel cells (PEM-FCs).

In principle, gaseous H_2 is oxidized at the anode and its protons migrate through the electrolyte to the cathode where they finally react with O_2 under uptake of four electrons to form two water molecules.

Despite the apparently simple reaction mechanism shown in Table 1.1, the exact reaction mechanism and thus the fundamental reaction steps of the ORR are still not fully understood. Indeed, this reaction is highly complex since it occurs in a multicomponent environment and is influenced by various environmental parameters: temperature, pressure, and electrode potential.

Before discussing the electrochemical ORR, we will investigate the Pt-catalyzed water formation out of gaseous O_2 and H_2, which is the closest surface science analogue to the ORR. After this, the generalization to electrocatalysis will be undertaken, for which only a few additional processes have to be included.

Table 1.1 The oxygen reduction reaction potentials.

Anode :	$2H_2$	$\rightarrow$	$4H^+ + 4e^-$	$E° = 0\,V$
Cathode :	$O_2 + 4H^+ + 4e^-$	$\rightarrow$	$2H_2O$	$E° = 1.229\,V$
Net :	$2H_2 + O_2$	$\rightarrow$	$2H_2O$	$E° = 1.229\,V$

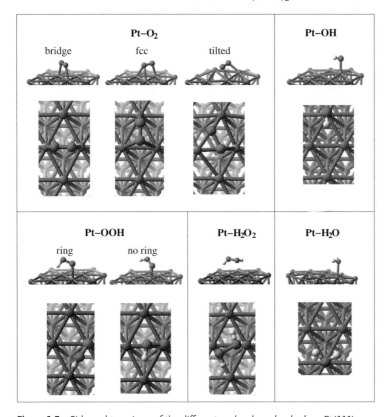

Figure 1.7 Side and top views of the different molecules adsorbed on Pt(111).

The fundamental steps for both surface science water formation and electrochemical ORR were calculated with DFT.[20] We considered adsorbed and nonadsorbed binding energies for all possible intermediates separately: H, H_2, O, O_2, OH, OOH, H_2O_2, and H_2O. We determined stable surface sites and binding energies for each intermediate on Pt(111) (see Figure 1.7). We then calculated barriers for all adsorbed molecule dissociation processes [50, 51]. We optimized transition states by a series of constrained optimizations along their reaction paths. Table 1.2 summarizes the corresponding binding energies and dissociation barriers. This includes values for the systems in gas phase, solvated in water, and finally under ambient conditions, that is, free energies at $T = 298$ K.

20) DFT calculations on the water formation mechanisms used the Jaguar code [49] spin restricted density DFT with the B3LYP gradient-corrected exchange–correlation functional and included zero-point energy (ZPE) corrections and implicit solvation when noted. The 60 core electrons in each Pt atom (1s-4f) were treated with the Hay and Wadt core-valencer elativistic effective core potential (ECP), leaving 18 valence electrons to be treated with the LACVP** basis set. The other elements (H and O) were described with the all-electron 6-31G** basis set. To represent the Pt(111), a 35-atom Pt cluster was used, which ensured cluster-size converged energies.

Table 1.2 Calculated binding energies and dissociation barriers for all investigated intermediates.

System	Adsorption site	Binding energies (eV)			Bond	Dissociation energies (eV)		
		ΔE^{gas}	ΔE^{solv}	ΔG^{solv}_{298K}		ΔE^{gas}	ΔE^{solv}	ΔG^{solv}_{298K}
$H_2^{g,aq}$	—	—	—	—		4.84	4.81	4.22
$O_2^{g,aq}$	—	—	—	—		4.95	5.46	5.05
$OH^{g,aq}$	—	—	—	—		4.57	4.83	4.31
$OOH^{g,aq}$	—	—	—	—	OO–H	2.79	2.69	2.12
	—	—	—	—	O–OH	3.17	3.32	2.86
$HOOH^{g,aq}$	—	—	—	—	HOO–H	3.83	4.00	3.44
	—	—	—	—	HO–OH	2.43	2.48	1.99
$H_2O^{g,aq}$	—	—	—	—		5.24	5.39	4.57
Pt–H	Top	2.73	3.09	2.67		—	—	—
	Bridge	2.64	3.43	3.07		—	—	—
Pt–O	fcc	3.24	4.40	4.04		—	—	—
	hcp	3.03	—	—		—	—	—
Pt–O_2	Bridge	0.49	1.31	0.81		1.34	0.81	0.90
	fcc	0.31	1.64	1.15		1.03	1.11	1.19
	Tilted	0.06	0.85	0.36		0.22	−0.10	−0.11
Pt–OH	Top	2.06	3.03	2.61		1.90	1.15	0.99
Pt–OOH	No ring	1.03	2.18	1.57	OO–H	1.03	0.72	0.57
	Ring	0.75	2.07	1.37	O–OH	0.74	0.62	0.59
					OO–H	0.36	0.81	0.71
Pt–HOOH	Bridge	0.41	1.36	0.52	HOO–H	0.94	0.96	0.78
					HO–OH	0.46	0.43	0.31
Pt–H_2O	Top	0.60	0.83	0.56		1.29	0.86	0.73

Energies are given for the compounds in gas-phase, solvated in water, and under ambient conditions (including thermal corrections).

This data affords a picture of different pathways (see Figure 1.8). In general, three major pathways are readily distinguishable. First, O_2^{ad} can dissociate on the surface generating two O^{ad} atoms and then react with H^{ad} atoms to form water. Second, the O_2^{ad} molecule can react with hydrogen to first form OOH^{ad} or then form $HOOH^{ad}$. After the O–O bond in these species breaks, the remaining species further reacts to the final product water. All pathways appear to be strongly influenced by the presence of the electrolyte.

1.3.1
Water Formation from Gaseous O_2 and H_2

Since many reactions and concepts of heterogeneous catalysis in surface science have analogues in electrocatalysis, we first discuss the Pt-catalyzed water formation from gaseous hydrogen and oxygen. The three major reaction pathways are separately described, and then we draw general conclusions on the overall mechanism.

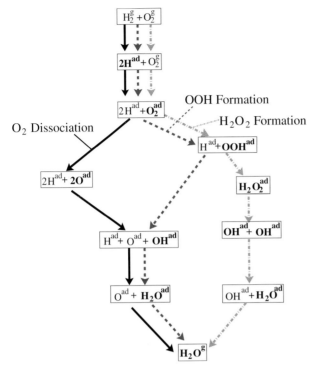

Figure 1.8 Schematic showing three main reaction pathways for the formation of water out of gaseous H_2 and O_2.

1.3.1.1 O_2 Dissociation

When O_2 adsorbs on the Pt(111) surface, three stable binding geometries are found: bridge (BE = 0.49 eV), fcc (BE = 0.31 eV), and tilted (BE = 0.06 eV) (see Figure 1.7). The most stable configuration is O_2^{ad} bound at a bridge position, where both oxygens use a doubly occupied p-orbital to form donor–acceptor bonds to the surface. The second stable structure corresponds to dioxygen above a fcc[21] surface-site, such that one oxygen binds on top of a Pt atom and the other oxygen at a bridge position. Finally, the last structure is somewhat comparable to bridge-bound O_2^{ad}, except that the molecule is tilted toward the surface such that the O=O π bond can form a donor–acceptor bond to an adjacent Pt atom.

After adsorption, O_2^{ad} may dissociate with one of the three dissociation barriers corresponding to those from the bridge (1.34 eV), the fcc (1.03 eV), or the tilted configuration (0.22 eV). Although, tilted O_2 forms the weakest surface bond, this adsorbate structure has the lowest dissociation barrier. Binding energy alone suggests that O_2^{ad} resides at a bridge surface site. However, the tilted configuration is 0.90 eV lower than the bridge-site dissociation barrier. Thus, O_2^{ad} may change its

---

21) The fcc-site is a threefold position where there is no Pt atom in the second layer below.

structure to the tilted configuration (which does not require much reorganization) and then dissociate via the O_2-tilted route. By doing so, the overall dissociation barrier reduces from 1.34 eV to only 0.65 eV, which is in much better agreement with the value of 0.38 eV measured by Ho and coworkers [52].

The products of O_2^{ad} dissociation are two O^{ad} atoms located in threefold sites: O_{fcc}/O_{fcc}, O_{hcp}/O_{hcp}, or O_{fcc}/O_{hcp}. Our simulations show dissociation results in two O^{ad} atoms at nonadjacent threefold positions (separated by two lattice constants), also in excellent agreement with the scanning tunneling microscopy (STM) experiments by Ho [53, 54]. When multiple O_2^{ad} dissociations are considered to occur across the entire surface, the final structure corresponds to a $p(2 \times 2)$ overlayer, a result we also obtained by evaluating the surface phase diagram of various adlayer configurations [55, 56] and which had also been observed experimentally [57]. Since the O−O interactions are already rather small, for coverages ≤ 0.25 ML it is justified to simulate O^{ad} atoms as independent adsorbates. Therefore, we neglected coverage effects in our simulations. The barrier to hop from a hexagonal close packed (hcp) to a fcc site via a bridge position is only 0.24 eV, so we expect most atomic oxygen will eventually migrate to the most thermodynamically stable fcc sites.

The next reaction step along the O_2 dissociation pathway is the formation of OH^{ad}, a process whose barrier is 1.25 eV. In contrast to O_2^{ad} dissociation, no alternative trajectory with a lower dissociation barrier was found. This process leads to adsorbed but mobile H^{ad}, partially hcp-bound O_{fcc}^{ad}, and OH^{ad}. Hydrogenation of OH^{ad} leads to a water molecule that can then desorb as a gas-phase water, the final product of the ORR reaction. Both H_2O^{ad} and OH^{ad} adsorb on Pt(111) with their oxygens at on-top surface sites. The binding energy to form H_2O^{ad} from OH^{ad} and H^{ad} is the sum of breaking the Pt−OH covalent bond (2.06 eV), forming the Pt−OH$_2$ donor–acceptor bond (−0.60 eV), breaking the surface Pt−H bond (2.73 eV), and forming the OH−H bond (−5.24 eV). This leads to a lowering of the system energy by −1.05 eV. The barrier connected with the water formation out of OH^{ad} and H^{ad} is only 0.24 eV. Finally, water desorption requires breaking the Pt−H$_2$O surface bond (0.60 eV), a value comparable to the experimental value of 0.52 eV [58].

The complete O_2 dissociation pathway in gas phase is shown in Figure 1.9. Energies reported so far *do not* include zero-point energy corrections or thermal contributions, thus represent the case of 0 K temperature. All reaction steps are exothermic except water desorption from the surface. The overall calculated reaction enthalpy at 0 K is 2.50 eV (per water molecule), a value comparable to the Nernst equation enthalpy value[22] of 2.46 eV. However, the latter value is under ambient conditions and includes the electrolyte effects we will discuss in later sections.

Only three steps of this entire process have barriers. The highest barrier (1.25 eV) is the $O_{fcc}^{ad} + H^{ad} \rightarrow OH^{ad}$ reaction. Dissociation of O_2^{gas} in the absence of the heterogeneous Pt catalyst is 4.92 eV, but dissociation of O_2^{ad}, usually considered the rate-determining process for the ORR reaction, is 0.65 eV. The last reaction barrier

---

22) Two electrons are transferred: $2e \times 1.23 V = 2.46$ eV.

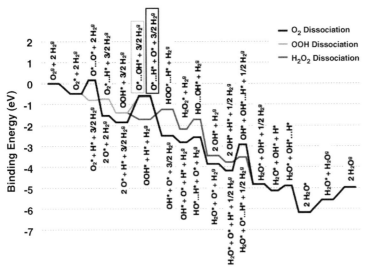

Figure 1.9 Gas-phase Langmuir–Hinshelwood reactions.

(0.24 eV) is formation of water on the surface from $OH^{ad} + H^{ad}$. This calculation data supports that the rate-determining step (RDS) should be the OH^{ad} formation. Experimental values for the overall ORR activation range from 0.1 to 1.0 eV; however, the value also depends on the experimental system and measurement conditions [59]. The discrepancy between our calculated barrier indicates that this reaction process may not be a relevant mechanism for the ORR.

1.3.1.2 OOH Formation

Instead of dissociating after adsorption, O_2^{ad} might first undergo a hydrogenation step to form adsorbed OOH^{ad}, which then may dissociate to form OH^{ad} and O^{ad}. As in the O_2 dissociation pathway discussed in the previous section, the generated OH^{ad} molecule could then react with another H^{ad} to from water, which finally desorbs from the surface. The energies along this OOH formation pathway have also been added to the diagram shown in Figure 1.9.

As adsorbed hydrogen is extremely mobile on the Pt(111) surface,[23] several different processes for the OOH formation mechanism should be considered. H^{ad} could approach O_2^{ad} roughly perpendicular to the O–O direction and then bind to an oxygen. Electronic rearrangement from an O–O π bond to the stronger O–H σ bond causes the other O atom to form a stronger covalent surface connection.

The barrier to form OOH^{ad} along this process is 0.43 eV. However, when H^{ad} approaches O_2^{ad} along the Pt–Pt bridge direction, another stable OOH^{ad} structure forms, in which OOH^{ad} and two surface Pt atoms form a five-membered ring structure. A strong Pt–H interaction weakens the adjacent Pt–O donor–acceptor bond, resulting in a 0.38 eV lower barrier for the $H^{ad} + O_2^{ad} \rightarrow OOH^{ad}$ process. The

---

23) Binding energies at different surface sites vary only by < 0.10 eV.

ring structure is unstable and can easily tautomerize into the nonring OOH^{ad} structure (see Figure 1.7). Therefore, we consider only the nonring OOH^{ad} structure as relevant.

Once adsorbed OOH has been formed, dissociation to OH^{ad} + O^{ad} has a barrier of 0.74 eV. The products are OH^{ad} at an on-top site and O^{ad} in a threefold site. The O—OH bond breaks along the O—O direction, and the single O atom must migrate over a top site before moving into its preferred threefold site. Along the dissociation, there is no calculated preference for O^{ad} to move into an adjacent fcc or hcp site. However, as discussed in the previous section, O^{ad} at an hcp position should easily migrate over a bridge position (0.24 eV barrier) to a fcc site. Finally, water formation and water desorption steps are the same as those described in the previous section.

Just like the O$_2^{ad}$ dissociation mechanism, every step of the OOH formation pathway is exothermic except for the last reaction step, that is, desorption of water. The OOH^{ad} mechanism involves three activation barriers: formation of OOH^{ad} (0.05 eV), the dissociation of an oxygen from OOH^{ad} (0.74 eV), and the final water surface desorption (0.24 eV). Thus, we find the O—OH dissociation is rate determining for the OOH^{ad} formation pathway, and due to an overall lower energy barrier it is a likelier ORR mechanism than the O$_2^{ad}$ dissociation (at least under gas-phase conditions).

1.3.1.3 HOOH Formation

The third mechanism considers OOH^{ad} undergoing another hydrogenation step to form HOOH^{ad} on the surface with an energy barrier of 0.47 eV. Two OH^{ad} adsorbates then form after HO—OH bond dissociation (0.46 eV). The remaining intermediates have already been discussed in the previous section. The highest barrier along the HOOH^{ad} pathway is the formation of HOOH^{ad} in gas phase (0.47 eV), thus being the rate-determining step. Consequently, the barrier for the RDS along the H$_2$O$_2$ formation pathway is 0.26 eV below the RDS of the OOH formation and 0.51 eV lower than that of the O$_2$ dissociation mechanism. Both the OOH formation and the HOOH formation pathways, which have energy barriers for the RDS of 0.74 eV and 0.47 eV, respectively, now show a much better agreement with experimental observations than the O$_2$ dissociation with an RDS barrier of 1.25 eV. Drawing this conclusion was possible only after investigating all three mechanisms individually. Furthermore, we have not yet included solvation or thermal effects, and either might certainly influence overall reaction kinetics. Therefore, we now will repeat the previous studies in the presence of water, which is the main constituent of the electrolyte.

1.3.2
Simulations Including Water Solvation

Electrocatalysis involves surface reactions under wet conditions. Compared to the gas-phase water formation, which had been discussed in the previous section, inclusion of an electrolyte will certainly alter energetics along the ORR and may even modify the preferred reaction mechanism. Besides these purely electronic

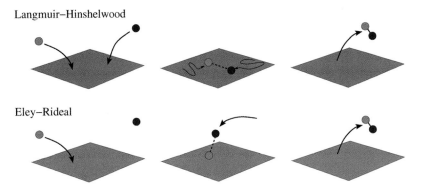

Figure 1.10 Schematic showing the Langmuir–Hinshelwood (*top*) and the Eley–Rideal (*bottom*) reaction mechanism.

effects, the solvent may also play an explicit role in the reaction steps. Therefore, two classes of reactions should be considered when an electrolyte is present. Besides Langmuir–Hinshelwood reactions, which have already been presented, Eley–Rideal reaction mechanisms rely on the electrolyte as a source of hydrogen atoms (see Figure 1.10).

We already mentioned that solvation presents a formidable problem for theory. Although results from gas-phase calculations are sometimes used to interpret experiments performed in solution, we believe that at least some treatment of the water solvent is required to obtain relevant results. Full quantum mechanics simulations alone would not provide completely accurate comparisons to experimental observations, but molecular dynamics simulations updated with quantum chemical forces (via *ab initio* molecular dynamics) usually provide reliable accuracy relevant to extended timescales. Unfortunately, the scaling of such methods quickly makes simulations too large and complex to simulate, especially for those with many intermediates and barriers such as the ORR reaction. As described in Section 2.5.5, an alternative approach is to treat solvation using a dielectric continuum. Although this method lacks dynamical information, qualitatively correct electrostatic behavior is attainable.

A self-consistent reaction field description of the water solvent augmented our studies on the ORR reactions. The binding energies and reaction barriers of the previous mechanisms were recalculated in the presence of water environment and used to generate Figure 1.11. Comparing gas-phase results to solvated results often shows large stabilizations of adsorbed species. Large stabilizations are certainly due to a partial charge transfer between the adsorbate and the surface, which was not observable without a polarizing environment. This results in a positive partial charge ($\delta+$) for each hydrogen and a slight negative partial charge ($\delta-$) for each oxygen. These charges interact with the water dipoles, polarizing the solvent, and thus further stabilize adsorbates.

We will now first discuss Langmuir–Hinshelwood-type reactions, describing the main electronic effects introduced by the water environment, and then consider

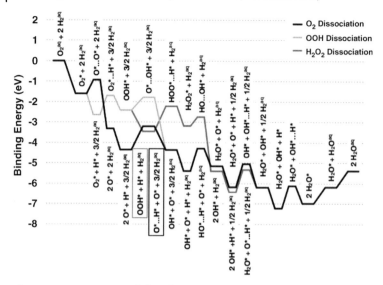

Figure 1.11 Langmuir–Hinshelwood reactions in water solvent.

Eley–Rideal-type reactions, where the protonations occur from the electrolyte directly.

1.3.2.1 Langmuir–Hinshelwood Mechanisms

- **O_2 Dissociation:** O_2^{ad} dissociation leads to two O^{ad}, which is now -1.71 eV downhill in solvent (water) rather than -1.05 eV in gas phase. The barrier for this dissociation, however, hardly changes from 0.65 to 0.68 eV. As mentioned before, this is certainly due to the higher degree of polarization of O^{ad} compared to O_2^{ad}. After dissociation of O_2^{ad}, O^{ad} reacts with a surface hydrogen to form OH^{ad}. Now, the reactants and products in this reaction are isoenergetic, instead of products being favored by 0.65 eV in gas phase. The barrier to form OH^{ad} drops from 1.25 eV in gas phase to 1.14 eV in solution. We observe a similar barrier (1.10 eV) to form H_2O^{ad}; however, this barrier was 0.24 eV in gas phase. Despite these changes, water solvation causes little change to the overall O_2^{ad} dissociation pathway. We find that the formation of OH^{ad} is still the rate-determining step for this mechanism, but this again does not conform to experimental expectations [60]. Water desorption from the solvated surface requires 0.83 eV, a value that is comparable to the binding energy of a water molecule within an entire water bilayer network.[24] This shows that application of an SCRF model in surface catalysis simulations is a relatively inexpensive approach to reproduce the qualitative behavior from experiment.

---

24) In the bilayer network, the Pt–H_2O bond energy is 0.38 eV, and the two hydrogen bonds bring 0.28 eV each. In total, removing H_2O^{ad} from a water bilayer network on top of the Pt(111) surface requires 0.94 eV [61–64].

- **OOH Formation:** For the OOH^{ad} formation pathway, the initial steps are again equivalent to the O$_2$ dissociation, but OOH^{ad} formation is now 0.22 eV *uphill* in water solvent compared to −0.60 eV *downhill* in gas phase. The OOH^{ad} formation barrier is greatly increased from 0.05 eV in gas phase to 0.94 eV in solvent. The solvent substantially destabilizes both the transition state and OOH^{ad}. Water has a small effect on the dissociation of OOH^{ad} (0.74 eV in gas phase and 0.62 eV in solvent). This reaction step is exothermic by −1.94 eV, which is −0.84 eV lower than in gas phase. Here again, OOH^{ad} formation should be the rate-determining process for this mechanism.
- **HOOH Formation:** Just as the OOH^{ad} formation step was heavily influenced by solvation, so is the HOOH^{ad} formation step. HOOH^{ad} formation in gas phase had a barrier of 0.47 eV, which increases to 1.23 eV in solvent. Similarly, due to the minor charge transfer between HOOH^{ad} and the Pt surface, HO−OH dissociation is hardly affected by solvation just as O−OH was not (gas-phase barrier = 0.46 eV; solvent-phase barrier = 0.43 eV). Overall, HOOH^{ad} formation is the RDS for this mechanism.

Overall Langmuir–Hinshelwood reaction mechanisms appear to be drastically influenced by solvation effects, particularly when H^{ad} reacts with another adsorbate. In gas phase, we found that HOOH^{ad} formation has the lowest barriers, thus being the overall RDS (0.47 eV). In solvent, OOH^{ad} formation is the overall RDS (0.62 eV), and HOOH^{ad} formation is unfavorable. The key energy barriers of these ORR mechanisms (O$_2^{ad}$ dissociations, OOH^{ad} formation, and HOOH^{ad} formation) are influenced by solvent as much as 0.89 eV, greatly altering interpretations from available literature data on the ORR activation energy [59]. In summary, by including the electronic effects of the surrounding water solution, we found a preference against HOOH^{ad} formation; however, both the O$_2^{ad}$ dissociation and the OOH^{ad} formation reactions should be competitive under these simulation conditions. Therefore, compared to the gas-phase system, considering the water environment not only changed the energies and barriers but also led to a different reaction mechanism to be favorable.

1.3.2.2 Eley–Rideal Reactions

Eley–Rideal mechanisms are surface reactions where a surface intermediate reacts with the solvent. In terms of the ORR, Eley–Rideal mechanisms involve hydrogenations from protons in the electrolyte. We use the initial thermodynamic resting state of hydrogen gas in solution as referenced to hydrogen gas out of convenience. This reference eschews problems and complexity with treating the electronic structure or protons in aqueous solution, an especially problematic simulation for theoretical methods. Indeed, a rigorous simulation for a full electrochemical system should consider a kinetic chemical equilibrium between H^{ad} and H$^+$; however, thermodynamic energies can still be reported. In Eley–Rideal mechanisms, hydrogen enters the simulation in the transition state for the hydrogenation process. Intermediates found in Langmuir–Hinshelwood mechanisms are exactly the same.

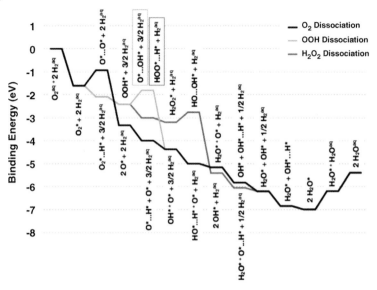

Figure 1.12 Eley–Rideal-type reactions in water solvent.

Simulations on the dissociation steps of O_2^{ad}, OOH^{ad}, and $HOOH^{ad}$ provide substantially different reaction details than those previously reported, where protons from the electrolyte first adsorb on the surface and then further react as H^{ad}. The rate-determining step for O_2^{ad} dissociation in solvent phase was OH^{ad} formation (1.14 eV); however, in an Eley–Rideal-type mechanism, this process has no barrier partly due to its reference at the highly acidic standard state (pH = 0) of the electrolyte, whereby protons should rapidly and easily protonate O^{ad} species. Thus, the resulting RDS of the O_2^{ad} dissociation pathway now becomes the dissociation of O_2^{ad} itself (0.68 eV), which agrees with the expectations from electrochemistry. For the OOH^{ad} formation pathway, the RDS still remains to be the O−OH dissociation (0.62 eV), and for the $HOOH^{ad}$ formation pathway also the HO−OH dissociation stays to be the RDS (0.43 eV). Overall, treatment of Eley–Rideal mechanisms shows that all three pathways are potential candidates to be possible reaction mechanisms for the ORR, with a slight preference for the $HOOH^{ad}$ formation pathway at zero potential (0.0 V). These results are summarized in Figure 1.12.

1.3.3
Including Thermodynamical Quantities

As we had shown in Section 2.5, quantum mechanical electronic energies alone should not be compared with experimental observables taken from measurements on macroscopic systems under ambient conditions. Zero-point energies and free energy contributions must be added to electronic energies of the solvated system. We use first-order approximations for these values obtained from statistical thermodynamics and the ideal gas approximation at room temperature $T = 298$ K.

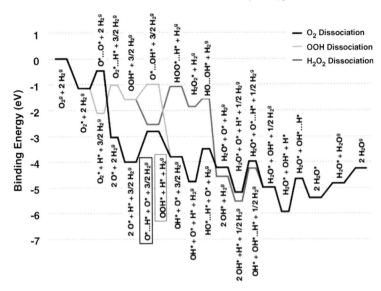

Figure 1.13 Langmuir–Hinshelwood reactions in water solvent under ambient conditions (including thermal contributions for $T = 298$ K).

1.3.3.1 Langmuir–Hinshelwood and Eley–Rideal Mechanisms

As expected, we do not find many notable differences due to free energy contributions in terms of barrier heights. However, we find that free energy contributions indeed make a very large difference when accounting for water formation at the end of the ORR reactions. In gas phase, the overall ORR reaction is exothermic by -5.00 eV (without E_{ZPE} corrections). In solvent, this value is even more exothermic, -5.36 eV. Including free energy contributions brings this value back even higher than the gas phase value, -4.65 eV. Thermal energy contributions also have a substantial role in favoring H^{ad} on the surface by 0.20 eV, generally shifting reaction intermediates that include H atoms slightly higher than were found with just solvation. We summarize this calculated data in Figure 1.13.

- **O_2 Dissociation:** In the O_2^{ad} dissociation mechanism, dissociation of O_2^{ad} was -1.71 eV downhill in solvent, but it is slightly lower when accounting free energies as well (-1.89 eV). The barrier for this process is calculated similar to the barrier in solvation (0.68 eV). OH^{ad} formation was energetically neutral in solvent with a barrier of 1.14 eV; however, free energy contributions make this value 0.20 eV uphill overall with a similar barrier (1.19 eV). A similar trend is seen with the final formation of H_2O^{ad}. The process is overall $+0.24$ eV in solvent with a 1.10 eV barrier. When accounting for free energies, the same reaction has an overall ΔG_{298K}^{solv} of $+0.55$ eV with a barrier of 1.27 eV.
- **OOH Formation:** In the OOH^{ad} formation mechanism, the energy required to form OOH^{ad} via the Langmuir–Hinshelwood-type reaction of O_2^{ad} and H^{ad} was

0.22 eV in solvent, but including thermal corrections this increases to 0.52 eV. The barrier for this process in solvent alone was 0.94 eV, but with free energy contributions it is slightly higher (1.09 eV). Finally, O–OH dissociation is essentially the same in both solvent (0.62 eV) and after additionally considering thermal corrections (0.59 eV).

- **HOOH Formation:** In the HOOH^{ad} formation mechanism, the energy to further hydrogenize OOH^{ad} via a Langmuir–Hinshelwood-type reaction required 0.27 eV in solvent and 0.68 eV with thermal corrections. Comparable to the formation of OOH^{ad}, the energy barrier to form HOOH^{ad} from OOH^{ad} and H^{ad} is slightly higher when including thermal corrections (1.09 eV compared to 0.94 eV with solvation only). Finally, the barrier for HOOH^{ad} dissociation is slightly lower than in our previous simulations where we included only the water solvent but neglected thermal corrections: 0.31 eV compared to 0.43 eV before.

It is clear that Langmuir–Hinshelwood mechanisms are affected by the motional freedom of H^{ad} on the surface. This stability rising from the presence of H^{ad} impacts relative thermodynamics by making other species containing H atoms relatively less stable by $\approx$ 0.2 eV. Reaction barriers for this mechanism are not greatly impacted by free energy effects, however.

Eley–Rideal-type reaction profiles with thermal corrections display the same trend shown as solvation-only simulations, which are shown in Figure 1.14. We note that our reliance on the standard state of protons (pH $=$ 0) strongly impacts these reaction profiles. After establishing the quantum mechanical energies for these processes, however, one can implement these barriers into a kinetic master equation that

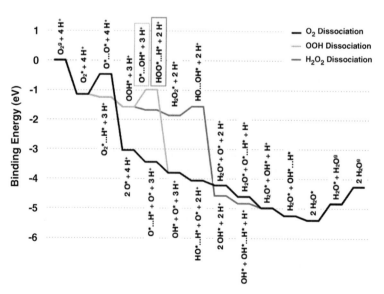

Figure 1.14 Eley–Rideal reactions in water solvent under ambient conditions (including thermodynamic contributions for $T = 298$ K).

depends on the concentration of protons. This type of analysis would be yet another step in moving up the hierarchy of multiscale modeling (see Figure 1.1).

1.3.4
Including an Electrode Potential

Realistic electrochemical systems (such as fuel cells) operate under the influence of an external electrostatic potential difference between the working and the counter electrode. In order to account for the presence of an electrode potential U, different attempts exist to model and understand the structure and properties of systems under electrochemical conditions. An overview can be found in the following reviews [65–68], articles [69–78], and references therein. For the different works mainly experimental input, semiempirical approaches, or quite simplified models are used. The presence of the electrode potential is either neglected or introduced by charging the electrode surface or applying an external electric field. While most of the theoretical studies disregard the presence of an electrode potential, some try to consider its influence on catalytic reactions. For instance, the group of Nørskov [74, 79] studied the hydrogen evolution reaction (HER) and oxygen reduction reaction on different electrodes whose Fermi energies were shifted by the value of the electrode potential. Focusing more on the atomistic structure of the interface, the group of Neurock [76] performed *ab initio* molecular dynamics simulations on charged electrodes surrounded by water. For compensation, a countercharge was located at a certain distance from the electrode surface, trying to mimic the potential profile within the interfacial region.

In order to account for the electrode potential, in the following this effect is approximated by shifting energy levels by a constant value $+e \cdot U$ for every process where a hydrogen (or proton) is dissociated from or attached to a surface species. This influences not only the energies of particular intermediates but also their corresponding transition states, that is, the dissociation or association barriers. Based on this approach, which should be capable to reproduce the overall behavior, Figure 1.15 shows the most favorable (lowest ΔG_{298K} barriers) Langmuir–Hinshelwood and Eley–Rideal processes.

The energy plot makes apparent that both classes of reactions (LH and ER) should be possible at electrode potentials near 1.23 eV, the reduction potential of the ORR established by the Nernst equation. While individual Eley–Rideal reaction barriers appear to be lower than Langmuir–Hinshelwood barriers, the energies for the O^{ad} and OH^{ad} species, which are influential for both classes of reactions, are almost identical and can therefore be expected to be competitive depending on the environmental conditions within the electrochemical system.

In summary, it appears that ORR reaction processes can be reduced to elementary forms of Langmuir–Hinshelwood and Eley–Rideal mechanisms, and treated with high-quality quantum mechanical approaches to obtain relevant thermodynamic stabilities of all species. Further studying the extent that these processes are coupled kinetically is the next logical step to provide insight into the complex ORR mechanism.

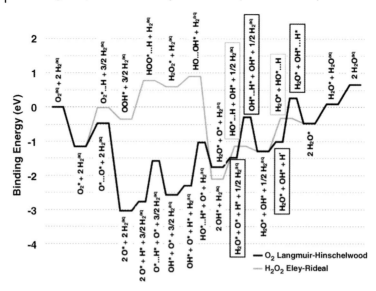

Figure 1.15 Selected (most favorable) reactions under ambient conditions (including solvent and thermal corrections) for an electrode potential of 1.23 V (versus RHE).

1.4
Conclusions

The enormous variety of possible surface reactions reveals many interesting intricacies regarding the ORR mechanisms. Using density functional theory to investigate different mechanisms of the ORR, in this chapter we showed the possibilities of modern *ab initio* modeling and tried to make the readers aware that conclusions might change after including environmental effects and thus should be drawn with special care. Among the three main ORR mechanisms (i.e., O_2^{ad} dissociation, OOH^{ad} formation, and $HOOH^{ad}$ formation), we found that $HOOH^{ad}$ formation is the preferred process in gas phase. When including water solvation as an environmental effect, the reaction paths were modified, leading to drastic changes in the energetics and a nearly identical preference for the O_2^{ad} dissociation and the OOH^{ad} formation mechanisms, but with a blocking of the hydrogen peroxide pathway. Interestingly, inclusion of solvent permits different classes or reaction mechanisms centered around electron transfers and protonations at different electrode potentials. Eley–Rideal variants of the previously investigated mechanisms are all substantially lower in energy at an electrode potential of 0 V. However, inclusion of thermal energy contributions due to ambient conditions, as well an approximate influence of the electrode potential, resulted in a picture showing that both Langmuir–Hinshelwood and Eley–Rideal mechanisms could be at play. Evaluation of the kinetics of all paths should further elucidate the complicated nature of the ORR mechanism.

Acknowledgments

The authors gratefully acknowledge support from the Alexander von Humboldt Foundation (AvH) and the Deutsche Forschungsgemeinschaft (DFG) within the framework of the Emmy-Noether Program.

References

1. Born, M. and Oppenheimer, J.R. (1927) *Ann. Phys.*, **84**, 457.
2. Jensen, F. (1999) *Introduction to Computational Chemistry*, John Wiley & Sons, Inc., New York.
3. Groß, A. (2003) *Theoretical Surface Science*, Springer, Berlin.
4. Hartree, D.R. (1927) *Proc. Cambridge Phil. Soc.*, **84**, 457.
5. Fock, V.A. (1930) *Z. Phys.*, **61**, 126.
6. Slater, J.C. (1929) *Phys. Rev.*, **34**, 1293.
7. Dawydow, A.S. (1978) *Quantenmechanik*, Deutscher Verlag der Wissenschaften, Berlin.
8. Dirac, P.A.M. (1930) *Proc. Cambridge Phil. Soc.*, **26**, 376.
9. Bloch, F. (1929) *Z. Phys.*, **57**, 545.
10. Slater, J.C. (1951) *Phys. Rev.*, **81**, 385.
11. Slater, J.C. (1972) *Advances in Quantum Chemistry*, vol. 6, Academic Press, New York.
12. Gaspar, R. (1954) *Acta Phys. Akad. Sci. Hungaria*, **3**, 263.
13. Schwarz, K. (1972) *Phys. Rev. B*, **5**, 2466.
14. Szabo, A. and Ostlund, N. (1989) *Modern Quantum Chemistry*, Dover, New York.
15. Bartlett, R.J. (1989) *J. Chem. Phys.*, **93**, 1697.
16. Møller, C. and Plesset, M.S. (1934) *Phys. Rev.*, **46**, 618.
17. Hohnberg, P. and Kohn, W. (1964) *Phys. Rev.*, **136**, 846.
18. Kohn, W. and Sham, L. (1965) *Phys. Rev.*, **140**, 1133.
19. Thomas, L.H. (1927) *Proc. Cambridge Phil. Soc.*, **23**, 542.
20. Fermi, E. (1927) *Rend. Accad. Lincei*, **6**, 602.
21. Tao, J., Perdew, J.P., Staroverov, V.N., and Scuseria, G.E. (2003) *Phys. Rev. Lett.*, **91**, 146401.
22. Perdew, J.P., Ruzsinszky, A., Tao, J., Staroverov, V.N., Scuseria, G.E., and Csonka, G.I. (2005) *J. Chem. Phys.*, **123**, 062201.
23. Perdew, J.P., Burke, K., and Ernzerhof, M. (1996) *Phys. Rev. Lett.*, **77**, 3865.
24. Slater, J.C. (1974) *Quantum Theory of Molecules and Solids, Vol. 4. The Self-Consistent Field for Molecules and Solids*, McGraw-Hill, New York.
25. Becke, A.D. (1988) *Phys. Rev. A*, **38**, 3098.
26. Vosko, S.H., Wilk, L., and Nusair, M. (1980) *Can. J. Phys.*, **58**, 1200.
27. Lee, C., Yang, W., and Parr, R.G. (1988) *Phys. Rev. B*, **37**, 785.
28. Press, W.H., Teukolsky, S.A., Vetterling, W.T., and Flannery, B.P. (1996) *Numerical Recipes in Fortran 77: The Art of Scientific Computing*, 2nd edn, Cambridge University Press.
29. Gao, J. (1996) *Acc. Chem. Res.*, **29**, 298.
30. Gao, J. (1996) *Rev. Comput. Chem.*, **7**, 119.
31. Car, R. and Parrinello, M. (1985) *Phys. Rev. Lett.*, **55**, 2471.
32. Field, M.J., Bash, P.A., and Karplus, M. (1990) *J. Comput. Chem.*, **11**, 700.
33. Gao, J. and Xia, X. (1992) *Science*, **258**, 631.
34. Dupuis, M., Schenter, G.K., Garrett, B.G., and Arcia, E.E. (2003) *J. Mol. Struct. Theochem.*, **632**, 173.
35. Moriarty, N.W. and Karlstrom, G. (1996) *J. Phys. Chem.*, **100**, 17791.
36. Moriarty, N.W. and Karlstrom, G. (1997) *J. Chem. Phys.*, **106**, 6470.
37. Tomasi, J. and Persico, M. (1994) *Chem. Rev.*, **94**, 2027.
38. Cramer, C.J. and Truhlar, D.G. (1999) *Chem. Rev.*, **99**, 2161.
39. Cossi, M., Mennucci, B., and Cammi, R. (1996) *J. Comput. Chem.*, **17**, 57.
40. Sanner, M.F., Olson, A.J., and Spehner, J.C. (1996) *Biopolymers*, **38**, 305.
41. Eisenberg, D. and McLachlan, A.D. (1986) *Nature*, **319**, 199.

42 Still, W.C., Tempczyk, A., Hawley, R.C., and Hendrickson, T. (1990) *J. Am. Chem. Soc.*, **112**, 6127.

43 Qiu, D., Shenklin, P.S., Hollinger, F.P., and Still, W.C. (1997) *J. Phys. Chem. A*, **101**, 3005.

44 Payne, M.C., Teter, M.P., Allan, D.C., Arias, T.A., and Joannopoulos, J.D. (1992) *Rev. Mod. Phys.*, **64** (4), 1045.

45 Monkhorst, H.J. and Pack, J.D. (1976) *Phys. Rev. B*, **13**, 5188.

46 Jacob, T., Muller, R.P., and Goddard, W.A., III (2003) *J. Phys. Chem. B*, **107**, 9465.

47 Jacob, T., Anton, J., Fritzche, S., Sepp, W.-D., and Fricke, B. (2002) *Phys. Lett. A*, **300**, 71.

48 Jacob, T., Geschke, D., Fritzsche, S., Sepp, W.-D., Fricke, B., Anton, J., and Varga, S. (2001) *Surf. Sci.*, **486**, 194.

49 Jaguar 4.2/5.0, Schrödinger Inc., Portland (2000/2002).

50 Jacob, T. (2006) *Fuel Cells*, **6**, 159.

51 Jacob, T. and Goddard, W.A., III (2006) *Chem. Phys. Chem.*, **7**, 992.

52 Stipe, B.C., Rezaei, M.A., Ho, W., Gao, S., Mersson, M., and Lundquist, B.I. (1997) *Phys. Rev. Lett.*, **78**, 4410.

53 Ho, W. (1998) *Science*, **279**, 1907.

54 Ho, W. (1998) *Acc. Chem. Res.*, **31**, 567.

55 Venkatachalam, S. and Jacob, T. (2009) *Density Functional Theory Applied to Electrocatalysis* in *Handbook of Fuel Cells: Advances in Electrocatalysis, Materials, Diagnostics and Durability*, vol. 5 & 6 (eds W. Vielstich, H.A. Gasteiger, and H. Yokokawa), John Wiley & Sons Ltd., Chichester, UK, pp. 133–151.

56 Keith, J.A. and Jacob, T. (2009). *Modeling Electrocatalysis from First Principles* in *Modern Aspects of Electrochemistry, Number 46: Advances in Electrocatalysis* (eds P. Balbuena and V. Subramanian), Springer Heidelberg, Germany.

57 Parker, D.H., Bartram, M.E., and Koel, B.E. (1989) *Surf. Sci.*, **217**, 489.

58 Thiel, P.A. and Madey, T.E. (1987) *Surf. Sci. Rep.*, **7**, 211.

59 Neyerlin, K.C., Gu, W., Jorne, J., and Gasteiger, H.A. (2006) *J. Electrochem. Soc.*, **153**, A1955.

60 Olsen, R.A., Kroes, G.J., and Baerends, E.J. (1999) *J. Chem. Phys.*, **111**, 11155.

61 Jacob, T. and Goddard, W.A., III (2004) *J. Am. Chem. Soc.*, **126**, 9360.

62 Meng, S., Xu, L.F., Wang, E.G., and Gao, S. (2002) *Phys. Rev. Lett.*, **89**, 176104.

63 Ogasawara, O., Brena, B., Nordlund, D., Nyberg, M., Pelmenschikov, A., Petterson, L.G.M., and Nilsson, A. (2002) *Phys. Rev. Lett.*, **89**, 276102.

64 Ruscic, B., Wagner, A.F., Harding, L.B., Asher, R.L., Feller, D., Dixon, D.A., Peterson, K.A., Song, Y., Qian, X.M., Ng, C.Y., Liu, J.B., Chen, W.W., and Schwenke, D.W. (2002). *J. Phys. Chem. A*, **106**, 2727.

65 Schmickler, W. (1996) *Chem. Rev.*, **96**, 3177.

66 Schmickler, W. (1999) *Annu. Rep. Prog. Chem., Sect. C*, **95**, 117.

67 Koper, M.T.M., van Santen, R.A., and Neurock, M. (2003) *Catalysis and Electrocatalysis at Nanoparticle Surfaces* (eds E. Savinova, C. Vayenas, and A. Wieckowski), Marcel Dekker, New York.

68 Koper, M.T.M. (2004) *Ab Initio Quantum-Chemical Calculations in Electrochemistry* in *Modern Aspects of Electrochemistry No. 36* (eds C. Vayenas, B. Conway, R. White, and M. Gamboa-Adelco), Springer, New York.

69 Nazmutdinov, R.R. and Shapnik, M.S. (1996) *Electrochim. Acta*, **41**, 2253.

70 Halley, J.W., Schelling, P., and Duan, Y. (2000) *Electrochim. Acta*, **46**, 239.

71 Vassilev, P., Hartnig, C., Koper, M.T.M., Frechard, F., and van Santen, R.A. (2001) *J. Chem. Phys.*, **115**, 9815.

72 Haftel, M.I. and Rosen, M. (2003) *Surf. Sci.*, **523**, 118.

73 Feng, Y.J., Bohnen, K.P., and Chan, C.T. (2005) *Phys. Rev. B*, **72**, 125401.

74 Kitchin, J.R., Nørskov, J.K., Barteau, M.A., and Chen, J.G. (2004) *J. Chem. Phys.*, **120**, 10240.

75 Gunnarsson, M., Abbas, Z., Ahlberg, E., and Nordholm, S. (2004) *J. Coll. Interf. Sci.*, **274**, 563.

76 Taylor, C.D., Wasileski, S.A., Filhol, J.S., and Neurock, M. (2006) *Phys. Rev. B*, **73**, 165402.

77 Jacob, T. (2007) *Electrochim. Acta*, **52**, 2229.

78 Jacob, T. (2007) *J. Electroanal. Chem.*, **607**, 158.

79 Rossmeisl, J., Nørskov, J.K., Taylor, C.D., Janik, M.J., and Neurock, M. (2006) *J. Phys. Chem. B*, **110**, 21833.

2
Dynamics of Reactions at Surfaces
Axel Groß

2.1
Introduction

Chemical reactions correspond to dynamical events involving bond making and bond breaking processes [1]. In the previous chapter, we have seen how the energetics of catalytic reactions can be determined from first principles using electronic structure theory. Due to the constant improvement in the computer power and the development of efficient algorithms for electronic structure calculations, mainly based on density functional theory (DFT), it has become possible to map out entire potential energy surfaces (PES) of complex catalytic reactions [2, 3]. However, this static information is often not sufficient to really understand how a reaction proceeds. Furthermore, in the experiment the potential energy surface is never directly measured but just reaction rates and probabilities that are a consequence of the interaction potential.

Thus for a thorough understanding of reaction mechanisms, dynamical simulations can be very helpful. Calculating the time evolution of processes also allows a genuine comparison between theory and experiment since experimentally accessible quantities such as reaction or adsorption probabilities can be directly derived from the simulations. Thereby, dynamical simulations also provide a reliable check on the accuracy of the calculated PES on which the dynamical simulations are based.

In Figure 2.1, a schematic two-dimensional PES is shown as a function of the distance of the reactants from a catalyst surface and of some molecular coordinate. It provides an illustration how a catalyst works. A reaction might be hindered by a relatively large barrier in the gas phase or in solution. However, for the adsorbed reactants, the barrier can be much lower. Thus, the catalyst provides a detour in the multidimensional configuration space with a lower barrier that can be much more easily traversed.

It should be noted that the catalyst not only provides reaction routes with smaller barriers but also acts as a thermal bath that can provide and dissipate energy. Therefore, many details of heterogeneous catalytic reactions can be understood on the basis of concepts derived from equilibrium thermodynamics. For example, rate constants can

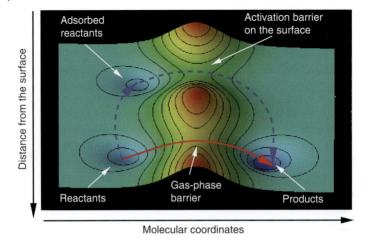

Figure 2.1 Schematic illustration of the role of a catalyst employing a two-dimensional representation of the potential energy surface. A catalyst provides a detour in the multidimensional PES (dashed line) with a lower activation barrier for the adsorbed species than in the gas phase (or in solution).

be estimated with transition state theory [4] where the assumption of strong friction is crucial [5]. This also means that the activity and selectivity of a heterogeneous catalytic reactions depend much more strongly on activation barrier heights, which enter the reaction rate constants exponentially, than on dynamical effects. This means that the length and the curvature of the detour in the multidimensional configuration space illustrated in Figure 2.1 hardly matters, it is just the reduction in the activation barrier height on the catalyst surface that accelerates the reaction.

Furthermore, activated processes usually correspond to rare events so that their dynamical description would involve the simulation of many unsuccessful events which is rather time consuming and thus not reasonable. For such processes, coarse-grained dynamical methods such as kinetic Monte Carlo simulations [6, 7] that are also presented in this book are much more appropriate.

Still, there are certain processes in catalytic reactions whose probability cannot be determined using transition state theory. The most prominent example is atomic and nondissociative molecular adsorption: here the adsorption probability strongly depends on the dissipation of the kinetic energy of the impinging atom or molecule that can be determined only in dynamical simulations [8].

In this chapter, I will show what kind of information about heterogeneous catalytic reactions can be gained by studying the dynamics of reactions at surfaces. I will first discuss the theoretical and computational methods required to perform dynamical simulations. Most of the dynamical simulations are based on the Born–Oppenheimer or *adiabatic* approximation taking advantage of the difference in the mass between nuclei and electrons. Furthermore, in principle the atomic motion should be described by a quantum mechanical treatment, but it turns out that classical mechanics is often sufficient.

In order to perform dynamical simulations, the potential energy surface and its gradients are needed at arbitrary points of the configuration space. However, electronic structure calculations yield energies only at discrete points of the configuration space. Hence, either one uses reliable interpolation methods for a continuous representation of the PES on which the dynamical simulations are performed or one employs methods where the gradients are calculated "on the fly." Such *ab initio* molecular dynamics (AIMD) simulations are rather time consuming, but I will show that it has become possible to determine a sufficient number of trajectories in order to obtain statistically significant results [9].

Most of the examples I will show are related to adsorption processes and simple reactions on surfaces. As mentioned above, such processes can usually be described employing the Born–Oppenheimer approximation. However, there are important examples of catalytically relevant reactions where this approximation breaks down. The dynamical treatment of electronically nonadiabatic reactions at surfaces is rather complex, but I will show that such reactions can also be treated from first principles nowadays.

2.2
Theoretical and Computational Foundations of Dynamical Simulations

In the realm of chemistry and solid-state physics, there is only one basic interaction that is relevant, namely, the electrostatic interaction between the charged nuclei and the electrons. Thus, together with the kinetic energy of the nuclei and the electrons only the nucleus–nucleus, nucleus–electron, and electron–electron electrostatic interaction energy enter the Hamiltonian describing catalytic systems:

$$H = T_{\text{nucl}} + T_{\text{el}} + V_{\text{nucl-nucl}} + V_{\text{nucl-el}} + V_{\text{el-el}}. \tag{2.1}$$

Relativistic effects are usually negligible except for the heaviest elements where the high charge of the nucleus can accelerate the electrons to velocities close to the speed of light. Thus, typically catalytic systems are described by the nonrelativistic Schrödinger equation

$$H\Phi(\mathbf{R}, \mathbf{r}) = E\Phi(\mathbf{R}, \mathbf{r}), \tag{2.2}$$

where $\mathbf{R}$ and $\mathbf{r}$ denote *all* nuclear and electronic coordinates, respectively. In principle, one is ready here because a complete knowledge about the system can be gained by solving the Schrödinger equation and determining the eigenfunctions of the many-body Hamiltonian taking into consideration the proper quantum statistics of the particles. Unfortunately, the solution of the many-body Schrödinger equation in closed form is not possible. Thus, a hierarchy of approximations is needed in order to make the solution feasible.

The first approximation that is typically used is the Born–Oppenheimer or adiabatic approximation [10]. Its central idea is the separation in the timescale of processes involving electrons and atoms because of their large mass mismatch.

Typically, at the same kinetic energy electrons are 10^2–10^3 times faster than the nuclei. Hence, one assumes that the electrons follow the motion of the nuclei instantaneously.

In practice, one splits up the full Hamiltonian and defines the electronic Hamiltonian H_{el} for fixed nuclear coordinates **R** as follows:

$$H_{el}(\{\mathbf{R}\}) = T_{el} + V_{nucl-nucl} + V_{nucl-el} + V_{el-el}. \tag{2.3}$$

In Eq. (2.3), the nuclear coordinates {**R**} do not act as variables but as parameters defining the electronic Hamiltonian. The Schrödinger equation for the electrons for a given fixed configuration of the nuclei is then

$$H_{el}(\{\mathbf{R}\})\Psi(\mathbf{r};\{\mathbf{R}\}) = E_{el}(\{\mathbf{R}\})\Psi(\mathbf{r};\{\mathbf{R}\}). \tag{2.4}$$

This is the basic equation of quantum chemistry that is solved by electronic structure codes, using either wave function-based methods or density functional theory, as described in the previous chapter. In the Born–Oppenheimer approximation, the eigenenergies $E_{el}(\{\mathbf{R}\})$ of the electronic Schrödinger equation as a function of the nuclear coordinates {**R**} define the potential for the nuclear motion. $E_{el}(\{\mathbf{R}\})$ is therefore called the Born–Oppenheimer energy surface. Minima of the Born–Oppenheimer surface correspond to stable and metastable configurations of the system, for example, energy minimum structures of molecules or adsorption sites on a surface, whereas saddle points are related to activation barriers for chemical reactions or diffusive motion.

The validity of the Born–Oppenheimer approximation is hard to prove. Still it has been very successful in providing theoretical description of chemical reactions. Qualitatively, two regimes can be identified in which the Born–Oppenheimer approximation should be justified. If there is a large energy gap between the highest occupied molecular orbital (HOMO) and the lowest unoccupied molecular orbital (LUMO) in the case of molecules or between the valence and the conduction band in the case of solids, then electronic transitions will be rather improbable and the system is likely to stay in the electronic ground state. If, on the other hand, there is no bandgap, like in the case of metals, but there are many coupled electronic states allowing electronic transitions with arbitrarily small excitation energies, then the strong coupling of the electronic states in the broad conduction band will lead to short lifetimes of excited states and thus to a fast quenching of these states. The Born–Oppenheimer approximation breaks down if either electronic states are directly excited like in photochemistry or if there are few weakly coupled electronic states so that there is a small probability for the system to relax to the electronic ground state. Such systems require a special treatment, as will be shown in Section 2.8.

Within the Born–Oppenheimer approximation, the quantum dynamics can be determined by solving the atomic Schrödinger equation

$$\{T_{nucl} + E_{el}(\mathbf{R})\}\Lambda(\mathbf{R}) = E_{nucl}\Lambda(\mathbf{R}), \tag{2.5}$$

where $E_{el}(\mathbf{R})$ acts as the potential. However, often quantum effects in the atomic motion can be neglected and the classical equation of motion are solved for the atomic motion:

$$M_I \frac{\partial^2}{\partial t^2} \mathbf{R}_I = -\frac{\partial}{\partial \mathbf{R}_I} E_{\text{el}}(\mathbf{R}). \quad (2.6)$$

This is the basis for molecular dynamics (MD) simulations. The interaction potential $E_{\text{el}}(\mathbf{R})$ does not necessarily have to be derived from first principles. In particular for biological systems, there are parameterized interaction potentials available, the so-called *force fields* that are often derived from a combination of experimental and theoretical data. However, such *classical* potentials are usually not well suited to describe bond making and bond breaking processes, as they occur in heterogeneous catalytic reactions. Hence, I will here discuss only dynamical simulations that are based on first-principles calculations.

2.3
Interpolation of Potential Energy Surfaces

In order to perform *quantum* dynamical (QD) simulations, a continuous presentation of the potential energy surface is needed since the wave functions are delocalized and always probe a certain extended area of the PES at any time. On the other hand, classical molecular dynamics simulations on a suitable analytical representation of a potential energy surface can be extremely fast. First-principles total energy calculations, however, just provide total energies for discrete configurations of the nuclei. Hence, it is desirable to adjust the first-principles energies to an analytical or numerical continuous representation of the PES. This is a not an easy task. The representation should be flexible enough to accurately reproduce the *ab initio* input data, yet it should have a limited number of parameters so that it is still controllable. Furthermore, a good parameterization should not only accurately interpolate between the actually calculated points but also give a reliable extrapolation to regions of the PES that have actually not been determined by the first-principles calculations.

A straightforward approach is to assume a certain analytical form of the interaction potential $V(\mathbf{R})$ depending on a certain number of parameters [11, 12] and then adjust these parameters in such a way that the root mean square error (RMSE)

$$\Delta E_{\text{RMSE}} = \sqrt{\frac{1}{N}\sum_i^N (V(\mathbf{R}_i) - E_{\text{el}}(\mathbf{R}_i))^2} \quad (2.7)$$

is minimal, where $\mathbf{R}_i$ are N configurations for which the energy has been evaluated by first-principles calculations. Such an approach has, for example, been used to interpolate the PES of the interaction of H_2 with a Pd(100) surface covered by a quarter monolayer of sulfur within a S(2 × 2)/Pd(100) geometry [13, 14]. Two so-called elbow plots of this PES, that is, two-dimensional cuts through the multidimensional PES as a function of the interatomic H−H distance and the H_2 center of

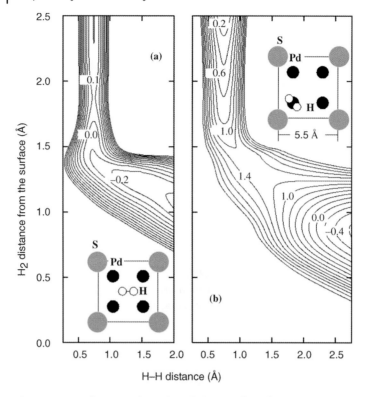

Figure 2.2 Two-dimensional cuts through the potential energy surface of $H_2/S(2 \times 2)/Pd(100)$ derived from DFT calculations as a function of the H–H distance and the H_2 distance from the surface. The insets illustrate the molecular orientation and lateral center of mass position. The contour spacing in (a) is 0.1 eV, while it is 0.2 eV in (b). Adapted from Ref. [13].

mass distance from the surface for different lateral positions and orientations of the H_2 molecule, are shown in Figure 2.2. Typically, these interaction potentials are strongly corrugated and anisotropic, that is, they depend on the lateral position and orientation of the molecule, even for smooth metal surfaces [15].

If more than the molecular degrees of freedom should be considered in a parameterization of an *ab initio* PES, analytical forms become very complicated and cumbersome. Furthermore, sometimes analytical expression introduce some artificial symmetries into the description of the PES that can have quite a significant influence on dynamical results [9]. There are some numerical schemes that avoid the restrictions of analytical expansions. Neural networks can fit, in principle, any real-valued, continuous function to any desired accuracy. They have already been successfully applied to interpolate *ab initio* potential energies describing chemical reactions [16–20]. Another approach is to use a modified Taylor expansion of the PES around the points calculated from first principles [21]. One disadvantage of these numerical schemes is that there is no physical insight used as an input in this parameterization. Hence, the parameters of the expansions do not reflect any physical or chemical property which usually leads to a large number of unknowns in the parameterization.

There are some interpolation methods that are not purely numerical. For example, the corrugation reducing procedures [22–25] take advantage of the observation that most of the corrugation in molecular potential energy surfaces can be derived from the atom–surface interaction. In this method, the interaction of both the atomic and the molecular species with the surface is determined by first-principles calculations. Then, a three-dimensional reference function is constructed from the atomic data that is subtracted from the molecular potential energy surface. The remaining function is much smoother than the original potential energy surface and therefore much easier to fit, and furthermore, the interpolated PES also reflects the correct symmetry of the system. This method has been used successfully for quite a number of interaction potentials [22–24, 26, 27]. Still, this method cannot be easily extended to include surface degrees of freedom.

As an intermediate method between purely numerical schemes and full first-principles calculations, the representation of the *ab initio* interaction potential by a tight binding (TB) formalism was suggested [28]. Tight binding methods are more time consuming than an analytical representation or a neural network since they require the diagonalization of matrices. However, due to the fact that the quantum mechanical nature of bonding is taken into account [29], a smaller number of *ab initio* input points is needed in order to perform a good interpolation [30]. Furthermore, tight binding schemes can even be used for the extrapolation of first-principles results since the parameters of the tight binding scheme, the Slater–Koster integrals [31], have a well-defined physical meaning. This makes, for example, the inclusion of lattice vibrations possible [32, 33], as will be shown in this chapter.

Finally, it should be emphasized that it is not trivial to judge the quality of the fit to a first-principles PES. Using just the integrated value of the root mean square error Eq. (2.7), is often not sufficient since some points of the potential energy surface, for example, activation barriers, are more important than others. This can be taken into account by introducing weighting factors in the formulation of ΔE_{RMSE}.

2.4
Quantum Dynamics of Reactions at Surfaces

In order to treat the quantum dynamics of reactions at surfaces, either the time-dependent Schrödinger equation

$$i\hbar \frac{\partial}{\partial t} \Psi(\vec{R}, t) = H\Psi(\vec{R}, t) \tag{2.8}$$

or the time-independent Schrödinger equation

$$H\Psi(\vec{R}) = E\Psi(\vec{R}, t) \tag{2.9}$$

may be solved. Both approaches are equivalent and should lead to the same results. The time-dependent Schrödinger equation is typically solved on a numerical grid using the wave-packet formalism [34–37]. In the time-independent formulation, the wave function is usually expanded in some suitable set of eigenfunctions leading to the so-called *coupled channel equations* [38, 39]. High-dimensional quantum dynamical simulations are computationally rather expensive. Still it is nowadays possible to describe the interaction dynamics of H_2 with metal surfaces including all six degrees of freedom of the H_2 molecule dynamically [13, 40–43], even at precovered [13, 44] and stepped [45] surfaces. However, the computational effort grows exponentially with the number of degrees of freedom considered, and hence it is almost impossible to include substrate degrees of freedom realistically in these quantum simulations.

One of these high-dimensional quantum studies was devoted to the H_2 dissociation on sulfur-precovered surfaces [13, 44]. This study is relevant for heterogeneous catalysis since sulfur is known to act as a catalyst poison, that is, its presence reduces the catalytic activity of a substrate. DFT calculations have shown that the poisoning of sulfur is due to a combination of direct and indirect effects [14, 46]. Since sulfur is strongly bound to Pd(100), it first of all blocks sites at which no further reaction can take place. But then, it also modifies the electronic structure of the Pd substrate. The interaction of sulfur with Pd leads to a downshift of the Pd d-states. As a consequence, in the vicinity of sulfur atoms on Pd the H_2 dissociation is no longer nonactivated, as on the clean Pd surface [47], but a dissociation barrier is built up. Such a barrier of height 0.1 eV is shown in Figure 2.2a above the hollow position. When the H_2 molecule is moved toward the sulfur atom, the direct repulsion between H_2 and sulfur leads to a dramatic increase in the dissociation barrier height. While this barrier above the top site of clean Pd(100) has a value of 0.15 eV [47], it rises to more than 1.2 eV above Pd atoms close to the adsorbed sulfur atoms (see Figure 2.2b). This results in a strongly corrugated PES.

In order to study the dynamical consequences of the sulfur poisoning on the hydrogen adsorption dynamics, a six-dimensional quantum dynamical study was performed [13, 44]. Figure 2.3 shows calculated sticking probabilities as a function of the initial kinetic energies for initially nonrotating and rotating molecules. In addition, results according to the so-called *hole model* [48] are plotted. This is the integrated barrier distribution $P_b(E)$ that is the fraction of the configuration space for which the barrier toward dissociation is less than E, and it would correspond to the sticking probability if the impinging H_2 molecules are not redirected and deflected upon approaching the surface.

The fact that the calculated sticking probabilities are all significantly larger than the results according to the hole model indicates that there are considerable dynamical effects during the dissociative adsorption in this system. As already mentioned, the PES of H_2 interacting with $S(2 \times 2)/Pd(100)$ is strongly corrugated. In addition, it is also highly anisotropic. As a consequence, molecules directed toward high barriers are very efficiently steered to configurations and sites with lower barriers [40]. This process is illustrated in Figure 2.4 where the calculated center-of-mass traces of H_2 molecules scattering and adsorbing on $S(2 \times 2)/Pd(100)$ are plotted.

The high anisotropy of the PES is also reflected in the strong dependence of the sticking probability on the initial rotational state of the molecules. When the

2.4 Quantum Dynamics of Reactions at Surfaces

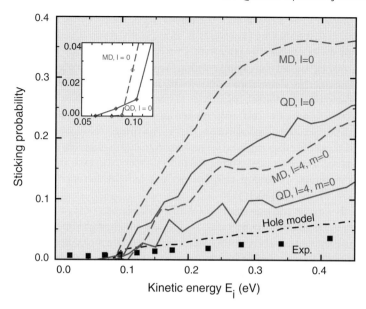

Figure 2.3 Quantum dynamical (QD) and classical (MD) dissociative adsorption probability of hydrogen on sulfur-covered Pd (100) as a function of the initial kinetic energy for initially nonrotating ($l=0$) and rotating H_2 molecules ($l=4$). (Adapted from Ref. [44]). In addition, experimental results [49] and the integrated barrier distribution (denoted by "hole model" [48]) are included.

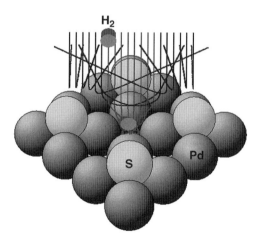

Figure 2.4 Illustration of the redirection of H_2 molecules upon approaching the highly corrugated $S(2 \times 2)/Pd(100)$ surface. The lines correspond to the calculated classical trajectories tracing the H_2 center of mass. Both scattering and adsorption events are depicted. Upon adsorption, not only the H_2 center of mass is redirected but also the molecular orientation is changed.

molecules are initially in a rotational state corresponding to the rotational quantum number $l = 4, m = 0$, the sticking probability is significantly reduced compared to the sticking probability for initially nonrotating molecules ($l = 0$). This rotational hindering is caused by the fact that rapidly rotating molecules rotate out of molecular orientations favorable for dissociation during the interaction with the surface. Invoking the principle of detailed balance or microscopic reversibility between adsorption and desorption, the rotational hindering in adsorption is reflected by the so-called rotational cooling in desorption, that is, the mean rotational energy of desorbing molecules is less than expected for molecules in thermal equilibrium at the surface temperature. This rotational cooling was indeed observed in experiments and reproduced in quantum dynamical calculations, not only at sulfur-covered Pd (100) [50] but also at clean Pd(100) [51].

There are further dynamical aspects of the interaction dynamics of H_2 at $S(2 \times 2)$/Pd(100) that are quite general and have also been found in other systems [52]. Not only the rotational motion but also the orientation of the rotating molecules matters. Molecules rotating in the so-called helicopter mode with their molecular axis parallel to the surface have a higher dissociation probability than molecules rotating in the cartwheel mode for which the *rotational* axis is parallel to the surface [44]. This is caused by the fact that molecules in the cartwheel mode have a high probability of hitting the surface in an upright configuration that is rather unfavorable for dissociative adsorption [14, 46]. The stereodynamical consequences can again be best observed in desorption where this anisotropy leads to rotationally aligned molecules [44, 53]. In addition, initial vibrational motion enhances the sticking probability and leads to vibrational heating in desorption [50]. However, the vibrational effects are much less pronounced than in the so-called late barrier systems, such as H_2/Cu [37, 41, 52, 54, 55] where in contrast to the $H_2/S(2 \times 2)$/Pd(100) PES shown in Figure 2.2 the barrier for dissociation is located after the curved region of the reaction path. Such a topology of the PES leads to a very efficient coupling between translational and vibrational motion.

Although qualitatively all experimentally observed dynamical aspects of the reaction dynamics of H_2 at $S(2 \times 2)$/Pd(100) are reproduced by the first-principles-based simulations, there are still significant quantitative differences, in particular as far as the sticking probability shown in Figure 2.3 is concerned. The experimental results are much smaller than the calculated ones. First of all, one has to note that the dynamical simulations are of only approximate nature. The calculated PES might not be fully correct due to problems associated with current DFT functionals [58]. Furthermore, the neglect of electronic excitations and the substrate motion in the dynamical simulations could have an influence on the accuracy of the results. However, it should also be noted that experimentally the preparation of an ordered (2×2) sulfur overlayer on Pd(100) is not trivial [49, 50]. Sulfur tends to form a $c(2 \times 2)$ overstructure on Pd(100) [50] corresponding to a higher coverage that would explain the low sticking probability observed in the experiment [49].

Finally, the difference between quantum dynamical results and classical results, which are also included in Figure 2.3, shall be discussed. There are basically two quantum effects that are absent in classical dynamics: tunneling and zero-point and quantization effects due to the localization of the wave function perpendicular to the reaction path, in particular in the so-called frustrated modes that are free in the gas phase. In fact, these two quantum effects have opposite consequences compared to classical results. Tunneling leads to additional particles that cross the barrier whereas zero-point effects typically lead to an effective increase in barrier heights that reduces the sticking probability. However, one has to note that upon dissociative adsorption the intramolecular vibration becomes softer leading to a decrease in the zero-point energy in this mode that can compensate for the increase in the zero-point energies in all other modes [59]. The fact that the quantum sticking probabilities are smaller than the classical sticking probabilities (see Figure 2.3) indicates that zero-point effects due to the frustrated modes are more dominant than tunneling in the quantum dynamics of $H_2/S(2 \times 2)/Pd(100)$. This is also true for H_2 dissociation on other metal surfaces [12].

Another quantum effect is the relatively strong structure of the quantum sticking probability as a function of the kinetic energy due to resonance effects and the opening up of new scattering channels with rising kinetic energies [12]. Such quantum oscillations have not been observed yet despite significant experimental efforts [60]; they are quickly washed out by imperfections of the substrate and thermal motion of the surface atoms [61]. Furthermore, it should be noted that in spite of the quantitative differences between quantum and classical results in Figure 2.3, qualitatively they show the same behavior as a function of the initial kinetic energy and the initial rotational states. This means that classical molecular dynamics simulations can be used to get qualitative trends in the reaction dynamics at surfaces even for the lightest element hydrogen; for heavier atoms, the quantum effects are actually less pronounced so that the results of classical molecular dynamics simulations become rather reliable.

2.5
Nondissociative Molecular Adsorption Dynamics

For the simulation of the dissociative adsorption dynamics of H_2 on metal surfaces, the recoil of the substrate atoms usually does not play a crucial role because of the large mass mismatch between metal and hydrogen atoms. After dissociation, the single hydrogen atoms will eventually accommodate at the surface due to the dissipation of their kinetic energy to the substrate, but the energy transfer to the substrate hardly influences the dissociation probability.

This situation is entirely different for atomic and nondissociative molecular adsorption. In order to stick to the surface, the impinging atoms and molecules have to transfer their initial kinetic energy to the substrate, that is, the sticking probability as a function of the initial energy E can be expressed as

$$S(E) = \int_E^\infty P_E(\varepsilon) \, d\varepsilon, \qquad (2.10)$$

where $P_E(\varepsilon)$ is the probability that an incoming particle with kinetic energy E will transfer the energy ε to the surface. This energy will be taken up by mainly substrate vibrations, but the excitation of electron–hole pairs at the surface can also contribute to the dissipation, as will be discussed below.

Typically, the sticking probability in atomic or nondissociative molecular adsorption decreases as a function of kinetic energy because the energy transfer to the substrate becomes less efficient at higher kinetic energies [39]. The qualitative features of nondissociative adsorption can be discussed within the so-called *hard cube model* (HCM) [56, 57] in which the impact of the atom on the surface is treated as a binary elastic collision between a gas phase atom (mass m) and a substrate atom (mass M_c) that is moving freely with a velocity distribution $P_c(v_c)$. Close to the surface, the particle becomes accelerated because of the adsorption well of depth E_{ad}. This model that is illustrated in Figure 2.5 can be solved analytically. Thus, it can be shown that the sticking probability becomes higher for larger well depth E_{ad} and greater mass ratio m/M_c because then the impact on the surface "cube" becomes stronger and consequently the energy transfer to the substrate is enlarged.

The HCM model has been used to model the energy transfer to the surface in atomic or molecular adsorption [56, 57], for example, for the sticking of O_2 on Pt (111) [62]. However, it assumes that the molecule is a point-like object impinging on a flat surface. This assumption is rather crude and can lead to an erroneous interpretation of measured sticking probabilities, as has been shown especially for the O_2/Pt(111) system [32, 33]. This system is of particular importance for understanding the elementary processes occurring in the car exhaust catalyst. In spite of its seeming simplicity, this system is rather complex since oxygen can exist in different states on Pt(111). A weakly bound physisorbed species exists at surface temperatures up to 30 K [63, 64]. Chemisorbed peroxo-like (O_2^{-2}) and superoxo-like (O_2^-) molecular species are found at surface temperatures below about 100 K [65, 66]. For higher surface temperatures, oxygen adsorbs dissociatively [67].

The chemisorbed O_2 species on Pt(111) have also been identified in DFT total energy calculations [68, 69]. Two elbow plots of the calculated O_2/Pt(111) PES are shown in Figure 2.6. The superoxo molecular precursor state corresponds to the minimum in Figure 2.6a above the bridge site with a binding energy of 0.72 eV [68, 69]. The other chemisorption state, the peroxo state that is energetically almost

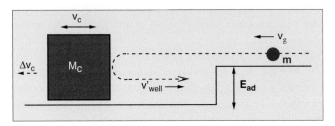

Figure 2.5 Schematic illustration of the hard cube model [56, 57]. An atom or molecule with mass m is impinging on an attractive potential with well depth E_{ad} on a surface modeled by a cube of effective mass M_c. The surface cube is moving with a velocity v_c given by a Maxwellian distribution.

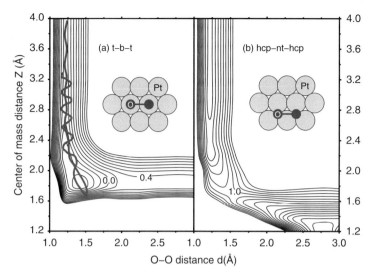

Figure 2.6 Two-dimensional cuts through the potential energy surface of O_2/Pt(111) determined by an *ab initio*-derived tight binding Hamiltonian as a function of the O–O distance and the O_2 distance from the surface [32]. The insets in (a) and (b) illustrate the molecular orientation and lateral center of mass position corresponding to a top–bridge–top (t-b-t) and a hcp hollow–near-top–hcp hollow (hcp-nt-hcp), respectively. The contour spacing is 0.2 eV per O_2 molecule. In (a) a trajectory of an O_2 molecule with an initial kinetic energy of 0.6 eV scattered at Pt(111) is also plotted.

degenerate with the superoxo state, is located above the threefold hollow sites with its axis slightly tilted from the parallel orientation (not shown in Figure 2.6). Both chemisorption states can be accessed from the gas-phase without encountering any barrier, that is, the adsorption in nonactivated.

Figure 2.6b demonstrates the strong corrugation of the O_2/Pt(111) PES. Although the molecule is shifted only by about 1 Å from the bridge position of Figure 2.6a to a near-top position, there is no longer any chemisorption well present but rather a large barrier of about 1 eV toward dissociative adsorption. This barrier becomes even larger for the molecule directly at the on-top site. Furthermore, the PES is also strongly anisotropic: for molecules in an upright orientation, the PES is repulsive. In fact, the majority of adsorption pathways are hindered by barriers; direct nonactivated access to the adsorption states is possible for only a small fraction of initial conditions.

Interestingly, molecular beam experiments showed the surprising result that oxygen molecules do not dissociate on cold Pt surfaces below 100 K [62, 70, 71], even at kinetic energies above 1 eV. As Figure 2.6a indicates, this energy is much greater than the dissociation barrier that is on the order of 0.2 eV with respect to the gas-phase energy of O_2. Apparently, in the system O_2/Pt(111) the kinetic energy of the impinging molecules is not operative in surmounting the dissociation barrier.

To simulate the adsorption dynamics of O_2/Pt(111) represents, in fact, a significant theoretical and computational challenge. On the one hand, a realistic PES is needed that reliably describes both the molecular and the dissociative adsorption channels.

On the other hand, molecular trapping processes can be reproduced only if the energy dissipation to the platinum substrate is properly taken into account. Using empirical classical potentials, almost arbitrarily many trajectories can be computed; however, there are no reliable interaction potentials available treating reactions on the surface and the surface recoil upon impact on an equal footing.

Ab initio molecular dynamics simulations represent a method that is well suited for this task. For H_2, AIMD simulations of the adsorption dynamics on precovered surfaces were indeed already successfully performed (see the next two sections). However, periodic DFT calculations involving oxygen are computationally much more expensive: on the one hand, the strong localization of the oxygen wave functions close to the nucleus requires a larger plane wave basis, on the other hand, oxygen has to be described in a spin-polarized manner because of its atomic and molecular triplet states. As a compromise, the results of static DFT calculations [68, 69] were used in order to adjust a TB Hamiltonian [28, 30, 32, 33]. This approach combines a quantum mechanical description of the molecule–surface interaction with the numerical efficiency of tight binding calculations that are about three orders of magnitude faster than DFT calculations. The elbow plots shown in Figure 2.6 were, in fact, produced using a TB Hamiltonian with its parameter adjusted to reproduce *ab initio* calculations.

Figure 2.7 shows sticking probabilities derived from tight binding molecular dynamics (TBMD) simulations. The results were obtained by averaging over a sufficient number of trajectories with random initial lateral positions and orientations of the O_2 molecules that were started 4 Å above the surface. The classical equations of motion were integrated using the Verlet algorithm [73] with a time step of 1 fs within the microcanonical ensemble. The statistical error ΔS of the calculated sticking probability S can be estimated by

$$\Delta S = \sqrt{S(1-S)}/\sqrt{N}, \tag{2.11}$$

where N is the number of calculated trajectories. Already for a relatively small number of trajectories such as $N = 200$, the statistical uncertainty of the calculated sticking probability is below 3.5% that is usually sufficient if the sticking probabilities lie in the range between 0.1 and 1.

Calculated and measured [71, 73] sticking probabilities of $O_2/Pt(111)$ as a function of the kinetic energy are compared in Figure 2.7a. There is a semiquantitative agreement between theory and experiment that is quite satisfactory with respect to the numerous approximations entering the calculations. The TBMD simulations also reproduced the surprising experimental finding that O_2 does not directly dissociate upon adsorption, even at kinetic energies of 1.1 eV.

This peculiar behavior can, in fact, be understood by inspecting the topology of the elbow plots of $O_2/Pt(111)$. At the energetically most favorable adsorption paths, O_2 is first attracted to the surface toward the molecular chemisorption states. Because it is a nondissociative molecular state, the O–O bond length hardly changes when the molecule enters the chemisorption well. On the other hand, during the dissociation the oxygen atoms remain at roughly the same distance from the

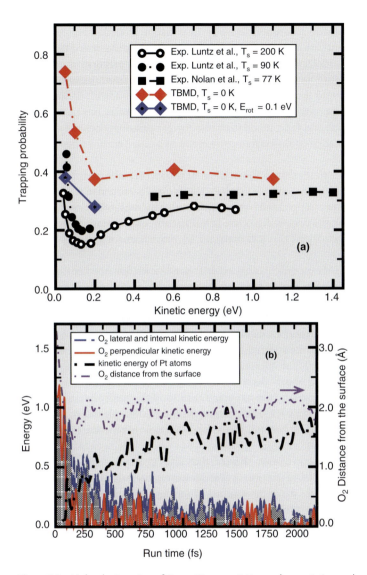

Figure 2.7 Molecular trapping of O_2 on Pt(111). (a) Trapping probability of O_2/Pt(111) as a function of the kinetic energy for normal incidence. Results of molecular beam experiments for surface temperatures of 90 and 200 K (Luntz et al. [72]) and 77 K (Nolan et al. [71]) are compared with tight binding molecular dynamics simulations for the surface initially at rest ($T_s = 0$ K) for initially rotating and nonrotating molecules. (b) Calculated energy redistribution and O_2 center of mass distance from the surface for an O_2 molecule impinging on Pt(111) with an initial kinetic energy of 1.1 eV, as a function of time. The lateral and internal kinetic energy and the perpendicular kinetic energy curves are indicated by the blue- and red-shaded areas, respectively.

surface. As a consequence, there is no smoothly curved energy minimum reaction path toward dissociative adsorption, as for H_2 dissociation plotted in Figure 2.2, but rather a sharply bent path, as shown in Figure 2.6a. This is also illustrated by a typical trajectory of an O_2 molecule directly aimed at the molecular precursor state that is included in the figure. Its initial kinetic energy of 0.6 eV is much higher than the dissociation barrier so that the molecule could, in principle, adsorb dissociatively. However, it rather becomes accelerated by the attractive potential, hits the repulsive wall of the potential, and is scattered back into the gas phase. It does not enter the dissociation channel since this would correspond to a sharp turn on the potential energy surface.

This does not mean that direct dissociation of O_2/Pt(111) is impossible, but it is very unlikely. Hence, dissociation of O_2 on Pt(111) is usually a two-step process. First, the molecule becomes trapped and is accommodated in the molecular chemisorption state and subsequently does it dissociate at sufficiently high surface temperatures due to thermal fluctuations that will make the O_2 molecules enter the dissociation channel.

Furthermore, STM experiments showed that after dissociation upon heating the Pt substrate, the two oxygen atoms are found on average two Pt lattice units apart from each other [67]. Kinetic Monte Carlo simulations [74] showed that this spatial distribution cannot result from two atomic jumps, but rather from a hot atom movement after dissociation, as will be discussed in detail in Section 2.7.

We will now focus on the sticking probability as a function of the kinetic energy. At low kinetic energies, it first strongly decreases and then it levels off after passing a minimum. This strong initial decrease in the sticking probability was initially interpreted as being caused by the trapping into a shallow physisorption state with a well depth of about 0.12 eV [62, 71]. However, such a shallow physisorption well is not present in the TB-PES, but still this strong decrease is reproduced in TBMD simulations.

An analysis of the trajectories showed that also at low kinetic energies the molecules enter the chemisorption wells. However, the sticking probability at these low kinetic energies is not determined by the energy transfer to the substrate *per se* but rather by the probability to access the entrance channels toward the chemisorption states. All molecules that find their way to the molecular chemisorption state at low kinetic energies do, in fact, remain trapped. At low kinetic energies, a high fraction of the impinging O_2 molecules is indeed steered to the attractive paths toward the chemisorption states (such a steering is illustrated in Figure 2.4). However, this steering effect becomes strongly suppressed at higher kinetic energies [40, 75], and this is the reason for the initial decrease in the sticking probability. This steering effect becomes, in fact, also suppressed by additional rotational motion. This is reflected in Figure 2.7a by the strong reduction in the calculated TBMD sticking probability for O_2 molecules that have initially a rotational energy of $E_{rot} = 0.1$ eV.

The leveling off of the sticking probability at higher kinetic energies is also a surprising result. According to the hard cube model, no O_2 molecule with a kinetic energy of 1.0 eV would stick at a surface with a chemisorption well depth of 0.7 eV.

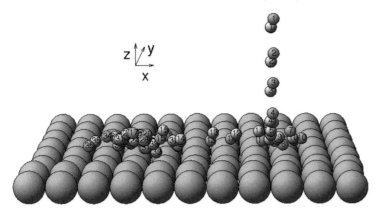

Figure 2.8 Illustration of the conversion of the molecular kinetic energy perpendicular to the surface into internal degrees of freedom (rotations, vibrations) and lateral movement across the surface leading to a dynamical precursor state. The snapshots along the trajectory are numbered consecutively.

However, the hard cube model assumes the impinging of a point-like particle on a flat surface; it does not take into account the corrugation and anisotropy of the PES.

In Figure 2.7b, the energy redistribution along the trajectory of an O_2 molecule impinging on Pt(111) with an initial kinetic energy of 1.1 eV is shown; in addition, the O_2 center of mass distance from the surface is plotted. When the molecule hits the surface for the first time after about 70 fs, less than 0.4 eV is transferred to the substrate vibrations. This transfer would not be sufficient for the molecule in order to remain at the surface. Still, almost 1 eV is transferred to lateral and internal (rotations, vibrations) energy of the molecule. The molecule starts to rotate, vibrate, and move parallel to the surface due to the corrugation and anisotropy of the PES. This dynamical trapping process is illustrated in Figure 2.8. In this so-called dynamical precursor state [76, 77], the molecule does not have sufficient kinetic energy perpendicular to the surface to scatter back in the gas phase. It bounces back and forth with respect to the substrate, and with every bounce it dissipates further energy to the substrate until after about 2 ps it has accommodated at the surface. The dynamical trapping of O_2 on Pt(111) depends relatively weakly on the initial kinetic energy that causes the leveling off of the trapping probability at higher kinetic energies.

It is important to realize that the dependence of the sticking probability of O_2 on Pt(111) as just discussed can be understood by only using dynamical simulations and concepts. Static information alone is not sufficient to derive these results.

2.6
Adsorption Dynamics on Precovered Surfaces

In Section 2.4, we have already seen that precovering a surface can significantly influence the adsorption and reaction dynamics on a surface. However, during the course of a reaction in heterogeneous catalysis, the reactants themselves remain for a certain period of time on the surface and block adsorption sites. Furthermore, their presence can also influence the electronic structure of the substrate and thus the reactivity of the catalyst [78]. Last but not least, the recoil of the substrate and thus the dissipation of the energy of impinging molecules will be modified.

To model coverage effects in the adsorption dynamics corresponds to a high-dimensional problem in which several inequivalent atoms have to be realistically described. An analytical or numerical interpolation of such a high-dimensional interaction potential is rather cumbersome but can be managed [24]. Fortunately, due to the increase in computer power and the development of efficient electronic structure codes, it is now possible to perform AIMD simulations of complex systems in which the forces necessary to integrate the equations of motion are determined "on the fly" by first-principles calculations. While some years ago AIMD studies were restricted to a small number of trajectories [79, 80], now a statistically meaningful number of AIMD trajectories can be determined, as has already been demonstrated [9, 24, 25, 81, 82].

The interest in the adsorption at precovered surface was fueled by a recent STM study [83, 84] showing that cold H_2 molecules impinging on an almost completely hydrogen-covered Pd(111) surface do not adsorb dissociatively in a hydrogen dimer vacancy. Rather, aggregates of three or more vacancies are required to dissociate hydrogen. A subsequent DFT study [85] demonstrated that the presence of the hydrogen overlayer has a poisoning effect; it leads to the formation of energetic barriers making the dissociative adsorption on hydrogen-precovered Pd(111) to an activated process, but the dissociative adsorption of H_2 in a hydrogen dimer vacancy is still exothermic.

These findings were confirmed in an AIMD study [9]: at low initial kinetic energies of 0.02 eV, as used in the experiments [83, 84], H_2 molecules are not able to overcome the dissociation barrier above a dimer vacancy on H-covered Pd(111) surface within a (3 × 3) geometry corresponding to a surface coverage of $\Theta_H = 7/9 \approx 0.78$. At higher kinetic energies, however, H_2 molecules can penetrate the dimer vacancies on hydrogen-covered Pd(111) and adsorb dissociatively.

This is shown in Figure 2.9 where the sticking probabilities on both Pd(111) and Pd(100) are plotted as a function of the hydrogen coverage for different arrangements of the preadsorbed hydrogen normalized to the value at the corresponding clean Pd surfaces for a kinetic energy of 0.1 eV. The results were obtained by averaging over at least 200 trajectories for each coverage. Hydrogen dimer and trimer vacancies were modeled within a (3 × 3) surface periodicity corresponding to hydrogen coverages of 7/9 and 2/3, respectively. As far as the dimer vacancy on Pd(111) is concerned (denoted by 2V in Figure 2.9), a small but nonvanishing relative adsorption probability of 0.05 is obtained. At the trimer vacancies on Pd(111) centered either around a hollow site ($3V_H$) or around a Pd top site ($3V_T$), the adsorption probability is more than twice as large as on the dimer vacancy (2V). Since the area of the trimer vacancy is only 50% larger than the one of the dimer vacancies, this indicates that it is not only the area of the vacancies that determines the adsorption probability but also indirect poisoning and dynamical effects.

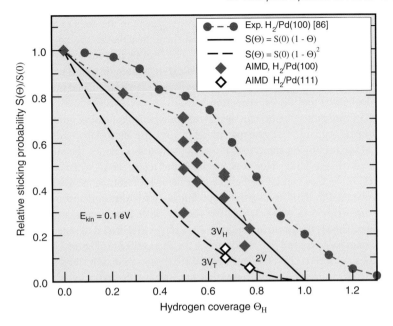

Figure 2.9 Relative dissociative adsorption probability of hydrogen on hydrogen-covered Pd(100) determined through AIMD simulations as a function of hydrogen coverage θ_H involving different arrangements of hydrogen. The initial kinetic energy is 0.1 eV. 2V, $3V_H$ and $3V_T$ denote the dimer vacancy and the trimer vacancy centered around a Pd hollow and a Pd top site on Pd(111), respectively. The experimental results [86] were obtained for a temperature of 170 K (Adapted from Ref. [82]).

These dynamical effects on the adsorption at precovered surfaces have been studied in greater detail at the more open Pd(100) surface. As a reference, two curves corresponding to $S(\Theta_H) = S(0)(1-\Theta_H)$ and $S(\Theta_H) = S(0)(1-\Theta_H)^2$ are included in Figure 2.9 that would correspond to the sticking probability if it was determined by pure site blocking requiring one or two empty sites. For Pd(111), the AIMD results are close to those predicted for site blocking requiring one empty site. However, for the more open and reactive Pd(100) surface where the adsorption into a dimer vacancy is nonactivated, the sticking probability is significantly larger than predicted from a simple site blocking picture, in particular for $\Theta_H = 0.5$. This is somewhat surprising since the hydrogen coverage leads to a small poisoning effect that should reduce the sticking probability. Yet, because there is no mass mismatch between the impinging hydrogen molecules and the adsorbed hydrogen atoms, there should be a larger energy transfer to the substrate. In order to check the dynamical role of the surface atoms, additional AIMD runs with fixed substrate atoms were performed [9]. This showed that it is indeed the energy transfer from the impinging H_2 molecule to the substrate that leads to this enhanced sticking probability compared to pure site blocking. Still, also the rearrangement of the substrate atoms when the incoming H_2 molecules strike the strongly corrugated surface contributes to the high sticking probability. In Figure 2.9, experimental results [86] are also included. These are obtained for lower kinetic energies than the ones used in the AIMD simulations, hence the experimental and theoretical results are not directly comparable. It should also be noted that the theoretical results for each coverage should, in principle, be averaged over a statistical ensemble of different surface structures with the same coverage. Still, it is gratifying that the calculated maximum sticking probabilities trail the experimental results.

The AIMD simulations do not only yield statistically reliable sticking probabilities but also provide valuable microscopic insights into the dissociative adsorption dynamics. Many impinging molecules either directly dissociate on the surface or scatter back into the gas phase. However, in some cases it takes some while before the adsorbing H_2 molecules end up in the energetically most favorable hollow sites on hydrogen-covered Pd. This is illustrated in Figure 2.10 where snapshots of an adsorption trajectory at a H(3 × 3)/Pd(100) surface with a hydrogen coverage of 5/9 are shown. The H_2 molecule first hits the Pd surface close to a Pd top site where it becomes dynamically trapped [59, 76, 77] above the Pt top site (see Figure 2.10a): the molecule does not directly find the pathway toward dissociation, but starts rotating, vibrating, and moving laterally, and due to the conversion of the initial kinetic energy into internal and lateral degrees of freedom, the H_2 molecule cannot escape back into the gas phase. This molecular precursor state above the top sites is stabilized due to

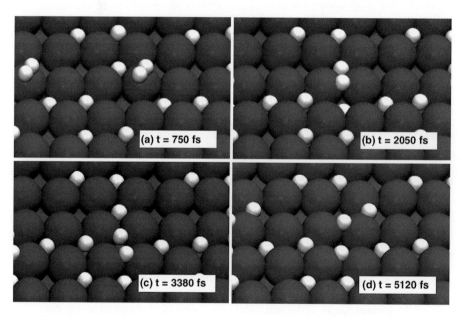

Figure 2.10 Snapshots of a trajectory calculated in *ab initio* molecular dynamics simulations [9] within a (3 × 3) surface periodicity illustrating details of the adsorption dynamics at a hydrogen-covered Pd(100) surface. (a) Molecular precursor, (b) dissociation into hydrogen atoms in the fourfold hollow and the bridge site, (c) exchange diffusion of the bridge-bonded hydrogen atom, and (d) final adsorption in fourfold hollow site by an exchange mechanism.

the poisoning effects of the preadsorbed hydrogen atoms, very similar to the one already identified at the hydrogen-covered stepped Pd(210) surface [87, 88]. It corresponds to a PdH$_2$ complex that is known in the gas phase [89] with the H$_2$ molecule moving relatively freely around the on-top position. This state has not been identified experimentally yet; however, it should be detectable at low surface temperatures by, for example, isotope exchange experiments since it is bound by 0.1 eV.

After about 2 ps, one of the two hydrogen atoms of the impinging H$_2$ molecule enters the energetically most favorable adsorption site, the fourfold hollow site (Figure 2.10b). The associated energy gain of about 0.5 eV is transferred to the hydrogen overlayer resulting in large vibrational amplitudes of the hydrogen atoms. This is visible as the large displacements of some of the hydrogen atoms from their equilibrium sites. However, the other hydrogen atom does not directly find another fourfold hollow site but becomes trapped at a bridge-bonded site.

The additional bridge-site hydrogen atom is not immobile and can actually move along the surface in an exchange mechanism [90]: after 3.4 ps it replaces one of the adsorbed hydrogen atoms at the fourfold hollow sites that is then pushed up to an adjacent bridge site (Figure 2.10c). In fact, several of these exchange processes occur until finally after 5 ps the bridge-bonded hydrogen atom pushes an adsorbed hydrogen atom to another empty fourfold hollow site (Figure 2.10d). Still it takes some time before the hydrogen atoms dissipate the energy gained upon entering the adsorption wells. This is reflected in some atomic jumps between adjacent atomic hydrogen adsorption sites (not shown in Figure 2.10) and will be discussed in the next section.

It is important to realize that some of the processes just described are not at all obvious. Thus, by following the time evolution of the adsorption dynamics obtained by AIMD simulations, interesting and unexpected insights into the reaction dynamics at surfaces can be gained.

2.7
Relaxation Dynamics of Dissociated H$_2$ Molecules

In the last section, we have focused on the dissociation of H$_2$ on hydrogen-covered Pd surfaces. We followed the H$_2$ dissociation dynamics on Pd, but we did not really consider the fate of hydrogen atoms after dissociation. However, immediately after dissociation the atoms gain a significant amount of energy when they enter the atomic adsorption wells, which amounts to about 1 eV for the two H atoms together on Pd surfaces. Since it takes some time before this excess kinetic energy is dissipated to the substrate, the energy gain leads to the formation of "hot" atoms, that is, atoms with energies much larger than thermal energies. These atoms can use their kinetic energy in order to propagate along the surface. The mean free path of these hot atoms is, for example, relevant for catalytic reactions on surfaces since it determines whether adjacent reactants can react directly after the dissociative adsorption with another species or whether some diffusive motion is required before any further reaction can occur.

There are some theoretical studies that modeled the motion of single atoms with initial velocities considered to be typical for dissociation fragments directly after the bond breaking process [91, 92] in order to address the relaxation of hot atoms after dissociation. However, in such simulations the interaction of the two fragments after the dissociation is not taken into account. This approximation has been avoided in AIMD simulations of the dissociation of H_2 on clean Pd(100) based on periodic DFT calculations [81]. In order to minimize the interaction of the hot hydrogen atoms with their periodic images, a large (6 × 6) surface unit cell was chosen.

Figure 2.11 displays one specific AIMD trajectory that was run for 2 ps. The trajectory shows that the single hydrogen atoms visit several surface sites before they come to rest. In this particular trajectory, the hydrogen atoms approach each other again after an initial increase in the interatomic distance. At a certain time, they are moving toward adjacent adsorption sites before they separate again. This shows that the mutual interaction can be important for the hot atom movement.

The displacement of the hydrogen atom and their energy redistribution as a function of time averaged over 100 AIMD trajectories is plotted in Figure 2.12. The initial kinetic energy was chosen to be 0.2 eV in order to avoid molecular trapping events. Running trajectories with other initial kinetic energies shows that the hot atom movement is only weakly dependent on the initial kinetic energy since most of the kinetic energy is provided by the energy gain upon dissociative adsorption.

In detail, the mean H−H distance, the mean displacement of the single H atoms and the H_2 center of mass, and the total kinetic energy of the hydrogen and the palladium atoms are plotted in Figure 2.12. Immediately after hitting the surface, the H_2 molecules dissociate and gain on average about 700 meV when hydrogen atoms enter adjacent atomic adsorption wells. Because of this high kinetic energy, hydrogen

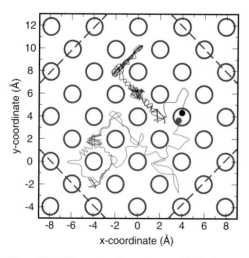

Figure 2.11 Illustration of a trajectory of the hydrogen atoms after dissociation on clean Pd(100) within a (6 × 6) surface unit cell. The initial kinetic energy was 0.2 eV. The total run time was 2 ps. The surface unit cell of the simulations is indicated by the dashed line.

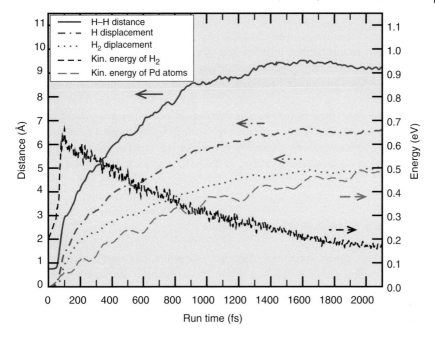

Figure 2.12 Calculated hot atom dynamics after H_2 dissociation on clean Pd(100) within a (6 × 6) surface unit cell. H−H distance, the displacement of the single H atoms and the H_2 center of mass from their initial lateral position, and the kinetic energy of the H_2 molecules and of the Pd substrate are plotted as a function of the run time averaged over 100 AIMD trajectories with an initial H_2 kinetic energy of 200 meV.

atoms make further jumps and increase their separation. At the same time, they constantly lose kinetic energy that is transferred to the substrate.

After about 1 ps, hydrogen atoms lose half their kinetic energy, and their joint mean kinetic energy is about 0.3 eV. The mean separation of the two hydrogen atoms then levels off at a value of about 9 Å that corresponds to about three Pd lattice units. However, there is a large variance in the separation, as single trajectories show a maximum H−H distance of more than 20 Å. The single hydrogen atoms move two lattice sites on average and the H_2 center of mass is displaced by about 5 Å from its initial lateral position. After 2 ps, the atoms have less than 0.2 eV kinetic energy left, and their mean displacement does not change any more. It is interesting to note that this kinetic energy is still larger than the diffusion barrier of hydrogen atoms on Pd (100) that is about 30 meV, but the irregular motion of hydrogen atoms does not lead to a further net displacement.

Further AIMD trajectories have been determined for a fixed substrate with and without rescaling of the velocities. In addition, AIMD runs were carried out in which the two single hydrogen atoms after the initial dissociation moved separately [81]. It turned out that although any single trajectory depends on the specific conditions, on average it is just the energy dissipation that determines the final mean distance of the two dissociation products; the mutual interaction of the two fragments plays a minor role, at least for H_2 dissociation on Pd(100).

Hydrogen is the lightest element and thus should show the smallest energy transfer to substrate atoms because of the large mass mismatch. Hence, the results for the hot atom motion of hydrogen should represent an upper limit for other heavier atoms such as oxygen or nitrogen that dissipate their energy more quickly to metal substrates. Thus, the hot atom motion should lead to mean displacements of at most three lattice sites. Wider separations [93] should be due to other processes, such as the so-called cannonball mechanism [94] where one atom in dissociative adsorption is emitted again and reimpinges on the surface at a laterally distant site.

2.8
Electronically Nonadiabatic Reaction Dynamics

So far, we have always assumed that the Born–Oppenheimer approximation is valid for the description of the reaction dynamics on surfaces. As long as there is a satisfactory agreement between electronically adiabatic simulations and experiment, there is no need to invoke electronic excitations in the dynamics [42]. Still there are dynamical processes where electronic excitations play an important role. One has to distinguish between two kinds of electronic excitations in the reaction dynamics at surfaces, delocalized excited states of the surface and localized excitations at the reactant or the adsorbate–surface bond. The first type corresponds to electron–hole pair excitations that exhibit a continuous spectrum, in particular at metal surfaces, and which can often be described using a friction formalism. Using a *molecular dynamics with electronic friction* approach [95] based on Hartree–Fock cluster calculations, it was shown that the excitation of electron–hole pairs upon the adsorption of CO on Cu(100) has only a minor influence on the sticking probability [96]. This is illustrated in Figure 2.13, taking into account energy dissipation to electron–hole pairs increases the sticking probability only slightly since the main channel for energy transfer is the excitation of substrate vibrations.

On the other hand, using thin polycrystalline metal and semiconductor films deposited on n-type Si(111) as a Schottky diode device, the so-called chemicurrent due to nonadiabatically generated electron–hole pairs upon both atomic and molecular chemisorption can be measured [97–99]. For the NO adsorption on Ag, it has been estimated that one quarter of the adsorption energy of about 1 eV is dissipated to electron–hole pairs.

The generation of the chemicurrent upon adsorption has been addressed in several theoretical studies [100–104]. Based upon time-dependent density functional theory (TDDFT), the excitation of electron–hole pairs in the atomic hydrogen adsorption on Al(111) was studied from first principles [103, 104]. The dynamics of the nuclei were treated in the mean field approximation, whereas the time evolution of the electron system was determined by integrating the time-dependent Kohn–Sham equations. However, because of the light mass of the electrons, a rather short time step had to be chosen that was on the order of 0.002 0.003 fs [103, 104], which is about three orders

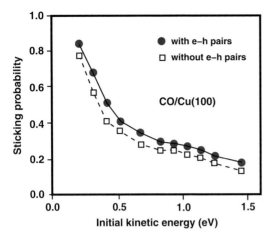

Figure 2.13 Sticking probability for CO/Cu(100) under normal incidence calculated without and with the consideration of electronic friction illustrating the minor role of e–h pairs in the molecular adsorption. Adapted from Ref. [96].

of magnitude smaller than the time step required for electronically adiabatic molecular dynamics simulations. Therefore, only very few selected trajectories could be calculated.

These simulations showed that the excitation of electron–hole pairs depends sensitively on the impact point of the hydrogen atom. If the hydrogen atom impinges at the Al(111) fcc hollow site, it penetrates into the Al crystal and couples strongly to the substrate electrons so that the energy loss to electron–hole pairs and phonons becomes comparable. At the on-top site, on the other hand, the energy dissipated to electron–hole pairs is much smaller, below 0.1 eV, whereas there is still a significant energy transfer to the phonons [103].

Note that there are also nonadiabatic effects as far as the spin of the hydrogen atom is concerned. Far away from the surface, the H atom is spin-polarized since there is only one electron with a specific spin, while it becomes spin-unpolarized upon adsorption on Al(111). The TDDFT-AIMD simulations showed [104], in agreement with simulations based on the Newns–Andersen model [102], that the hydrogen atom approaching the surface does not follow the adiabatic spin ground state because of the weak coupling between the spin states. These nonadiabatic spin dynamics lead to additional electronic dissipation.

Spin effects are also believed to be crucial in the system O_2/Al(111). The sticking of O_2 on Al(111) has been a puzzle for a long time. On the one hand, molecular beam experiments yield a vanishing sticking probability at low kinetic energies and thus suggest that the adsorption is hindered by a small adsorption barrier [93, 105]. On the other hand, adiabatic electronic structure calculations using DFT yield a potential energy surface with large purely attractive portions [107–109] so that the dissociation probability for all kinetic energies should be close to one. This is demonstrated in Figure 2.14 where the experimentally measured sticking probability is contrasted with the MD results based on the adiabatic DFT-PES [110].

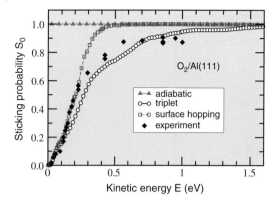

Figure 2.14 Sticking probability of O_2/Al(111) as a function of the kinetic energy for normal incidence computed by MD simulations on the adiabatic (triangle) and triplet (open circle) PES, as well as by a surface hopping (SH) method including both the triplet and the singlet PES (open square). The experimental data (diamond) are taken from Ref. [105]. Adapted from Ref. [106].

To resolve this puzzle, it was proposed that spin selection rules play an important role in understanding the dissociation dynamics of O_2/Al(111) [110–112]. Upon adsorption, oxygen changes its spin state from the gas-phase $^3\Sigma_g^-$ triplet state to the singlet state. Because of the low density of states of aluminum at the Fermi level, the probability for the triplet-to-singlet transition is rather small. Hence, O_2 molecules do not follow the adiabatic potential energy curve but rather stay in the triplet state that becomes repulsive close to the surface.

The shape of the spin-dependent potential energy curves of O_2/Al(111) is illustrated in Figure 2.15. The triplet curve is obtained within a constrained DFT formalism [110–112] in which an auxiliary magnetic field was introduced in order to keep O_2 in its triplet state. The first excited state of the system is modeled by an

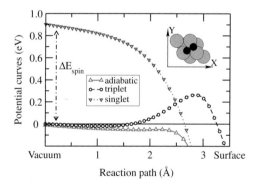

Figure 2.15 Minimum energy pathway in the triplet state and corresponding adiabatic and singlet potentials for dissociation over the fcc site, with the molecular axis aligned parallel to the surface, as derived from constrained DFT calculations [110]. ΔE_{spin} indicates the singlet–triplet splitting in the gas phase. The inset illustrates the molecular configuration.

adiabatic spin-unpolarized DFT calculation, leading to an approximate description of the singlet state. In the triplet state, O_2 adsorption is no longer nonactivated but hindered by a small barrier, whereas for the singlet state there are purely attractive adsorption paths.

Classical MD simulations were performed on the triplet PES interpolated by a neural network [20] with the Al substrate atoms kept fixed. The resulting sticking probability is also included in Figure 2.14 denoted as triplet. Due to the presence of a minimum adsorption barrier, the triplet results are in good agreement with the experiment.

Still, in these simulations triplet–singlet transitions were completely neglected. Since the adiabatic, triplet, and singlet PES are known, the coupling matrix elements between singlet and triplet states can be derived by inverting the diagonalization of the diabatic Hamiltonian. Using these coupling elements, molecular dynamics simulations including electronic transitions were performed with the fewest switches algorithm [113–115]. In this surface hopping method, the nuclear degrees of freedom are integrated classically on one potential energy surface at each time step. Simultaneously, the density matrix including all electronic states is calculated by integrating the time–dependent Schrödinger equation along this trajectory. Transitions from one PES to another are introduced in such a way that for a large number of trajectories the occupation probabilities given by the density matrix are achieved within the smallest number of switches possible [113, 115].

Including triplet–singlet transitions in the adsorption dynamics of O_2 on Al(111) leads to an increase in the sticking probability at intermediate energies, as the surface hopping results included in Figure 2.14 demonstrate [106, 116]. Molecules with low kinetic energies are scattered at the repulsive triplet potential before they can reach the region where the triplet–singlet transitions occur with a significant probability. However, O_2 molecules with intermediate energies can reach this region, where the transition from the activated triplet PES to the attractive singlet PES facilitate the dissociative adsorption and thus increase the sticking probability.

Still it should be noted that the inclusion of triplet–singlet transitions changes the sticking probability with respect to the pure triplet results only quantitatively, not qualitatively. Furthermore, the experimental sticking curve could also be reproduced in a purely electronically adiabatic framework on a slightly changed PES. It might well be that an improved description of exchange and correlation effects in the adiabatic DFT calculations would lead to the presence of an activation barrier [117, 118]. Hence, the sticking probability of triplet O_2 does not allow an unequivocal assessment of the role of spin transitions in the O_2–Al interaction dynamics.

It was therefore suggested to consider the adsorption of singlet O_2 molecules on Al (111) [106, 116]. This is experimentally feasible, as was just demonstrated for the adsorption of O_2 on small Al_n clusters [119]. If all impinging O_2 molecules stayed in the singlet state, the sticking probability would be unity, as shown in Figure 2.16 where the calculated sticking probability for O_2 molecules on the singlet PES is plotted. Allowing spin transitions, however, leads to a significant reduction in the singlet sticking probability. This is due to the fact that some of the impinging singlet molecules that suffer a spin transition near the crossing seam of the singlet and

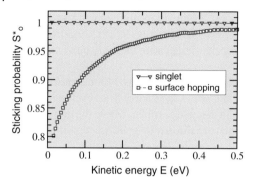

Figure 2.16 Sticking probability for singlet oxygen molecules ($^1\Delta_g$) impinging on Al(111) as computed by molecular dynamics simulations on the singlet PES (inverted triangle) and by the fewest switches surface hopping algorithm (open square) based on first-principles calculations. Adapted from Ref. [106].

triplet PES are backscattered on the triplet PES that indeed exhibits some barriers larger than 1 eV. Furthermore, upon the singlet–triplet transition, the possible energy gain is not only transferred into the propagation along the reaction path but also into other molecular degrees of freedom that also reduces the sticking probability. These effects lead to a measurable reflection probability for singlet molecules approaching Al(111) with a low kinetic energy [106].

Even if the experimental molecular beam does not completely consist of singlets, it would still be possible to discriminate the nonadiabatically reflected molecules. All reflected singlet molecules leave the surface in the triplet state. However, this also means that the difference between the O_2 singlet and the triplet energy ΔE_{spin} (indicated in Figure 2.15) gets released during the scattering event. Due to the redistribution of this excess energy, the reflected molecules become translationally, vibrationally, and rotationally hot, that is, upon reflection they gain a significant amount of energy in all molecular degrees of freedom. Their energy distribution thus provides an unequivocal signature for the nonadiabatic spin transition.

2.9
Conclusions

Due to the improvement in the computer power and the development of efficient algorithms, it is now possible to reliably simulate the dynamics of reactions at surfaces from first principles, that is, without invoking any empirical parameter. Thus, detailed insight into reaction mechanisms in heterogeneous catalysis can be gained. However, the simulations are not only limited to explanatory purposes but also gain a predictive power, as demonstrated for the case of spin effects in the interaction of singlet O_2 with Al(111) in this chapter.

Still, most of the reactants addressed in the presented dynamical simulations are rather simple, as mainly diatomic molecules were considered. In particular in *ab initio* molecular dynamics simulations, it is in principle no problem to treat more complex molecules. Hence, we will certainly see dynamical studies including more complex catalytically relevant molecules such as methanol in the future.

On the other hand, dynamical simulations are typically restricted to a rather short timescale only up to pico- or at most nanoseconds. This means that they can often only treat specific steps of a catalytic reaction. In particular, rare processes hindered by large activation barriers are usually out of reach for dynamical simulation. Nevertheless, within a multiscale approach, involving either horizontal coupling (hybrid methods) or vertical coupling (parameter inheritance), as described elsewhere in this volume, dynamical simulations are an indispensable tool for a complete understanding of catalytic reactions at surfaces.

Acknowledgments

It is a pleasure to acknowledge all my coworkers and friends without whom many of the studies on reaction dynamics at surfaces presented in this chapter would not have been possible. These are in particular Christian Bach, Jörg Behler, Wilhelm Brenig, Christian Carbogno, Arezoo Dianat, Andreas Eichler, Jürgen Hafner, Michael Mehl, Christian Mosch, Dimitrios Papaconstantopulos, Karsten Reuter, Matthias Scheffler, Ching-Ming Wei, and Steffen Wilke. I sincerely enjoyed working with them.

Some of the calculations presented in this chapter have been made possible by computational resources provided by the John von Neumann Institute for Computing in Jülich and by the bwGRiD project of the Federal State of Baden-Württemberg, Germany, which is gratefully acknowledged.

References

1 Kroes, G.-J. (2008) *Science*, **321**, 794.
2 Groß, A. (2008) *Theoretical Surface Science: A Microscopic Perspective*, 2nd edn, Springer, Berlin.
3 Nørskov, J.K., Abild-Pedersen, F., Studt, F., and Bligaard, T. (2011) *Proc. Natl. Acad. Sci.*, **108**, 937–943.
4 Hänggi, P., Talkner, P., and Borkovec, M. (1990) *Rev. Mod. Phys.*, **62**, 251.
5 Cucchetti, A. and Ying, S.C. (1996) *Phys. Rev. B*, **54**, 3300–3310.
6 Fichthorn, K.A. and Weinberg, W.H. (1991) *J. Chem. Phys.*, **95**, 1090.
7 Sendner, C., Sakong, S., and Groß, A. (2006) *Surf. Sci.*, **600**, 3258.
8 Groß, A. (2003) Chapter 1, in *The Chemical Physics of Solid Surfaces*, vol. 11 (ed. D.P. Woodruff), Elsevier, Amsterdam.
9 Groß, A. and Dianat, A. (2007) *Phys. Rev. Lett.*, **98**, 206107.
10 Born, M. and Oppenheimer, J.R. (1927) *Ann. Phys.*, **84**, 457.
11 Wiesenekker, G., Kroes, G.-J., and Baerends, E.J. (1996) *J. Chem. Phys.*, **104**, 7344.
12 Groß, A. and Scheffler, M. (1998) *Phys. Rev. B*, **57**, 2493.
13 Groß, A., Wei, C.M., and Scheffler, M. (1998) *Surf. Sci.*, **416**, L1095.
14 Wei, C.M., Groß, A., and Scheffler, M. (1998) *Phys. Rev. B*, **57**, 15572.
15 Hammer, B., Scheffler, M., Jacobsen, K., and Nørskov, J. (1994) *Phys. Rev. Lett.*, **73**, 1400.
16 Blank, T.B., Brown, S.D., Calhoun, A.W., and Doren, D.J. (1995) *J. Chem. Phys.*, **103**, 4129.

17 No, K.T., Chang, B.H., Kim, S.Y., Jhon, M.S., and Scheraga, H.A. (1997) *Chem. Phys. Lett.*, **271**, 152.
18 Lorenz, S., Groß, A., and Scheffler, M. (2004) *Chem. Phys. Lett.*, **395**, 210.
19 Lorenz, S., Scheffler, M., and Groß, A. (2006) *Phys. Rev. B*, **73**, 115431.
20 Behler, J., Lorenz, S., and Reuter, K. (2007) *J. Chem. Phys.*, **127**, 014705.
21 Crespos, C., Collins, M.A., Pijper, E., and Kroes, G.-J. (2003) *Chem. Phys. Lett.*, **376**, 566.
22 Busnengo, H.F., Salin, A., and Dong, W. (2000) *J. Chem. Phys.*, **112**, 7641.
23 Kresse, G. (2000) *Phys. Rev. B*, **62**, 8295.
24 Lozano, A., Groß, A., and Busnengo, H.F. (2010) *Phys. Rev. B*, **81**, 121402(R).
25 Lozano, A., Groß, A., and Busnengo, H.F. (2009) *Phys. Chem. Chem. Phys.*, **11**, 5814.
26 McCormack, D.A., Olsen, R.A., and Baerends, E.J. (2005) *J. Chem. Phys.*, **122**, 194708.
27 Salin, A. (2006) *J. Chem. Phys.*, **124**, 104704.
28 Mehl, M.J. and Papaconstantopoulos, D.A. (1996) *Phys. Rev. B*, **54**, 4519.
29 Goringe, C.M., Bowler, D.R., and Hernández, E. (1997) *Rep. Prog. Phys.*, **60**, 1447.
30 Groß, A., Scheffler, M., Mehl, M.J., and Papaconstantopoulos, D.A. (1999) *Phys. Rev. Lett.*, **82**, 1209.
31 Slater, J.C. and Koster, G.F. (1954) *Phys. Rev.*, **94**, 1498.
32 Groß, A., Eichler, A., Hafner, J., Mehl, M.J., and Papaconstantopoulos, D.A. (2003) *Surf. Sci.*, **539**, L542.
33 Groß, A., Eichler, A., Hafner, J., Mehl, M.J., and Papaconstantopoulos, D.A. (2006) *J. Chem. Phys.*, **124**, 174713.
34 Newton, R. (1982) *Scattering Theory of Waves and Particles*, 2nd edn., Springer, New York.
35 Fleck, J.A., Morris, J.R., and Feit, M.D. (1976) *Appl. Phys.*, **10**, 129.
36 Tal-Ezer, H. and Kosloff, R. (1984) *J. Chem. Phys.*, **81**, 3967.
37 Kroes, G.-J. (1999) *Prog. Surf. Sci.*, **60**, 1.
38 Groß, A. (1998) *Surf. Sci. Rep.*, **32**, 291.
39 Groß, A. (2002) *Theoretical Surface Science: A Microscopic Perspective*, Springer, Berlin.
40 Groß, A., Wilke, S., and Scheffler, M. (1995) *Phys. Rev. Lett.*, **75**, 2718.
41 Kroes, G.-J., Baerends, E.J., and Mowrey, R.C. (1997) *Phys. Rev. Lett.*, **78**, 3583.
42 Nieto, P., Pijper, E., Barredo, D., Laurent, G., Olsen, R.A., Baerends, E.-J., Kroes, G.-J., and Farias, D. (2006) *Science*, **312**, 86.
43 Diaz, C., Pijper, E., Olsen, R.A., Busnengo, H.F., Auerbach, D.J., and Kroes, G.J. (2009) *Science*, **326** (5954), 832–834.
44 Groß, A. and Scheffler, M. (2000) *Phys. Rev. B*, **61**, 8425.
45 Olsen, R.A., McCormack, D.A., Luppi, M., and Baerends, E.J. (2008) *J. Chem. Phys.*, **128**, 194715.
46 Wilke, S. and Scheffler, M. (1996) *Phys. Rev. Lett.*, **76**, 3380.
47 Wilke, S. and Scheffler, M. (1996) *Phys. Rev. B*, **53**, 4926.
48 Karikorpi, M., Holloway, S., Henriksen, N., and Nørskov, J.K. (1987) *Surf. Sci.*, **179**, L41.
49 Rendulic, K.D., Anger, G., and Winkler, A. (1989) *Surf. Sci.*, **208**, 404.
50 Rutkowski, M., Wetzig, D., Zacharias, H., and Groß, A. (2002) *Phys. Rev. B*, **66**, 115405.
51 Wetzig, D., Rutkowski, M., Zacharias, H., and Groß, A. (2001) *Phys. Rev. B*, **63**, 205412.
52 Kroes, G.-J., Groß, A., Baerends, E.J., Scheffler, M., and McCormack, D.A. (2002) *Acc. Chem. Res.*, **35**, 193.
53 Rutkowski, M., Wetzig, D., and Zacharias, H. (2001) *Phys. Rev. Lett.*, **87**, 246101.
54 Rettner, C.T., Auerbach, D.J., and Michelsen, H.A. (1992) *Phys. Rev. Lett.*, **68**, 1164.
55 Rettner, C.T., Michelsen, H.A., and Auerbach, D.J. (1995) *J. Chem. Phys.*, **102**, 4625.
56 Grimmelmann, E.K., Tully, J.C., and Cardillo, M.J. (1980) *J. Chem. Phys.*, **72**, 1039.
57 Kuipers, E.W., Tenner, M.G., Spruit, M.E.M., and Kleyn, A.W. (1988) *Surf. Sci.*, **205**, 241.
58 Cohen, A.J., Mori-Sánchez, P., and Yang, W. (2008) *Science*, **321**, 792.

59 Groß, A. and Scheffler, M. (1997) *J. Vac. Sci. Technol. A*, **15**, 1624.
60 Rettner, C.T. and Auerbach, D.J. (1996) *Chem. Phys. Lett.*, **253**, 236.
61 Groß, A. and Scheffler, M. (1996) *Phys. Rev. Lett.*, **77**, 405.
62 Rettner, C.T. and Mullins, C.B. (1991) *J. Chem. Phys.*, **94**, 1626.
63 Luntz, A.C., Grimblot, J., and Fowler, D.E. (1989) *Phys. Rev. B*, **39**, 12903.
64 Wurth, W., Stöhr, J., Feulner, P., Pan, X., Bauchspiess, K.R., Baba, Y., Hudel, E., Rocker, G., and Menzel, D. (1990) *Phys. Rev. Lett.*, **65**, 2426.
65 Steininger, H., Lehwald, S., and Ibach, H. (1982) *Surf. Sci.*, **123**, 1.
66 Puglia, C., Nilsson, A., Hernnäs, B., Karis, O., Bennich, P., and Martensson, N. (1995) *Surf. Sci.*, **342**, 119.
67 Wintterlin, J., Schuster, R., and Ertl, G. (1996) *Phys. Rev. Lett.*, **77**, 123.
68 Eichler, A. and Hafner, J. (1997) *Phys. Rev. Lett.*, **79**, 4481.
69 Eichler, A., Mittendorfer, F., and Hafner, J. (2000) *Phys. Rev. B*, **62**, 4744.
70 Nolan, P.D., Lutz, B.R., Tanaka, P.L., Davis, J.E., and Mullins, C.B. (1998) *Phys. Rev. Lett.*, **81**, 3179.
71 Nolan, P.D., Lutz, B.R., Tanaka, P.L., Davis, J.E., and Mullins, C.B. (1999) *J. Chem. Phys.*, **111**, 3696.
72 Luntz, A.C., Williams, M.D., and Bethune, D.S. (1988) *J. Chem. Phys.*, **89**, 4381.
73 Verlet, L. (1967) *Phys. Rev.*, **159**, 98.
74 Sendner, C. and Groß, A. (2007) *J. Chem. Phys.*, **127**, 014704.
75 Groß, A. and Scheffler, M. (1998) *Phys. Rev. B*, **57**, 2493.
76 Busnengo, H.F., Dong, W., and Salin, A. (2000) *Chem. Phys. Lett.*, **320**, 328.
77 Crespos, C., Busnengo, H.F., Dong, W., and Salin, A. (2001) *J. Chem. Phys.*, **114**, 10954.
78 Hammer, B. (2001) *Phys. Rev. B*, **63**, 205423.
79 Groß, A., Bockstedte, M., and Scheffler, M. (1997) *Phys. Rev. Lett.*, **79**, 701.
80 Ciacchi, L.C. and Payne, M.C. (2004) *Phys. Rev. Lett.*, **92**, 176104.
81 Groß, A. (2009) *Phys. Rev. Lett.*, **103**, 246101.
82 Groß, A. (2010) *ChemPhysChem*, **11**, 1374.
83 Mitsui, T., Rose, M.K., Fomin, E., Ogletree, D.F., and Salmeron, M. (2003) *Nature*, **422**, 705.
84 Mitsui, T., Rose, M.K., Fomin, E., Ogletree, D.F., and Salmeron, M. (2003) *Surf. Sci.*, **540**, 5.
85 Lopez, N., Lodziana, Z., Illas, F., and Salmeron, M. (2004) *Phys. Rev. Lett.*, **93**, 146103.
86 Behm, R.J., Christmann, K., and Ertl, G. (1980) *Surf. Sci.*, **99**, 320.
87 Schmidt, P.K., Christmann, K., Kresse, G., Hafner, J., Lischka, M., and Groß, A. (2001) *Phys. Rev. Lett.*, **87**, 096103.
88 Lischka, M. and Groß, A. (2002) *Phys. Rev. B*, **65**, 075420.
89 Dedieu, A. (2000) *Chem. Rev.*, **100**, 543.
90 Feibelman, P.J. (1990) *Phys. Rev. Lett.*, **65**, 729.
91 Engdahl, C. and Wahnström, G. (1994) *Surf. Sci.*, **312**, 429.
92 Pineau, N., Busnengo, H.F., Rayez, J.C., and Salin, A. (2005) *J. Chem. Phys.*, **122**, 214705.
93 Brune, H., Wintterlin, J., Behm, R.J., and Ertl, G. (1992) *Phys. Rev. Lett.*, **68**, 624.
94 Diebold, U., Hebenstreit, W., Leonardelli, G., Schmid, M., and Varga, P. (1998) *Phys. Rev. Lett.*, **81**, 405.
95 Head-Gordon, M. and Tully, J.C. (1995) *J. Chem. Phys.*, **103**, 10137.
96 Kindt, J.T., Tully, J.C., Head-Gordon, M., and Gomez, M.A. (1998) *J. Chem. Phys.*, **109**, 3629.
97 Gergen, B., Nienhaus, H., Weinberg, W.H., and McFarland, E.W. (2001) *Science*, **294**, 2521.
98 Gergen, B., Weyers, S.J., Nienhaus, H., Weinberg, W.H., and McFarland, E.W. (2001) *Surf. Sci.*, **488**, 123.
99 Nienhaus, H. (2002) *Surf. Sci. Rep.*, **45**, 1.
100 Trail, J.R., Graham, M.C., and Bird, D.M. (2001) *Comp. Phys. Comm.*, **137**, 163.
101 Trail, J.R., Graham, M.C., Bird, D.M., Persson, M., and Holloway, S. (2002) *Phys. Rev. Lett.*, **88**, 166802.

102 Mizielinski, M.S., Bird, D.M., Persson, M., and Holloway, S. (2005) *J. Chem. Phys.*, **122**, 084710.
103 Lindenblatt, M., van Heys, J., and Pehlke, E. (2006) *Surf. Sci.*, **600**, 3624.
104 Lindenblatt, M. and Pehlke, E. (2006) *Phys. Rev. Lett.*, **97**, 216101.
105 Österlund, L., Zorić, I., and Kasemo, B. (1997) *Phys. Rev. B*, **55**, 15452–15455.
106 Carbogno, C., Behler, J., Groß, A., and Reuter, K. (2008) *Phys. Rev. Lett.*, **101**, 096104.
107 Sasaki, T. and Ohno, T. (1999) *Surf. Sci.*, **433**, 172.
108 Honkala, K. and Laasonen, K. (2000) *Phys. Rev. Lett.*, **84**, 705.
109 Yourdshahyan, Y., Razaznejad, B., and Lundqvist, B.I. (2002) *Phys. Rev. B*, **65**, 075416.
110 Behler, J., Delley, B., Lorenz, S., Reuter, K., and Scheffler, M. (2005) *Phys. Rev. Lett.*, **94**, 036104.
111 Behler, J., Delley, B., Reuter, K., and Scheffler, M. (2007) *Phys. Rev. B*, **75**, 115409.
112 Behler, J., Reuter, K., and Scheffler, M. (2008) *Phys. Rev. B*, **77**, 115421.
113 Tully, J.C. (1990) *J. Chem. Phys.*, **93**, 1061.
114 Bach, C., Klüner, T., and Groß, A. (2004) *Appl. Phys. A*, **78**, 231.
115 Carbogno, C., Groß, A., and Rohlfing, M. (2007) *Appl. Phys. A*, **88**, 579.
116 Carbogno, C., Behler, J., Reuter, K., and Groß, A. (2010) *Phys. Rev. B*, **81**, 035410.
117 Mosch, C., Koukounas, C., Bacalis, N., Metropoulos, A., Groß, A., and Mavridis, A. (2008) *J. Phys. Chem. C*, **112**, 6924.
118 Bacalis, N.C., Metropoulos, A., and Groß, A. (2010) *J. Phys. Chem. A*, **114**, 11746–11750.
119 Burgert, R., Schnöckel, H., Grubisic, A., Li, X., Bowen, S.T.S.K.H., Ganteför, G.F., Kiran, B., and Jena, P. (2008) *Science*, **319**, 438.

3
First-Principles Kinetic Monte Carlo Simulations for Heterogeneous Catalysis: Concepts, Status, and Frontiers

Karsten Reuter

3.1
Introduction

Forming the basis for the production of virtually all everyday products, catalysis has always been the driving force for chemical industries. In the twenty-first century, the concomitant importance of catalysis research is even further increased by the rapidly growing demand worldwide for more efficient exploitation of energy and material resources. As in many other areas of materials science, modern computational science is becoming a key contributor in the quest to quantitatively understand the molecular mechanisms underlying the macroscopic phenomena in chemical processing, envisioned to ultimately enable a rational design of novel catalysts and improved production strategies. Of particular relevance are hierarchical approaches that link the insights modeling and simulation can provide across all relevant length- and timescales [1]. At the molecular level, first-principles electronic structure calculations unravel the making and breaking of chemical bonds. At the mesoscopic scale, statistical simulations account for the interplay between all elementary processes involved in the catalytic cycle, and at the macroscopic scale, continuum theories yield the effect of heat and mass transfer, ultimately scaling up to a plant-wide simulation.

A comprehensive control of catalytic processes requires addressing all these facets and will thus ultimately necessitate novel methodological approaches that integrate the various levels of theory into one multiscale simulation. With the focus on the surface chemistry, first-principles kinetic Monte Carlo (kMC) simulations for heterogeneous catalysis represent precisely one step in this direction. A proper evaluation of the surface kinetics also dictates to unite two distinctly different aspects and in turn two distinct methodologies [2]: The first important part is an accurate description of the involved elementary steps, typically comprising adsorption and desorption processes of reactants and reaction intermediates, as well as surface diffusion and surface reactions. When aiming at a material-specific modeling at its best of predictive quality, the computation of the corresponding kinetic parameters is

---

Modeling and Simulation of Heterogeneous Catalytic Reactions: From the Molecular Process to the Technical System,
First Edition. Edited by Olaf Deutschmann.
© 2012 Wiley-VCH Verlag GmbH & Co. KGaA. Published 2012 by Wiley-VCH Verlag GmbH & Co. KGaA.

the realm of electronic structure theories [3–5] that explicitly treat the electronic degrees of freedom and thus the quantum mechanical nature of the chemical bond.

Even though such a set of first-principles kinetic parameters constitutes an (even for the most simple model systems hitherto barely achieved) important intermediate goal and highly valuable result, it does not suffice for the description of the surface catalytic function. For this, the second key ingredient is the occurrence (and thus relevance) of the individual elementary processes, in particular of the different reaction mechanisms. An evaluation of this statistical interplay within the manifold of elementary processes obviously needs to go beyond a separate study of each microscopic process. Taking the interplay into account naturally necessitates the treatment of larger surface areas. Much more challenging than this size, however, is the fact that the surface catalytic system is of course "open" in the sense that reactants, reaction intermediates, and products continuously impinge from and desorb into the surrounding gas phase. Owing to the highly activated nature of many of the involved elementary steps, the correspondingly required evaluation of the chemical kinetics faces the problem of a so-called "rare event dynamics," meaning that the time between consecutive events can be orders of magnitude longer than the actual process time itself. Instead of the typical picosecond timescale on which, say, a surface diffusion event takes place, it may therefore be necessary to follow the time evolution of the system up to seconds and longer in order to arrive at meaningful conclusions concerning the statistical interplay.

Tackling such demanding simulation times is the objective of modern nonequilibrium statistical mechanics techniques, many of which rely on a master equation type of description that coarse-grains the time evolution to the relevant rare event dynamics [6, 7]. Kinetic Monte Carlo simulations[1] [8] fall within this category and distinguish themselves from alternative microkinetic approaches in that they explicitly consider the correlations, fluctuations, and spatial distributions of the chemicals at the catalyst surface. For given gas-phase conditions, the typical output of such simulations are then the detailed surface composition and occurrence of each individual elementary process at any time. Since the latter comprises the surface reaction events, this also gives the catalytic activity in the form of products per surface area, either time resolved, for example, during induction, or time averaged during steady-state operation.

Summarized in one sentence, the central idea of first-principles kMC simulations for heterogeneous catalysis is thus to combine an accurate description of the elementary processes with an account for their statistical interplay in order to properly evaluate the surface chemical kinetics. As such, the approach is the result of a twofold choice: the choice to employ first-principles electronic structure calculations, presently predominantly density functional theory (DFT), to obtain the kinetic parameters of the individual processes, and the choice to employ a master equation-based kMC algorithm to tackle the required long simulation times.

---

1) Another less frequently employed name for kinetic Monte Carlo is dynamic Monte Carlo, with all of the early conceptual papers typically just referring to Monte Carlo. See, for example, the excellent review by Voter [8] for a nice account of the historic development of this technique.

Although aiming to give an introduction to this technique, this chapter discusses in detail why such a combination of distinct theories is necessary and what can be expected from it, and also discusses its advantages and limitations in comparison to alternative approaches. The underlying concepts will be first discussed, mostly illustrating the ideas with a simple toy system. Then, the use of first-principles kMC for catalysis-related problems focusing on one showcase will be discussed. Emphasis is placed not only on highlighting the capabilities but also on highlighting the challenges to such modeling. In the last section, the current frontiers of this technique are discussed, and as always "there is lots of room for improvement at the bottom".

3.2
Concepts and Methodology

3.2.1
The Problem of a Rare Event Dynamics

Instead of a more formal derivation, let me develop the more detailed conceptual discussion on the basis of a simple toy system. Consider the diffusive surface motion of an isolated atomic reaction intermediate that chemisorbs more or less strongly to specific sites offered by a solid surface. At finite temperature T, the adsorbate will vibrate around its adsorption site with a frequency on picosecond timescale and diffuse (depending on its bond strength) about every microsecond between two neighboring sites of different stability. The actual diffusion process itself also takes only a picosecond or so, and in the long time in between the adsorbate does really nothing else than performing its random thermal vibrations. Obviously, to get a proper understanding of what is going on in this system, we should follow its time evolution at least over a time span that includes several of the rare hops to neighboring sites, that is, here on the order of several microseconds or more. Despite its simplicity, this toy model carries all characteristics of a rare event system, in which the relevant dynamics of interest (here the hops over the surface) proceeds by occasional transitions with long periods of time in between. This separation of timescales between thermal vibrations and actual diffusion events results from the necessity to break (or "activate") chemical bonds. Highly activated processes are quite typical for surface chemistry and catalysis [9–11], and correspondingly such a rare event dynamics is more of a norm than an exception.

An important concept to further analyze the dynamics of our toy system is the so-called potential energy surface (PES) [12]. While chemical bonds are the consequence of electronic interactions, a frequently justified approximation is to assume that the electron dynamics takes place on much faster timescales than the motion of the atomic nuclei. In this Born–Oppenheimer picture, the electrons therefore adapt adiabatically to every configuration of atomic positions $\{\mathbf{R}_I\}$, and, reciprocally, the atoms can be viewed as traveling on the PES landscape $E\{\mathbf{R}_I\}$ established by the electronic interactions. The forces acting on a given atomic configuration are

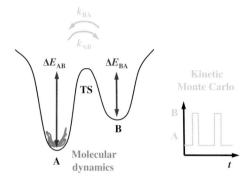

Figure 3.1 Schematic representation of the relevant degree of freedom of the PES underlying the surface diffusion problem discussed in the text: Two stable adsorption sites A and B of different stability are separated by a sizable barrier. An MD simulation of this problem would explicitly follow the dynamics of the vibrations around one minimum and would therefore most likely never escape to the other basin within affordable simulation times. KMC simulations coarse-grain this short-time dynamics into the rate constants, k_{AB} and k_{BA}, and consider the discrete state-to-state dynamics as prescribed by a Markovian master equation. Transition state theory is employed to derive the rate constants from the underlying PES, where in harmonic TST the barrier heights, ΔE_{AB} and ΔE_{BA}, as well as the vibrational modes at the minima and the transition state (TS) are required.

then the local gradient of the PES, and a (meta)stable atomic configuration corresponds to a (local) minimum of this landscape. In the language of a PES, the relevant degree of freedom of our surface diffusion problem can therefore schematically be described as shown in Figure 3.1: Two neighboring adsorption sites of different stability are represented by two minima of different depths, separated by a sizable barrier as a characteristic for the activated diffusion process.

A widely employed approach to follow the time evolution at the atomic level is molecular dynamics (MD) simulation that corresponds to a numerical integration of Newton's equations of motion [13]. Starting in any one of the two minima and using the forces provided by the PES gradient, an MD trajectory would explicitly track the entire thermal motion of the adsorbate. To accurately resolve the picosecond-scale vibrations around the PES minimum, this requires time steps in the femtosecond range. Surmounting the high barrier to get from one basin to the other is possible only if enough of the random thermal energy stored in all other degrees of freedom gets coincidentally united in the right direction. If this happens only every microsecond or so, as in our example, an MD simulation would have to first calculate on the order of 10^9 time steps until one of the relevant diffusion events can be observed. Even if the computational cost to obtain the forces would be negligible (which, as we will see later, is certainly not the case for PESs coming from first-principles calculations), this is clearly not an efficient tool to study the long-term time evolution of such a rare event system. In fact, such long MD simulations are presently computationally not feasible for any but the most simple model systems, and spending CPU time on any shorter MD trajectory would only yield insight into the vibrational properties of the system.

3.2.2
State-to-State Dynamics and kMC Trajectories

The limitations to a direct MD simulation of rare event systems are therefore the long time spans, in which the system dwells in one of the PES basins before it escapes to another one. Precisely this feature can, however, be seen as a virtue that enables a very efficient access to the system dynamics. Just because of the long time spent in one basin, it should not be a bad assumption that the system has forgotten how it actually got in there before it undergoes the next rare transition. In this case, each possible escape to another PES basin is then completely independent of the preceding basins visited, that is, of the entire history of the system. With respect to the rare jumps between the basins, the system thus performs nothing but a simple Markov walk [6].

Focusing on this Markovian state-to-state dynamics, that is, coarse-graining the time evolution to the discrete rare events, is the central idea behind a kMC simulation. Methodologically, this is realized by moving to a stochastic description that focuses on the evolution with time t of the probability density function $P_i(t)$ to find the system in state i representing the corresponding PES basin i. This evolution is governed by a master equation that, in view of the discussed system properties, is of a simple Markovian form [6]

$$\frac{dP_i(t)}{dt} = -\sum_{j \neq i} k_{ij} P_i(t) + \sum_{j \neq i} k_{ji} P_j(t), \qquad (3.1)$$

where the sums run over all system states j. The equation thus merely states that the probability to find the system in a given state i at any moment in time t is reduced by the probabilities to jump out of the present state i into any other basin j and is increased by the probabilities to jump from any other basin j into the present state i. These various probabilities are expressed in the form of rate constants k_{ij}, which give the average escape rate from basin i to basin j in units of inverse time. Because of the Markovian nature of the state-to-state dynamics, these quantities are also independent of the system history and thus an exclusive function of the properties of the two states involved. Assuming for now that all rate constants for each system state are known, the task of simulating the time evolution of the rare event system is in this description then shifted to the task of solving the master equation (3.1).

As will be discussed in the application example later, the typical number of states in models for a reactive surface chemistry is so huge that an analytical solution of the corresponding high-dimensional master equation is unfeasible. Following the usual stochastic approach of Monte Carlo methods [13, 14], the idea of a kMC algorithm is to achieve a numerical solution to this master equation by generating an ensemble of trajectories, where each trajectory propagates the system correctly from state to state in the sense that the average over the entire ensemble of trajectories yields probability density functions $P_i(t)$ for all states i that fulfill Eq. (3.1).

Since this ensemble aspect of kMC trajectories is quite crucial, let me return to our example of the diffusing adsorbate to further illustrate this point. Figure 3.2 shows a

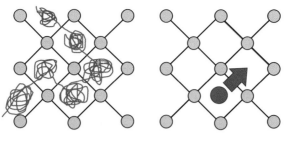

Molecular dynamics:
the whole trajectory

Kinetic Monte Carlo:
coarse-grained hops

Figure 3.2 Schematic top view explaining the differences between an MD (left panel) and a kMC (right panel) trajectory. Sketched is the path covered by an adsorbate that diffuses over the surface by rare hops to nearest-neighbor sites. While the MD trajectory resolves the short-time vibrational dynamics around the stable adsorption sites explicitly, this is coarse-grained into the rate constants in the kMC simulations so that the corresponding trajectory consists of a sequence of discrete hops from site to site.

schematic top view of our model surface sketching the path covered by the adsorbate in a time span covering a few of the rare diffusion events. The left panel shows the trajectory as it would have been obtained in an MD simulation. The rare event limitations to this simulation technique are obvious by the many wiggles around each one of the stable adsorption sites visited, representing the large time span of vibrational motion before the adsorbate achieves the rare transition to a neighboring site. The efficiency of a corresponding kMC trajectory, shown in the right panel of Figure 3.2, results from the fact that this vibrational short-term dynamics is eliminated from the simulation and (as we will see later) is instead appropriately accounted for in the rate constants k_{ij}. The trajectory consists of a mere series of discrete hops from one stable adsorption site to another, or in the language of the master equation, from state i to another state j. The sequence of sites and transition times of the proceeding kMC trajectory are randomly chosen by the kMC algorithm under appropriate consideration of the probabilities as contained in the rate constants k_{ij}. Contrary to the (for defined initial positions and momenta) deterministic MD trajectory, kMC trajectories are meaningful only in the sense that averaging over a sufficiently large number yields the correct probability with which the system is in any of the states i at any moment in time t.

While this must always be kept in mind, it is also worth noticing that computation of most of the quantities of interest would equally require some averaging over MD trajectories. If we take the diffusion coefficient as a typical target quantity within the area of our surface diffusion model, one possible approach to the involved computation of the average displacement per time would be, for example, to average over many hops observed during one sufficiently long MD trajectory or average over many shorter MD trajectories covering just one hop and starting with different initial momenta. As long as the Markov approximation is exactly fulfilled and the rate constants are accurate, the result obtained from an averaging over different kMC trajectories and the result obtained from the MD procedure would then be indistinguishable, only that the kMC approach is vastly more efficient and thus presumably the only one of the two that is computationally tractable.

3.2.3
kMC Algorithms: from Basics to Efficiency

A kMC trajectory consists of a sequence of discrete hops from one system state to another, where the random selection of which state is next visited and after which amount of time the corresponding hop occurs follows the probabilities prescribed by the master equation (3.1). Starting in any given state of the system, an algorithm generating such a trajectory must therefore appropriately determine in which state to jump next and what the corresponding escape time Δt_{escape} is. The system clock is then advanced by the escape time, $t \rightarrow t + \Delta t_{escape}$, and the procedure starts anew. As illustrated in Figure 3.1 for the case of our two-well toy system, the resulting timeline has a staircase-like structure where the system always remains in one state for the time t to $t + \Delta t_{escape}$ before it hops to the next state.

Obviously, Δt_{escape} is determined by the rate constants, and to derive the exact relationship required for the kMC algorithm, let us stay with the two-well example. In this system, there is always only one possibility where to go next. If in state A, there is the possibility to jump into state B with a probability expressed by the rate constant k_{AB}, and if in state B, there is the possibility to jump into state A with a probability prescribed by the rate constant k_{BA}. Now, one of the fundamental ideas behind the entire kMC approach is that during its thermal vibrational motion in one of the PES basins, the system loses the memory of its past history. Carrying this picture one step further, one may assume that this loss of memory occurs continuously, so that the system has the same probability of finding the escape path during each short increment of time it spends in the PES basin. This leads to an exponential decay statistics; that is, the probability that the system has, for example, escaped from state A into state B after a time Δt is $1 - \exp(-k_{AB}\Delta t)$. The connected probability distribution function for the time of first escape $p_{AB}(\Delta t)$ is just the time derivative of this and is a Poisson distribution,

$$p_{AB}(\Delta t) = k_{AB}\exp(-k_{AB}\Delta t), \qquad (3.2)$$

centered around the average time for escape given by $\overline{\Delta t_{escape}} = k_{AB}^{-1}$. When considering only this average time for escape, an executed jump from state A to state B would therefore simply advance the system clock by $t \rightarrow t + \overline{\Delta t_{escape}}$. Formally more correct is, however, to advance the system clock by an escape time that is properly weighted by the probability distribution function $p_{AB}(\Delta t)$ [15, 16]. Generating such an exponentially distributed escape time is numerically achieved through the expression

$$\Delta t_{escape} = -\frac{\ln(\varrho)}{k_{AB}}, \qquad (3.3)$$

where $\varrho \in]0, 1]$ is a random number. Equally, a jump from state B to A would advance the system clock by an exponentially distributed time $-\ln(\varrho)/k_{BA}$. If, as shown in

Figure 3.1, basin B has a lower stability, then the probability of jumping out of this shallower well will obviously be larger than jumping out of the deeper well A, that is, $k_{BA} > k_{AB}$. From Eq. (3.3), this means that the typical escape time out of A will be larger than out of B, as illustrated by the schematic time line shown in Figure 3.1. Averaged over a sufficiently long time, we therefore arrive at the expected result that the system spends more time in the more stable state.

Generalizing the escape time procedure to realistic systems with a manifold of different states i is straightforward [16]. Because of its loss of memory, the system has at any moment in time a fixed probability to find any one of the now many possible pathways out of the present minimum, with the fixed probabilities for the different pathways given by the different rate constants. Each pathway has thus its own probability distribution function for the time of first escape. Say, if we are in state i, we have for each possible pathway to another state j a Poisson distribution

$$p_{ij}(\Delta t) = k_{ij}\exp(-k_{ij}\Delta t). \tag{3.4}$$

Since only one event can be the first to happen, an intuitive generalization to propagate the kMC trajectory, also known as the first-reaction method [17], would be to draw a random number for each possible pathway and then determine for each one of them an exponentially distributed escape time through an expression of the form of Eq. (3.3). We then pick the pathway with the shortest escape time, move the system to the state reached by that pathway, advance the system clock by the corresponding shortest escape time, and begin again from the new state. Even though this is a perfectly valid kMC algorithm, it is clearly not particularly efficient, as we have to draw a lot of random numbers to generate all the different escape times and then discard all but one of them.

Among a variety of numerically efficient kMC algorithms suggested in the literature [18, 19], the one most commonly used is often attributed to Bortz et al. [20], even though one can clearly trace its idea further back[1]. As such, it is sometimes referred to as the BKL algorithm, with the "N-fold way," residence time algorithm, and Gillespie algorithm being other frequently used names. Figure 3.3 shows a flowchart of this algorithm. As necessarily required in any kMC procedure, it starts with the determination of all N possible processes (also known as escape pathways) out of the present system configuration or state. The corresponding N different rate constants are then summed to yield the total rate constant

$$k_{\text{tot}} = \sum_{p=1}^{N} k_p. \tag{3.5}$$

Executed is the process q, which fulfills the condition

$$\sum_{p=1}^{q} k_p \geq \varrho_1 k_{\text{tot}} \geq \sum_{p=1}^{q-1} k_p, \tag{3.6}$$

where $\varrho_1 \in]0, 1]$ is a random number. To understand the idea behind this condition, imagine for each process p a block of height k_p. If we then stack all these blocks on top

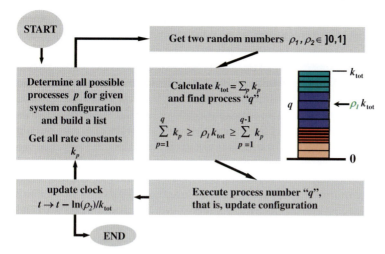

Figure 3.3 Flowchart illustrating the basic steps in the rejection-free BKL algorithm. The loop starts with the determination of all processes (and their rate constants) that are possible for the current system configuration. After the generation of two random numbers, the system is advanced according to the process selected by the first random number and the system clock increments according to the computed total rate constant and the second random number, as prescribed by an ensemble of Poisson processes. Thereafter, the loop starts anew, or the simulation is terminated, if a sufficiently long time span has been covered.

of each other as illustrated in Figure 3.3, we arrive at a stack of total height k_{tot}. Choosing a random height along this stack, that is, $0 < \varrho_1 k_{\text{tot}} \leq k_{\text{tot}}$, will point to one of the blocks and this is the process that is selected. Obviously, a process with a large rate constant, that is, a large block height, has a higher chance of being chosen in this way, and this probability-weighted selection is precisely what the partial sums in Eq. (3.6) achieve. By executing the selected process, the system is moved to the new configuration, and the system clock is advanced by

$$t \to t - \frac{\ln(\varrho_2)}{k_{\text{tot}}}, \qquad (3.7)$$

where $\varrho_2 \in]0, 1]$ is another random number, and the entire cycle starts anew from the new system state. Instead of drawing N different random numbers for the N possible pathways as in the intuitive first-reaction method, this algorithm thus needs only two in each cycle, which in case of realistic systems of the type discussed below makes a huge computational difference. Each cycle, furthermore, definitely propagates the system to a new state, which is also often viewed as an advantage of the corresponding class of "rejection-free" kMC algorithms.

An important aspect of the BKL algorithm is that the time by which the clock is advanced (cf. Eq. (3.7)) is independent of which process was actually chosen. To understand this, it is important to realize that the overall scale of Δt_{escape} is governed by the fastest process that can occur in a given configuration. This process with the highest rate constant has the shortest average escape time (cf. Eq. (3.4)), and even in the more intuitive first-reaction method, a slower process has a chance of getting selected only if its randomly drawn escape time is of this order of magnitude. Since the Poisson distribution of this slower process is centered around a longer average escape time, this happens only rarely and correspondingly the slow process occurs less often than the fast ones in the generated kMC trajectory as should be the case. In the BKL algorithm, this differentiation between slow and fast processes is instead achieved through the probability-weighted selection of Eq. (3.6). Nevertheless, since the overall magnitude of k_{tot} is predominantly determined by the high rate constant processes, the amount of time by which the clock is advanced after the execution of a process (cf. Eq. (3.7)) is equally reduced as soon as a fast process is possible. The same argument holds, of course, for increasing system sizes. Larger system size typically means more processes that can occur. With more terms in the sum in Eq. (3.5), k_{tot} gets larger, reducing the time increment achieved by the individual kMC steps and thereby limiting the total simulation times that can be reached.

In this respect, we see that the often-made statement that kMC enables simulation time up to seconds or longer is a bit sloppy. The total simulation time that can be reached depends instead on the system size and predominantly on the fastest process in the system, both of which dictate the increment in time typically achieved by one kMC step. If the system features a process that happens on a nanosecond timescale, it is this process and not the possibly more relevant less frequent ones that will almost always be executed. Each kMC step then advances the time also only on the order of a nanosecond, which may become a bottleneck for the simulation. When people hence refer to the macroscopic times that can be reached with kMC simulations, this simply means that the fastest processes in the particular problems and system sizes they studied allowed them to do so–or that they resorted to some of the known tricks to address this problem (see below). In fact, it depends on the physics of the problem what simulation times are required, and even when "only" achieving say microseconds, a kMC simulation may still be an enormous asset. This said, the limitations set by the presence of very fast processes are one of the current frontiers of this method, which will be discussed later.

3.2.4
Transition State Theory

From the preceding discussion, it is clear that the efficiency of a kMC simulation in tackling the long-time evolution of rare event systems arises from the fact that the short-time vibrational motion around the individual PES basins is appropriately coarse-grained into the rate constants k_{ij}. Up to now we have, however, simply assumed that these probabilities for the hops between the different system states are known. Obviously, if one only aspires a conceptual discussion, one could use here *ad hoc* values that fall purely from heaven or are (with more or less justification) believed to be somehow "characteristic" of the problem. In fact, the original development of kMC theory was done within this kind of approach [8, 15–17, 20, 21] and by now there is an entire bulk of literature where kMC simulations have been used in this sense. As stated in the introduction, the idea (and value) of modern first-principles kMC simulations is instead to be material specific, and in the philosophy of hierarchical models, to carry the predictive power of first-principles electronic structure theories to the mesoscopic scale [2]. For this type of modeling based on a proper microscopic meaning, an important aspect is to derive a stringent relationship between the rate constants k_{ij} and the PES information provided from the first-principles theories.

As the first step in this derivation, let us consider the constraints coming from the master equation for a system that has reached steady state. With a vanishing time derivative in Eq. (3.1), we arrive at one condition

$$\sum_{j \neq i} \left[k_{ij} P_i^* - k_{ji} P_j^* \right] = 0 \tag{3.8}$$

for every state i, where P_i^* and P_j^* are the time-independent probabilities that the steady-state system is in states i and j, respectively. This condition states that at steady state the sum of all transitions into any particular state i equals the sum of all transitions out of this state i. Since such a condition holds for every single state i in the system, the manifold of such conditions is, in general, fulfilled only if every term in the sum in Eq. (3.8) separately equals to zero:

$$\frac{k_{ij}}{k_{ji}} = \frac{P_j^*}{P_i^*}. \tag{3.9}$$

This is the detailed balance (or microscopic reversibility) criterion [6] that holds independently for the transitions between every pair of states i and j in the system. Furthermore, if the system has reached thermodynamic equilibrium with respect to the population of these two states, then the fractional population of states i and j on the right-hand side of Eq. (3.9) is simply proportional to the Boltzmann weighted difference in the free energies of these two states. For the case of our surface diffusion model and the two bound states of different stability, A and B, we would have the condition

$$\frac{k_{AB}}{k_{BA}} = \exp\left(-\frac{F_B(T) - F_A(T)}{k_B T}\right), \tag{3.10}$$

where k_B is the Boltzmann constant, and $F_A(T)$ and $F_B(T)$ are the free energies of states A and B, respectively, each comprising the total energy of the state and its vibrational free energy at temperature T.

The detailed balance relation places some constraints on the rate constants, and the PES information entering are the relative energies of the different states and their vibrational properties, that is, data about the (meta)stable minima. For the evaluation of equilibrium properties, no further specification of the rate constants is in fact needed. On the basis of the knowledge of the PES minima equilibrium MC algorithms such as the one by Metropolis [22] or Kawasaki [23] thus construct their transition probabilities according to detailed balance. In this respect, it is worth noting that in a dynamical interpretation, the resulting MC trajectories can therefore also be seen as numerical solutions to the master equation(3.1). This holds only for the equilibrated system though. While it is admittedly tempting to also assign some temporal meaning, for example, to the initial part of an MC trajectory when the system has not yet reached equilibrium, it is important to realize that an MC step in general does not correspond to a fixed amount of real time [14]. Only kMC algorithms propagate the system properly in time, and for this they need the absolute values of the rate constants, not just the relative specification as achieved by detailed balance [16]. The latter criterion is nevertheless an important constraint, in particular in light of the approximate determination of the rate constants to be discussed later. If (inadvertently) different and inconsistent approximations are made for the forward and backward rate constants between any two states, detailed balance is violated and the corresponding kMC simulations will never attain the correct thermodynamic limits.

At present, the most commonly employed approach to obtain the absolute values of the rate constants in first-principles kMC simulations in the area of surface chemistry and catalysis is transition state theory (TST) [24–26]. In TST, the rate constant for the transition from state A to state B in our two-well toy system is approximated by the equilibrium flux through a dividing surface separating the two states: Imagine creating an equilibrium ensemble by allowing a large number of replicas of this two-state system to evolve for a sufficiently long time that many transitions between the two states have occurred in each replica. If we then count the number of forward crossings through a dividing surface separating states A and B that occur per unit time in this ensemble and divide this by the number of trajectories that are on average in state A at any time, this yields the TST approximation to the rate constant k_{AB}^{TST}. Here, apart from the assumed equilibrium, another important implicit assumption is that successive crossings through the dividing surface are uncorrelated in the sense that each forward crossing takes the system indeed from state A to state B. In reality, there is, of course, the possibility that a trajectory actually recrosses the dividing surface several times before falling into state A or state B. Since this leads to k_{AB}^{TST} overestimating the real rate constant, the best choice of dividing surface is typically near the ridge top of the PES, where recrossings and in fact the entire equilibrium flux is minimized [26].

In this respect, it comes as no surprise that in harmonic TST (hTST) [27], the dividing surface is actually taken to be the hyperplane perpendicular to the reaction coordinate at the maximum barrier along the minimum energy path connecting the two states, with the understanding that the equilibrium flux occurs predominantly through this one transition state. With the additional approximation to the displacements sampled by the thermal motion, the PES around the minimum and at the saddle point (in the dimensions perpendicular to the reaction coordinate) is well described by a second-order expansion, that is, the corresponding vibrational modes are harmonic, this then leads to very simple expressions for the rate constants [26]. For the transition between the two bound states in Figure 3.1, one can, for example, derive the form [28]

$$k_{AB}^{TST}(T) = f_{AB}^{TST}(T) \left(\frac{k_B T}{h}\right) \exp\left(-\frac{\Delta E_{AB}}{k_B T}\right), \tag{3.11}$$

where h is the Planck constant and

$$f_{AB}^{TST}(T) = \frac{q_{TS(AB)}^{vib}}{q_A^{vib}}. \tag{3.12}$$

Here, ΔE_{AB} denotes the energy barrier between the two sites, $q_{TS(AB)}^{vib}$ the partition function at the transition state, and q_A^{vib} the partition function at the bound state A.

With a corresponding expression for the backward rate constant k_{BA}^{TST}, the form of Eq. (3.11) naturally fulfills the detailed balance condition (cf. Eq. (3.10)). Equally important is to realize that because k_{AB}^{TST} is an equilibrium property, there is no need to perform any actual dynamical simulations. Instead, only static PES information is required in the form of the energies at the initial state minimum and at the transition state, as well as the curvature around these two points for the harmonic modes. It is this efficiency of hTST that largely explains its popularity in first-principles kMC simulations, where as we will further discuss below every single energy and force evaluation "hurts." The other reason is that hTST tends to be a quite good approximation to the exact rate constant for the typical processes in surface chemical applications, such as diffusion or surface reaction events with a tight transition state. The making and breaking of strong covalent bonds involved in these processes seems to give rise to rather smooth PES landscapes, with deep troughs and simply structured ridges in between. These (anticipated) characteristics do not necessarily call for more elaborate reaction rate theories such as transition path sampling [29] or even more refined TST versions such as variational TST [30], in particular, as all these approaches require significantly more PES evaluations and thus come at a significantly higher computational cost in first-principles kMC.

In fact, with the vibrational properties of atomic or small molecular adsorbates at minimum and transition state often found to be rather similar, and hence $f_{AB}^{TST}(T) \sim 1$, practical work has, on the contrary, frequently dodged the vibrational calculations and resorted instead to the yet more crude approximation of setting the prefactor simply constant to $k_B T/h \approx 10^{13}$ s^{-1} for temperatures around room temperature. Especially with applications in surface chemistry and catalysis (involving larger molecules), one needs to stress in this respect that this procedure is not generally valid and certainly does not apply to, for example, unactivated adsorption processes, where the prefactor needs to account for the strong entropy reduction in going from the gas phase to the bound state at the surface. Assuming at a local partial pressure p_n of species n of mass m_n an impingement as prescribed by kinetic gas theory, the starting point for the calculation of the rate constant, for example, into site B, is then the expression [28]

$$k_{n,B}^{ads}(T, p_n) = \tilde{S}_{n,B}(T) \frac{p_n A_{uc}}{\sqrt{2\pi m_n k_B T}}, \tag{3.13}$$

where the local sticking coefficient $\tilde{S}_{n,B}(T)$ governs which fraction of these impinging particles actually sticks to a free site B, and A_{uc} is the area of the surface unit cell containing site B. The determination of the local sticking coefficient dictates, in principle, explicit dynamical simulations, which in practice may again be avoided by resorting to some approximate treatments. Directing the reader to Ref. [28] for a detailed discussion of this point, suffice it to repeat that in this case, or when in general resorting to approximations in the rate constant determination, particular care has to be taken that this does not infer a violation of detailed balance.

3.2.5
First-Principles Rate Constants and the Lattice Approximation

Within the hTST framework, the most important PES information entering the determination of a rate constant is apart from the (meta)stable minimum the location of the transition state. In the surface catalytic applications on which we focus here, the latter region of the PES corresponds typically to the situation where chemical bonds are made or broken. This fact imposes an important constraint when considering which methodology to use that would provide the required PES data. Electronic structure theories [3, 4], comprising both *ab initio* quantum chemistry and density functional methods, explicitly treat the electronic degrees of freedom and are therefore the natural base for such a modeling. However, in view of the high computational cost incurred by these techniques, and considering that for the rate constant determination only the ground-state energy and not the additionally provided detailed electronic structure data are necessary, an appealing approach would be to condense the information into a suitably parameterized interatomic potential. Such potentials (or force fields) then yield the energy and forces solely as a function of the nuclear positions and do so at a significantly reduced computational cost. While this approach appears to be highly successful, for example, in the area of biophysical applications and finds widespread use in applied materials research, it is crucial to realize that few, if any, of the existing atomistic potentials can deal with the complex charge transfer or changes in hybridization that are critical when bond breaking or bond making plays a role. When really aspiring a material-specific and predictive quality modeling of surface chemistry, the quantum mechanical nature of the chemical bonds needs to be explicitly treated, dictating the use of electronic structure theories to generate what is then commonly called first-principles rate constants.

For the extended metal or compound surfaces encountered in catalytic problems, DFT is the main workhorse among the electronic structure theories. Deferring to excellent textbooks for details [31–33], the major feature of this technique is that all complicated many-body effects of the interacting electron gas are condensed into the so-called exchange and correlation functional. The exact form of this functional remains elusive, necessitating the use of approximate functionals [34], with local (local density approximation, LDA) and semilocal (generalized gradient approximation, GGA) functionals still forming the most widely employed class in current applications [5]. While exact in principle, DFT in practice mostly does not even meet the frequent demand for "chemical accuracy" (1 kcal/mol $\approx$ 0.04 eV) in the energetics [35]. As studies of the kind illustrated in the next section only become computationally feasible by the unbeaten efficiency of DFT, this uncertainty is something that must be kept in mind at all times and we will return to a more detailed analysis of its consequences toward the end of this chapter. With expected typical errors in DFT-LDA or DFT-GGA barriers on the order of, say, ~0.2 eV, this also further justifies the use of a crude reaction rate theory such as hTST in most existing first-principles kMC work: With such an error entering the exponential in the rate constant expression, Eq. (3.11), a more elaborate determination, for example, of the prefactor may not be necessary. To put these critical remarks into perspective though, let me stress that DFT is at present the method of choice. With not many (semi-) empirical potentials on the market that are parameterized to deal with molecules at metal or oxide surfaces anyway, qualitatively wrong barriers with errors exceeding ~1 eV (and with that often completely wrong hierarchies in the process rate constants) would not be uncommon for any of them. If one thinks kMC based on DFT rate constants has problems, then simulations based on barriers from semi-empirical potentials are completely pointless.

Getting DFT energies and forces for reactions at extended metal or compound surfaces is typically achieved using so-called supercell geometries [36, 37]. With the technicalities of such calculations, for example, reviewed by Payne et al. [38], suffice it here to say that this automatically implies having to deal with rather large systems. In view of the huge computational cost connected to such calculations, one will consequently want to make sure that obtaining the required PES information and, in particular, locating the transition state can be done with as little energy and force evaluations as possible. In mathematical terms, locating the transition state means identifying a saddle point along a reaction path on a high-dimensional surface. Completely independent of the computational constraints, this is not at all a trivial problem, and the development of efficient and reliable transition state search algorithms is still a very active field of research [39, 40]. Among a variety of existing approaches, string algorithms such as climbing image nudged elastic band [41, 42] enjoy quite some popularity in surface reaction studies and are by now included in most of the modern DFT software packages. Despite this ready availability, caution is always advised with these algorithms as none of them is foolproof. In particular, if existing symmetries can be efficiently exploited to reduce the dimensionality of the problem, mapping the PES along a couple of suitable reaction coordinates might always be a worthwhile alternative that in addition provides more insight into the overall topology of the PES than just knowing where the saddle point is.

In either approach, identification of the transition state and the ensuing calculation of the rate constant for an individual pathway using DFT energies and forces is nowadays typically computationally involved, but feasible. However, the more different processes there are, the more calculations are necessary. The feasibility statement holds only if the total number of independent rate constant calculations, that is, primarily the expensive transition state searches, remains finite, if not small. In order to ensure this, first-principles kMC simulations up to now resort almost exclusively to a lattice mapping in conjunction with the thereby enabled exploitation of locality. Let me explain this concept again on the basis of the surface diffusion problem. Quite characteristic for many problems in surface catalysis, the adsorbate considered in this example shows a site-specific adsorption. In the case of a simple single-crystal surface, it is then straightforward to map the total diffusion problem on the periodic lattice formed by these sites, say, a cubic lattice in the case of adsorption into the hollow sites of an fcc(100) surface. On this lattice, a single diffusion event translates simply into a hop from one discrete site to another, and the advantage is that one has a unique specification of a given system state i just on the basis of the population on different lattice sites. If one now constructs a catalog of possible processes and their corresponding rate constants for every possible state on the lattice before the kMC simulation, one can check at every kMC step the configuration on the lattice, identify the state, and perform the random selection which process is executed at essentially zero computational cost by looking up the processes and their rate constants in the catalog [43].

The lattice mapping, together with the rate constant catalog, already suggests that the total number of rate constants that needs to be computed is finite. A significant reduction in this number that ultimately makes the simulations feasible is then achieved by considering the locality of most surface chemical processes. Assume in the simplest case that the diffusion process of the adsorbate would not be affected by the presence of other nearby adsorbates, apart from the fact that no diffusion could occur into any sites already occupied. Despite an astronomically large number of different system states corresponding each to a different spatial distribution of a given number of adsorbates on the lattice, only one single rate constant would need to be computed, namely, the rate constant for the pathway of a hop from one site to the next. Admittedly, such a situation with a complete absence of lateral interactions between the adsorbates is rarely realized. Nevertheless, nothing prevents us from extending the dependence on the local environment. Staying with the diffusion problem, imagine that the diffusion process is now affected only by the presence of other adsorbates in nearest-neighbor sites. For the illustrative example of just one adsorbate species on a cubic lattice, this means that–regardless of the occupation of more distant lattice sites–now five different rate constants need to be computed as illustrated in Figure 3.4: one for the diffusion process of an "isolated" adsorbate without any nearest neighbors, two with one nearest-neighbor site occupied, and one with two and one with three such sites occupied (if all four nearest-neighbor sites are occupied, the adsorbate obviously cannot move via a nearest-neighbor hop, so no rate constant calculation is required here). If the diffusion process depends further on how the neighbors are arranged around the diffusing adsorbate, further rate constants could be required, for example, to distinguish the situation with two neighbors, whether they are located on diagonal neighboring sites of the adsorbate as in Figure 3.4 or not. Thus, an increasing local environment dependence increases the number of inequivalent rate constants that need to be computed, and this number increases steeply when accounting for the constellation in the second, third, and so on nearest-neighbor shells. On the other hand, depending on the nature of the surface chemical bond, such lateral interactions also decay more or less quickly with distance, so a truncation or a suitable interpolation [44] is possible, in either case with only a quite finite amount of different DFT rate constant calculations remaining.

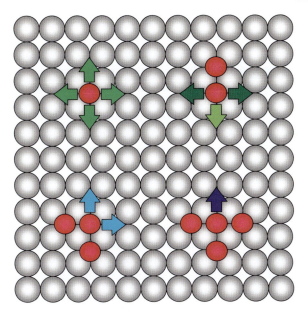

Figure 3.4 Schematic top view illustrating the effect of lateral interactions with the nearest neighbors in the surface diffusion problem with one atomic surface species and site-specific adsorption into the hollow sites of an fcc(100) lattice. Consideration of such lateral interactions requires the computation of five different rate constants, corresponding to the motions sketched with differently colored arrows, see text (small red spheres = adsorbate; large gray sphere = substrate atoms).

This leaves the problem of how to determine in the first place which lateral interactions are actually operative in a given system, keeping in mind that lateral interactions, especially at metal surfaces, are by no means necessarily restricted to just the pairwise interactions discussed in the example. For site-specific adsorption, the rigorous approach to this problem is to expand DFT energetics computed for a number of different ordered superstructures into a lattice gas Hamiltonian [2]. This cluster expansion technique [45–47] is to date primarily developed with the focus on lateral interactions between surface species in their (meta)stable minima [2, 44, 48–52], but generalizing the methodology to an expansion of the local environment dependence of the transition states is straightforward. Owing to the computational constraints imposed by the expensive transition state searches, existing expansions in first-principles kMC work in the field are often rather crude, truncating the dependence on the local environment often at the most immediate neighbor shells. In this situation, it has been suggested [53] that a less rigorous alternative could be to resort to semiempirical schemes such as the unity bond index-quadratic exponential potential (UBI-QEP) method [54] to account for the effects of the local environment. In either case, great care has again to be taken to ensure that applying any such approximations does not lead to sets of rate constants that violate the detailed balance criterion. Particularly in models with different site types and different surface species, this is anything but a trivial task.

3.3
A Showcase

Complementing the preceding general introduction to the underlying concepts, let me continue in this section with a demonstration of how first-principles kMC simulations are put into practice. Rather than emphasizing the breadth of the approach with a multitude of different applications, this discussion will be carried out using one particular example, namely, the CO oxidation at $RuO_2(110)$. With a lot of theoretical work done on the system, this focus enables a coherent discussion of the various facets, which is better suited to provide an impression of the quality and type of insights that first-principles kMC simulations can contribute to the field of heterogeneous catalysis (with Refs [55, 56], for example, providing similar compilations for epitaxial growth related problems). Since the purpose of the example is primarily to highlight the achievements and limitations of the methodological approach, suffice to say that the motivation for studying this particular system comes largely from the extensively discussed pressure gap phenomenon exhibited by "Ru" catalysts, with the $RuO_2(110)$ surface possibly representing a model for the active state of the catalyst under technologically relevant O-rich feeds (see, for example, Ref. [57] for more details and references to the original literature).

3.3.1
Setting up the Model: Lattice, Energetics, and Rate Constant Catalog

To some extent, the system lends itself to a modeling with first-principles kMC simulations, as extensive surface science experimental [58] and DFT-based theoretical studies [28] have firmly established that the surface kinetics predominantly takes place at two prominent active sites offered by the rutile-structured $RuO_2(110)$ surface, namely, the so-called coordinately unsaturated (cus) and the bridge (br) site. As illustrated in Figure 3.5, this naturally leads to a lattice model where these two sites are arranged in alternating rows and to consider as elementary processes the adsorption and desorption of O and CO at the bridge and cus sites, as well as

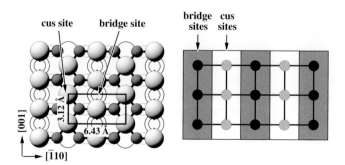

Figure 3.5 Top view of the $RuO_2(110)$ surface showing the two prominent active sites (bridge and cus). Ru = light, large spheres; O = dark small spheres. When focusing on these two site types, the surface can be coarse-grained to the lattice model on the right, composed of alternating rows of bridge and cus sites. Atoms lying in deeper layers have been whitened in the top view for clarity.

diffusion and surface chemical reactions of both reaction intermediates adsorbed at these sites. With a very small DFT-computed CO_2 binding energy to the surface [28], the surface reactions can further be modeled as associative desorptions; that is, there is no need to consider processes involving adsorbed CO_2.

Another benign feature of this system is its extreme locality in the sense that DFT-computed lateral interactions at the surface are so small that they can be neglected to a first approximation [28]. Considering the exhaustive list of noncorrelated, element-specific processes that can occur on the two-site-type lattice then leads to 26 different elementary steps, comprising the dissociative adsorption of O_2 resulting in three possible postadsorption states (two O atoms in neighboring cus sites, two O atoms in neighboring br sites, or two O atoms–one in a br site and one in a neighboring cus site), the associative desorption of O_2 from each of the three configurations of the O atoms, the adsorption of CO in cus or br sites, the desorption of CO from cus or br sites, the reaction of CO with O from four different initial states with the intermediates in neighboring sites (O in cus reacting with CO in cus, O in cus with CO in br, O in br with CO in br, and O in br with CO in cus), the corresponding four back-reactions dissociating gas-phase CO_2 into adsorbed CO and O, and the hops of O and CO from a site to the nearest site (for all possible site combinations).

The rate constants for all these processes were calculated using DFT-GGA to determine the energy barriers and TST expressions similar to those of Eqs (3.11) and (3.13), ensuring that all forward and backward processes obey detailed balance [28]. Figure 3.6 shows one of the correspondingly computed PES mappings

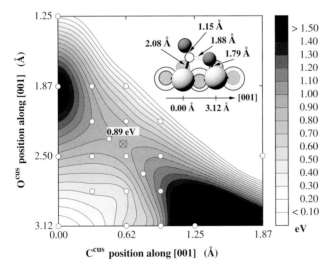

Figure 3.6 Calculated DFT-PES of one of the possible CO oxidation reaction mechanisms at the $RuO_2(110)$ model catalyst surface. The high-dimensional PES is projected onto two reaction coordinates, representing two lateral coordinates of the adsorbed O^{cus} and CO^{cus} (cf. Figure 3.5). The energy zero corresponds to the initial state at (0.00 Å, 3.12 Å), and the transition state is at the saddle point of the PES, yielding a barrier of 0.89 eV. Details of the corresponding transition state geometry are shown in the inset. Ru = light, large spheres; O = dark, medium spheres; and C = small, white spheres (only the atoms lying in the reaction plane itself are drawn as three-dimensional spheres). Reproduced from Ref. [59].

Table 3.1 Binding energies, E_b, for CO and O (with respect to $(1/2)O_2$) at bridge and cus sites (cf. Figure 3.5) and diffusion energy barriers, ΔE^{diff}, to neighboring bridge and cus sites, as used in the kMC simulations.

	E_b	ΔE^{des}			ΔE^{diff}	
		Unimol.	With O^{br}	With O^{cus}	To br	To cus
CO^{br}	−1.6	1.6	1.5	0.8	0.6	1.6
CO^{cus}	−1.3	1.3	1.2	0.9	1.3	1.7
O^{br}	−2.3	—	4.6	3.3	0.7	2.3
O^{cus}	−1.0	—	3.3	2.0	1.0	1.6

The desorption barriers are given for unimolecular and for associative desorption with either O^{cus} or O^{br}. This includes therefore surface reactions forming CO_2, which are considered as associative desorption of an adsorbed O and CO pair. All values are in eV. Reproduced from Ref. [28].

along high-symmetry reaction coordinates and Table 3.1 lists all the resulting energy barriers as used in the kMC simulations. To set the perspective, it is worth mentioning that it took half a million CPU hours at the time on Compaq ES45 servers to assemble this totality of energetic information required for the rate constant catalog.

The relevant physics that emerges at the level of this energetics is that there is a strong asymmetry in O binding to the two active sites, quite strong to the bridge sites (~ 2.3 eV/atom) and only moderate at the cus sites (~ 1.0 eV/atom). CO adsorption, on the other hand, has a rather similar strength on the order of ~ 1.5 eV/atom at both sites. From the established importance of the oxygen–metal bond breaking step in catalytic cycles and the Sabatier principle [9–11], one would thus expect the O^{cus} species to be mostly responsible for the high activity of this model catalyst at high pressures. In fact, this notion, as expressed by the well-known Brønsted–Evans–Polanyi-type relationships [60], is fully confirmed by the computed reaction barriers, with the two reactions involving the strongly bound O^{br} species exhibiting rather high barriers ($O^{br} + CO^{br}$:~ 1.5 eV, $O^{br} + CO^{cus}$:~ 1.2 eV) and the two reactions involving the moderately bound O^{cus} species exhibiting lower barriers ($O^{cus} + CO^{br}$:~ 0.8 eV, $O^{cus} + CO^{cus}$:~ 0.9 eV). From a mere inspection of these energetics, particularly the lowest barrier $O^{cus} + CO^{br} \rightarrow CO_2$ reaction appears most relevant for the catalysis, and one would imagine it to dominate the overall activity.

3.3.2
Steady-State Surface Structure and Composition

With the rate constant catalog established, the first important application area for kMC simulations is to determine the detailed composition of the catalyst surface under steady-state operation, that is, the spatial distribution and concentration of the reaction intermediates at the active sites. In the absence of mass transfer limitations in the reactor setup, the reactant partial pressures entering the surface impingement

(Eq. (3.13)) can be taken as equal to those at the inlet, without buildup of a significant product concentration in the gas-phase above the working surface. Furthermore, if one assumes that any heat of reaction is quickly dissipated, an approximation to steady-state operation can simply be achieved by performing the kMC simulations at constant reactant partial pressures p_{O_2} and p_{CO}, at constant global temperature T, and instantly removing any formed CO_2.

Figure 3.7 shows the time evolution of the actual and the time-averaged surface coverage for a corresponding constant (T, p_{O_2}, p_{CO}) run starting from an arbitrary initial lattice population (in this case, a fully O-covered surface) [28]. Despite notable fluctuations in the actual populations (characteristically determined by the size of the employed simulation cell, here a (20 × 20) lattice with 200 br and 200 cus sites), steady-state conditions corresponding to constant average values for all surface species are reached after some initial induction period. Although the dynamics of the individual elementary processes takes place on a picosecond timescale, this induction period can – depending on the partial pressures – take on the order of a tenth of a second even at this rather elevated temperature. At lower temperatures around room temperature, this becomes even more pronounced, and owing to the decelerated rate constants, the corresponding times (covered by an equivalent number of kMC steps) are orders of magnitude longer. These timescales are an impressive manifestation of the rare event nature of surface catalytic systems, considering that in the present example the evolution to the active state of the surface involves only the kinetics at two prominent adsorption sites without any

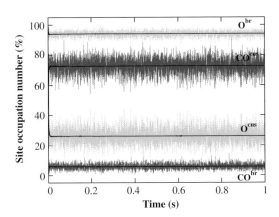

Figure 3.7 Time evolution of the site occupation by O and CO of the two prominent adsorption sites, bridge and cus (cf. Figure 3.5). The temperature and pressure conditions chosen ($T = 600$ K, $p_{CO} = 7$ atm, $p_{O_2} = 1$ atm) correspond to optimum catalytic performance (see below). Under these conditions, kinetics builds up a steady-state surface population in which O and CO compete for either site type at the surface, as reflected by the strong fluctuations in the site occupations within the employed (20 × 20) simulation cell. Note the extended timescale, also for the "induction period" until the steady-state populations are reached when starting from a purely oxygen-covered surface. Reproduced from Ref. [28].

complex morphological changes in the underlying substrate. It is clearly only the efficient time coarse-graining underlying kMC algorithms that makes it possible to reach such timescales, while still accounting for the full atomic-scale correlations, fluctuations, and spatial distributions at the catalyst surface.

Performing kMC runs starting from different initial lattice configurations and with different random number seeds allows to verify if the true dynamic steady state for a given (T, p_{O_2}, p_{CO}) environment is reached, and in the present system no indication for multiple steady states was found [28]. In this case, the surface populations under given gas-phase conditions obtained from averaging over sufficiently long time spans are then well defined in the sense that no further averaging over different kMC trajectories is required. Corresponding information about the concentration of the reaction intermediates at the active sites is therefore readily evaluated for a wide range of reactive environments, from ultrahigh vacuum to technologically relevant conditions with pressures on the order of atmospheres and elevated temperatures. One way of summarizing the obtained results is shown in the middle panel of Figure 3.8, which compiles the dominant surface species, also called most abundant reaction intermediates (MARIs), at constant temperature and as a function of the reactant partial pressures [28]. Note that the total computational time to obtain such a "kinetic phase diagram" [62] is typically insignificant compared to the aforementioned cost of assembling the first-principles rate constant catalog. In the present system, this is particularly pronounced, as the short correlation lengths resulting from the absence of lateral interactions enable the use of small simulation cells and there is no kMC bottleneck in the form of the discussed problem of a low-barrier process operating on a much faster timescale than all others.

At first glance, the overall structure of Figure 3.8 is not particularly surprising. In O-rich environments, the surface is predominantly covered with oxygen, in CO-rich

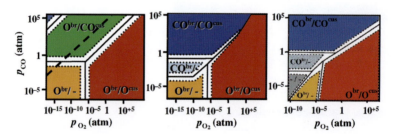

Figure 3.8 Steady-state surface composition of $RuO_2(110)$ in an O_2/CO environment at $T = 600$ K. In all nonwhite areas, the average site occupation is dominated (>90%) by one species, that is, O, CO, or empty sites (−). The labels correspondingly characterize the surface populations by indicating this majority species at the bridge and cus sites. In the white regions, there is a more complex coexistence of reaction intermediates and vacancies at the active sites. Compared are the results obtained by the first-principles kMC simulations (middle panel) to more approximate theories: *Constrained* atomistic thermodynamics (left panel) and microkinetic rate equations (right panel). Above the dashed line in the left panel, bulk RuO_2 is thermodynamically unstable against CO-induced decomposition; see text. Reproduced from Refs [28, 59, 61].

environments the surface is predominantly covered with CO, and in between there is a transition from one state to the other that coincides with O and CO, both being present at the surface in appreciable amounts. Such a transition from O-poisoned to CO-poisoned state depending on the partial pressure ratio is intuitive and conceptually already grasped, for example, by the early ZGB model [63]. The real advance brought by the first-principles kMC simulations is that this information is provided not only for a generic model but instead for the $RuO_2(110)$ system and without empirical input. Unlike in macroscopic engineering-type models, for example, through the MARI approximation, the dominant surface species are thus not assumed, but come out as a result of the proper evaluation of the statistical interplay of correctly described elementary processes at the microscopic level.

This rigorous solution of the master equation (cf. Eq. (3.1)) is also what distinguishes kMC simulations from more approximate theories that are otherwise equally built on a microscopic reaction mechanism as concerns a set of elementary processes and their rate constants. In prevalent microkinetic modeling, the master equation is simplified through a mean field approximation, leading to a set of coupled rate equations for the average concentrations of the different surface species [10, 11, 64]. *Constrained* atomistic thermodynamics, on the other hand, neglects the effect of the ongoing catalytic reactions and evaluates the surface populations in equilibrium with the reactive environment [59, 65]. Essentially, it thus corresponds to a kMC simulation where the reaction events are switched off, but because of its equilibrium assumption, it gets away with significantly less first-principles calculations. To better assess how these approximations can harm the theory, it is instructive to compare the results obtained with the three techniques when using exactly the same input in the form of the described set of elementary processes and DFT rate constants for the CO oxidation at $RuO_2(110)$. The corresponding "kinetic phase diagrams" are shown in Figure 3.8, which shows that in terms of the overall topology, the rate equation approach comes rather close to the correct kMC solution [61]. Quantitatively, there are, however, notable differences and we will return to this point in more detail later. In the case of constrained atomistic thermodynamics, the deviations under some environmental conditions are much more substantial, yet to be expected and easily rationalized [59, 65]. They concern prominently the presence of the strongly bound O^{br} species. For the thermodynamic approach, only the ratio of adsorption to desorption matters, and owing to its very low desorption rate, O^{br} is correspondingly stabilized at the surface even in highly CO-rich feeds. The surface reactions, on the other hand, are a very efficient means of removing this O^{br} species, and under most CO-rich conditions, they consume the surface oxygen faster than it can be replenished from the gas phase. Theories like kMC and the rate equation approach that explicitly account for the surface reactions yield a much lower average surface concentration of O^{br} than the thermodynamic treatment and, as a consequence, show an extended stability range of surface structures with CO^{br} at the surface (i.e., the $CO^{br}/-$ and CO^{br}/CO^{cus} regions in Figure 3.8).

Particularly the stability of these structures under rather CO-rich conditions has to be considered with care though. In such reducing environments, one would expect a CO-induced decomposition of the entire RuO_2 substrate to Ru metal, and the dashed line in the left panel of Figure 3.8 represents a thermodynamic estimate where this instability of the oxide bulk sets in [66]. Considering only the kinetics involving br and cus sites of an otherwise fixed lattice, neither kMC nor rate equations account for this instability, but yield at maximum a completely CO-covered surface. With its ability to quickly compare the stability of structures with completely different morphology, the constrained atomistic thermodynamics approach can in this respect be seen as a nice complement to the otherwise more accurate kinetic theories. Relevant for the ensuing discussion is also that the transition from O-poisoned to CO-poisoned surface, which as we will see below corresponds to catalytically most relevant environments, is quite far away from the oxide instability limit. Under such conditions, the kinetics determining the catalytic function concentrates on the br and cus sites, and the lattice model underlying the first-principles kMC simulations is fully justified.

With its explicit account of the reaction kinetics, it is needless to stress that kMC simulations do, of course, yield the correct temperature dependence. This is shown in Figure 3.9, which displays the "kinetic phase diagram" obtained at $T = 350$ K [28]. Shown is a region of much lower reactant partial pressures that, however, corresponds exactly to the same range of O_2 and CO chemical potentials as in the corresponding kMC diagram at $T = 600$ K in the middle panel of Figure 3.8. This is done to briefly address the general notion of thermodynamic scaling that expects equivalent surface conditions in thermodynamically similar gas phases and is thus often employed to relate results from surface science studies performed under ultrahigh vacuum conditions and low temperatures to catalytically relevant environments at ambient pressures and elevated temperatures. If such a scaling applies, the topology of the two diagrams at the two temperatures would exactly be the same, with only the width of the white coexistence regions varying according to the changing configurational entropy. Comparing the two kMC diagrams in Figures 3.8 and 3.9, it is clear that scaling is indeed largely present in the sense that the transition from O-poisoned to CO-poisoned state occurs roughly at similar chemical potentials.

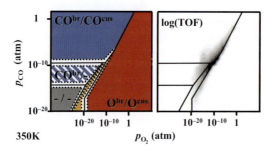

Figure 3.9 (Left panel) Equivalent plot of the first-principles kMC determined surface composition as in the middle panel of Figure 3.8, yet at $T = 350$ K (see Figure 3.8 for an explanation of the labeling). (Right panel) Map of the corresponding turnover frequencies in 10^{15} cm^{-2} s^{-1}: white areas have a TOF $< 10^3$ cm^{-2} s^{-1} and each increasing gray level represents one order of magnitude higher activity. Thus, the black region corresponds to TOFs above 10^{11} cm^{-2} s^{-1}, while in a narrow (p_{CO}, p_{O_2}) region the TOFs actually peak over 10^{12} cm^{-2} s^{-1}. Reproduced from Ref. [28].

Nevertheless, in detail notable differences due to the surface kinetics can be discerned, cautioning against a too uncritical use of thermodynamic scaling arguments and emphasizing the value of explicit kinetic theories like kMC to obtain the correct surface structure and composition at finite temperatures.

3.3.3
Parameter-Free Turnover Frequencies

Besides the surface populations, another important group of quantities that is straightforward to evaluate from kMC simulations of steady-state operation are the average frequencies with which the various elementary processes occur. Apart from providing a wealth of information on the ongoing chemistry, this also yields the catalytic activity as the sum of the averaged frequencies of all surface reaction events. Properly normalized to the surface area, a constant (T, p_{O_2}, p_{CO}) first-principles kMC run thus provides a parameter-free access to the net rate of product formation or turnover frequency (TOF) (measured in molecules per area and time).

A corresponding TOF plot for the same range of gas-phase conditions as discussed for the "kinetic phase diagram" at $T = 350$ K is also included in Figure 3.9 [28]. The catalytic activity is narrowly peaked around gas-phase conditions corresponding to the transition region where both O and CO are present at the surface in appreciable amounts, with little CO_2 formed in O-poisoned or CO-poisoned state. On a conceptual level, this is not particularly surprising and simply confirms the view of heterogeneous catalysis as a "kinetic phase transition" phenomenon [59, 62, 63], stressing the general importance of the enhanced dynamics and fluctuations when the system is close to an instability (here the transition from O-covered to CO-covered surface). Much more intriguing is the quantitative agreement achieved with existing experimental data [67] as shown in Figure 3.10. Recalling that the calculations do not rely on any empirical input, this is quite remarkable. In fact, considering the multitude of uncertainties underlying the simulations, in particular the approximate DFT-GGA energetics entering the rate constants, such an agreement deserves further comment and this will be discussed in more detail later.

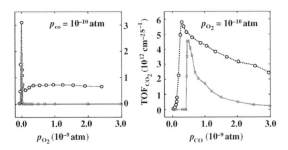

Figure 3.10 Comparison of the steady-state TOFs at $T = 350$ K from first-principles kMC (solid line) and the experimental data from Wang et al. [67] (dotted line). Shown is the dependence with p_{O_2} at $p_{CO} = 10^{-10}$ atm (left), and the dependence with p_{CO} at $p_{O_2} = 10^{-10}$ atm (right) (cf. the overall TOF plot in Figure 3.9). Reproduced from Ref. [28].

Figure 3.11 Snapshot of the steady-state surface population under optimum catalytic conditions at $T = 600$ K ($p_{O_2} = 1$ atm, $p_{CO} = 7$ atm). Shown is a schematic top view, where the substrate bridge sites are marked by gray stripes and the cus sites by white stripes (cf. Figure 3.5). Oxygen adatoms are drawn as light gray circles and adsorbed CO molecules as dark gray circles. Reproduced from Ref. [28]

In the catalytically most active coexistence region between O-poisoned and CO-poisoned state, the kinetics of the ongoing surface chemical reactions builds up a surface population in which O and CO compete for either site type at the surface. This competition is reflected by the strong fluctuations in the site occupations visible in Figure 3.7, leading to a complex spatial distribution of the reaction intermediates at the surface. In the absence of lateral interactions in the first-principles kMC model, there is no thermodynamic driving force favoring a segregation of adsorbed O and CO. Nevertheless, as shown in Figure 3.11, a statistical analysis of the surface population reveals that they are not distributed in a random arrangement [28]. Instead, they tend to cluster into small domains that extend particularly in the direction along the bridge rows and cus trenches. This tendency arises out of the statistical interplay of all elementary processes, but emerges at the considered ambient pressures primarily from diffusion limitations and the fact that the dissociative adsorption of oxygen requires two free neighboring sites and the unimolecular adsorption of CO requires only one free site.

The resulting complex, and fluctuating spatial arrangement has important consequences for the catalytic function. As shown in Figure 3.12, under these conditions of highest catalytic performance, it is not the reaction mechanism with the highest rate constant, that is, the lowest barrier reaction $O^{cus} + CO^{br} \rightarrow CO_2$ that dominates the overall activity [28]. Although the process itself exhibits very suitable properties for the catalysis, it occurs too rarely in the full concert of all possible processes to decisively affect the overall functionality. Even under most active conditions, for example, at $T = 600$ K, for $p_{O_2} = 1$ atm and $p_{CO} = 7$ atm, it contributes only around 10% to the total TOF, while essentially all of the remaining activity is due to the $O^{cus} + CO^{cus} \rightarrow CO_2$ reaction mechanism [28]. This finding could not have been obtained on the basis of the first-principles energetics and rate constants alone, and it emphasizes the importance of the statistical interplay and the novel level of understanding that is provided by first-principles kMC simulations. How critical it is that this evaluation of the interplay is done rigorously is again nicely illustrated by comparing to microkinetic modeling using mean field rate equations. As shown in Figure 3.12, although using exactly the same set of elementary processes and first-principles rate constants, this theory incorrectly predicts an almost equal contribution of the two competing low-barrier reaction mechanisms to the total TOF

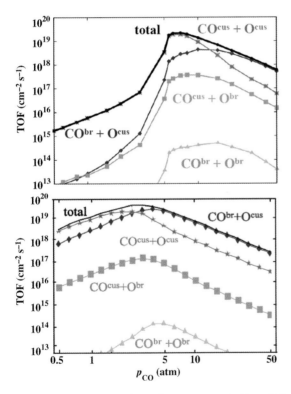

Figure 3.12 Variation in the catalytic activity with CO partial pressure at $T = 600$ K and $p_{O_2} = 1$ atm. The displayed gas-phase conditions comprise the most active state and the transition from O-poisoned to CO-poisoned surface (sketched as white region in Figure 3.8). Shown is the total TOF, as well as the contribution of the four different reaction mechanisms. (Upper panel) First-principles kMC. (Lower panel) Microkinetic rate equations. Reproduced from Ref. [61]

at peak performance [61]. This peak performance (at the corresponding transition from O-poisoned to CO-poisoned surface) is furthermore obtained at slightly shifted pressure conditions as was also apparent from the "kinetic phase diagrams" in Figure 3.8. The neglect of the spatial inhomogeneity in the mean field approach thus leads to severe shortcomings, underscoring the virtue of first-principles kMC simulations to explicitly resolve the locations of all reaction intermediates at the surface.

In this respect, it is intriguing to notice that the microkinetic modeling yields a TOF that is about three orders of magnitude in error particularly under conditions where one would intuitively expect the mean field approximation to be well fulfilled (the O-poisoned state at the lowest pressures shown in Figure 3.12 on the left). The reason for this deficiency even at an almost perfectly homogeneous surface coverage is that mean field assumes the random existence of independent vacant sites with the probability of neighboring divacancies (created by associative desorption of O_2 molecules or required for dissociative adsorption of gas-phase O_2) correspondingly being proportional to the vacancy concentration squared, θ_{vac}^2. However, due to severe diffusion limitations particularly along the cus site trenches, divacancies created through oxygen desorption persist for such a long time that they completely dominate the total vacancy concentration at the surface. Rather than going as θ_{vac}^2, the probability for a divacancy is thus close to half the probability of a single vacancy, $\approx \theta_{vac}/2$. If one patches the oxygen adsorption expression in the rate equations accordingly, a TOF virtually identical to the kMC result is obtained [68].

This trick works, of course, only for gas-phase conditions corresponding to the O-poisoned surface, and incorrect TOFs are then obtained, for example, in the CO-poisoned state. This nothing but exemplifies the well-known difficulties of effectively correcting a rate equation formulation to at least partly account for site correlations. KMC on the other hand, does not suffer from this deficiency, and fully includes all correlations and stochastic fluctuations at the active surface sites into the modeling. In this context, it is important to realize that another way of patching up deficiencies in the statistical modeling is to resort to effective kinetic parameters. If one considers all rate constants in the mean field rate equation model not to be fixed by the underlying first-principles calculations, but instead to be free parameters, it is possible to achieve a perfect fit to the kMC TOF profile shown in Figure 3.12 [61, 68]. However, the result is just an effective description, with the optimized rate constants no longer having any kind of microscopic meaning and deviating in their values from the true first-principles rate constants underlying the kMC TOF profile by several orders of magnitude. Not surprisingly, this effective description works only inside the parameterized range and is not transferable to gas-phase conditions outside those shown in Figure 3.12, where it predicts grossly wrong catalytic activities.

This highlights a crucial conceptual point: A hierarchical technique like first-principles kinetic Monte Carlo that builds on microscopically well-defined parameters is substantially more involved than existing empirical approaches not only because of the intense first-principles calculations but also because it typically requires a significantly improved description at all ensuing levels, here the solution of the statistical mechanics problem. Effective parameters, for example, rate constants fitted to experimental data, provide the possibility to (at least partly) cover up deficiencies in the modeling. In the present context, this means that one can get away with mean field rate equations despite existing site correlations or that one can get away with a significantly reduced number of kMC processes that then represent some unspecified lump-sum of not further resolved elementary processes. The price paid by such a seemingly simpler modeling is that it is typically neither transferable nor predictive, and most importantly lacks the reverse mapping feature much aspired in multiscale modeling, that is, to be able to unambiguously trace the correct microscopic origin of properties identified at the meso- or macroscopic level. All of this is possible in a theory like first-principles kMC, but one has to walk an extra mile in the form of an unprecedented level of detail and accuracy in the modeling. This level is at present mostly not matched in engineering-style approaches to heterogeneous catalysis, and is also the reason why existing first-principles kMC applications are hitherto still confined to well-defined model catalysts and simple reaction schemes.

3.3.4
Temperature-Programmed Reaction Spectroscopy

Properly tracking the system time, kMC simulations can, of course, not only address catalytic systems during steady-state operation. A prominent example for a time-dependent application in the field would be temperature-programmed desorption (TPD) or reaction (TPR) spectroscopies, which provide insight into the binding energetics of reaction intermediates by recording the amount of desorbing species while ramping the substrate temperature [69]. With typical experimental heating rates of a few Kelvin per second, modeling an entire TPD/TPR spectrum covering, say, a temperature range of a few hundred Kelvin leads again to timescales that are naturally tackled by kMC simulations. Always starting from a defined initial lattice configuration, the desorption or reaction rate as a function of temperature is there obtained from averages over an ensemble of kMC trajectories in which with progressing simulation time the temperature-dependent first-principles rate constants are adapted according to the applied heating ramp. As the kMC time does not evolve continuously, but in finite steps according to Eq. (3.7), there can be some technicalities with continuous temperature programs [70], which in practice can be addressed by larger simulation cells (leading to a larger k_{tot} and thus smaller time steps), finite temperature bins, or other system-specific solutions.

Another type of TPD/TPR quantity that is typically much less sensitive to this and some other problems like slight inaccuracies in the first-principles rate constants are integral yields, that is, the total amount of a species that has come off the surface during some extended temperature window. Such yields were, for example, measured for the $RuO_2(110)$ system from surfaces in which initially all bridge sites were always fully occupied by O^{br}, while the coverage of O^{cus} varied in the range $0 < \theta < 0.8$ monolayer (ML), where 1 ML corresponds to an occupation of all cus sites [71]. The remaining free cus sites were then each time saturated with CO, so that the initially prepared surfaces contained an amount of 1 ML O^{br}, θ ML O^{cus}, and

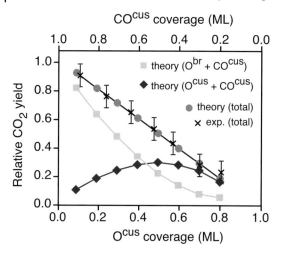

Figure 3.13 Total TPR CO_2 yield from $RuO_2(110)$ surfaces initially prepared with varying O^{cus} coverage θ. In all surfaces, all bridge sites are initially covered with O^{br} species, and the remaining $(1-\theta)$ cus sites not covered by O^{cus} are occupied by CO^{cus}. The CO_2 yield is given relative to the one obtained for the surface with zero O^{cus} coverage. Shown are the total simulated CO_2 yield and the contributions from the two dominant reaction mechanisms under these conditions, $O^{cus} + CO^{cus}$ and $O^{br} + CO^{cus}$. The experimental data are taken from Ref. [71]. Reproduced from Ref. [72].

$(1-\theta)$ ML CO^{cus}. The idea of this set of experiments was to evaluate which of the surface oxygen species, O^{cus} or O^{br}, is more reactive. With a constant population of O^{br} and a linearly decreasing amount of CO^{cus}, the expectation within a mean field picture was that in case of a dominant $O^{br} + CO^{cus}$ reaction the total CO_2 yield should show a linear $(1-\theta)$ dependence, whereas with a linearly varying amount of CO^{cus} a dominant $O^{cus} + CO^{cus}$ reaction would be reflected by a parabolic $\theta(1-\theta)$ variation. The measured linear dependence shown in Figure 3.13 was therefore taken as evidence for a much more reactive O^{br} species, which was difficult to reconcile both with the much lower DFT-GGA O^{cus} binding energy and with the lower $O^{cus} + CO^{cus}$ reaction barrier (cf. Table 3.1).

Only subsequent first-principles kMC simulations based exactly on this DFT-GGA energetics were able to resolve this puzzle as yet another consequence of the inhomogeneities in the adlayer caused by the specific arrangement of the active sites in conjunction with diffusion limitations of the reaction intermediates [72]. As apparent from Figure 3.13, the simulations perfectly reproduce the experimental data and even reveal that the contributions from the two competing reaction channels follow indeed the functional form expected from mean field (linear versus parabolic). Despite the much higher $O^{cus} + CO^{cus}$ rate constant, the share of this reaction mechanism is, however, largely suppressed, as under the conditions of the TPR experiments rows of strongly bound O^{br} species confine the reactive O^{cus} species to one-dimensional cus site trenches. With strong diffusion limitations inside these

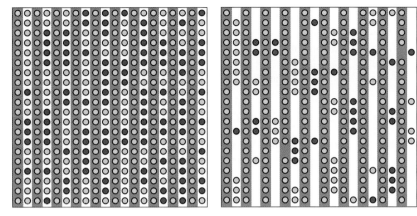

Figure 3.14 Snapshots of the surface population in the first-principles TPR kMC simulation for the surface covered initially with 1 ML O^{br}, 0.5 ML O^{cus}, and 0.5 ML CO^{cus}. Shown is a schematic top view as in Figure 3.11, where the substrate bridge sites are marked by gray stripes and the cus sites by white stripes. Oxygen adatoms are drawn as lighter gray circles and adsorbed CO molecules as darker gray circles. (Left panel) Initial population at $T = 170\,K$. (Right panel) Population at $T = 400\,K$. At this temperature, according to mean field >98% of all initially present CO^{cus} species should already have been reacted off by the low-barrier $O^{cus} + CO^{cus}$ reaction. Instead, 40% of them are still present, namely, essentially all those that were not initially adsorbed immediately adjacent to a O^{cus} species. Reproduced from Ref. [72].

trenches (cf. Table 3.1), the CO^{cus} molecules can access only a fraction of the O^{cus} species and instead react with their immediate O^{br} neighbors as illustrated in Figure 3.14.

As also evident from Figure 3.14, decreasing surface coverages toward the high-temperature end of the heating ramp are an inherent feature of TPD/TPR spectroscopy. With the concomitant increasing role of surface diffusion and a manifold of systems exhibiting rather low diffusion barriers, it is finally worth pointing out that TPD/TPR kMC simulations are particularly prone to the mentioned fastest-process bottleneck. As the gap between low and higher barrier rate constants opens up more and more with temperature (cf. Eq. (3.11)), it is especially at the high-temperature end that the kMC algorithm increasingly ends up just executing diffusion events at minuscule time increments. Approximate workarounds to this problem include either an artificial raising of the lowest barriers or a mixing of kMC with MC schemes [21]. Both approaches work on the assumption that on the timescale of the slower processes the system essentially equilibrates over the entire subset of states that can be reached via the fast processes, that is, that there is a separation of timescales. Raising the lowest barriers will slow down the fastest processes, which at an increased kMC efficiency will still yield an accurate dynamics if the fast processes are still able to reach equilibration even when they are slowed down. Alternatively, reducing the kMC algorithm to just the slow processes an equilibration over the subset of states reached by the fast processes can be achieved by performing some appropriate MC simulations after every kMC step. While quite some system-specific progress has been achieved along these lines, it is in general hard to know for sure that such approximations are not corrupting the dynamics – and the low-barrier problem prevails as one of the long-standing challenges to kMC simulations.

3.4
Frontiers

By now it has become clear that the crucial ingredients to a first-principles kMC simulation are the electronic structure (DFT) calculations to get the PES, the mapping of this PES information onto a finite set of elementary processes and rate constants, and the master equation-based kMC algorithm to evaluate the long-time evolution of the rare event system. Quite natural for a hierarchical approach spanning electronic to mesoscopic length and time scales, this reflects three complementary regimes of methodological frontiers: at the level of the electronic interactions, at the statistical mechanics level, and in the interfacing in between.

Necessarily representing the finest scale in any multiscale materials modeling, it is natural to start a short survey of open issues in the three regimes at the electronic structure level. Obviously, if the needed accuracy is lacking at this base, there is little hope that accurate predictions can be made at any level of modeling that follows. In this respect, the accuracy level at which first-principles rate constants can presently be computed for catalytic surface systems of a complexity as the $RuO_2(110)$ example is of course a major concern. As discussed in Section 3.2.5, the limitation is thereby predominantly in the PES uncertainty, and only to a lesser extent in the prevailing use of hTST as reaction rate theory. The problem here is that approaches with an accuracy superior to the present-day workhorse DFT-GGA, but still tractable computational demand, are not readily available. This is particularly pronounced for catalytically most relevant transition metal surfaces, where post Hartree–Fock quantum chemistry methods are ill-positioned and there is increasing evidence that the much acclaimed hybrid DFT functionals are hardly a major step forward [5, 73]. On top of this, the current focus of first-principles kMC work on the Born–Oppenheimer ground-state PES must not necessarily always convey the correct physics. Especially for core steps in the catalytic cycle like adsorption processes involving electron transfer or spin changes electronically nonadiabatic effects and the consideration of excited states may be essential [74].

With all these issues under active research, it is crucial to realize that while an accurate description of every elementary process is of course desirable, what matters at first in the context of first-principles kMC simulations is how the error contained in the rate constants propagates to the statistical mechanics level and affects the resulting ensemble properties. A very promising approach is therefore to employ sensitivity analyses to identify which processes critically control the targeted quantities. In the catalysis context, one possibility to identify such "rate-limiting steps" are degree of rate control (DRC) approaches, which essentially correspond to studying the linear response of the TOF to a change in one of the rate constants of the reaction network [75–78]. Such an analysis has recently been performed for the $RuO_2(110)$

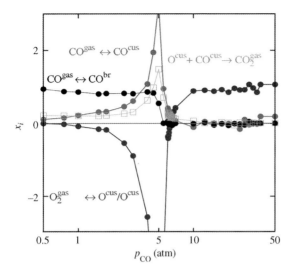

Figure 3.15 Degree of rate control x_i of different elementary processes in the first-principles kMC model of CO oxidation at RuO$_2$(110). Shown is the dependence on the gas-phase conditions of Figure 3.12, where at constant $T = 600$ K and $p_{O_2} = 1$ atm the variation of p_{CO} causes the transition from the CO-poisoned surface state (left end of shown pressure range) to the O-poisoned surface state (right end of shown pressure range) with the catalytically most active coexistence region around $p_{CO} = 7$ atm. The DRC values for all processes not shown are zero on the scale of this figure and modest errors contained in the corresponding first-principles rate constants will practically not affect the computed TOF. Reproduced from Ref. [79].

system and the corresponding results for the set of gas-phase conditions also discussed in Figure 3.12 are summarized in Figure 3.15 [79]. The most important conclusion from this work for the present discussion is that while particularly in the catalytically most active state there is no single but an entire group of rate-limiting steps, this number of processes controlling the overall CO$_2$ production is small. This indicates that if the rate constants of these processes are known accurately, the kMC procedure will produce correct results even if the other rate constants are inaccurate. Strikingly, some of the unimportant rate constants could, in fact, be varied over several orders of magnitude with no effect on the computed TOF [79]. Apart from providing a tool for analyzing the mechanism of a complex set of catalytic reactions, such a DRC sensitivity analysis thus tells for which aspects of the reaction mechanism, say, a potential DFT-GGA error in the rate constants is not much of a problem. On the other hand, it also tells which aspects must be treated accurately with corresponding errors in the rate constants directly carrying through to the mesoscopic simulation result. In this respect, the RuO$_2$(110) DRC work revealed that under the gas-phase conditions of the experiments, shown in Figure 3.10, the ruling processes are the dissociative adsorption of O$_2$ into a cus site pair and the desorption of CO^{cus}. The almost quantitative agreement with the experimental data reached by the first-principles kMC simulations suggests therefore that especially the microscopic quantities entering the corresponding rate constants, namely, the sticking coefficient of oxygen at a cus site pair and the CO^{cus} binding energy, are rather well described [79].

At the level of the interfacing between the electronic structure and the stochastic system description, the most pressing issue is without doubt the identification of and ensuing coarse-graining to the relevant elementary processes. As most transition-state search algorithms require knowledge of the final state after the process has taken place [40, 41], already quite some insight into the physics of an elementary process is needed just to determine its rate constant and include it in the list of possible kMC processes. Which processes are to be included in this list at all, and when it can be deemed complete, requires even more system-specific information. If the latter is not fully available, as is presently mostly the case, the modeler is forced to resort to chemical intuition. Unfortunately, it is by now well established that the real dynamical evolution of surface systems is full with nonobvious, unexpected, and complex pathways, with surface diffusion occurring via hopping or exchange mechanisms forming a frequently cited classic example [2, 80]. In this respect, the transition from MD to kMC is a paradigm for the no-free-lunch theorem: In an MD simulation, the system would tell us by itself where it wants to go, yet we would be unable to follow its time evolution for a sufficiently long time. In kMC, on the other hand, the latter can be achieved, but we have to explicitly provide the possibilities where the system can go and cannot be sure that we know them all.

The risk of overlooking a potentially relevant molecular process is especially consequential in first-principles kMC simulations. As discussed above, there is no possibility to cover up for a missing process as in traditional kMC simulations, where an unknown number of real molecular processes are in any case lumped together into a handful of effective processes with optimized rate constants. Instead, in a theory based on parameters with correct microscopic meaning such as first-principles kMC simulations, omission of an important process means just one thing: a wrong result. Even if in the position of knowing all relevant processes, the problem frequently arises that with increasing system complexity this number of processes just about explodes, making a direct computation of all required rate constants from first principles at present intractable. Examples of this are multisite lattices or multispecies problems or simply, as already indicated in Section 3.2.5, the existence of far-reaching significant lateral interactions. And this is not to mention the degree of complications that come in if the system cannot be easily mapped onto a lattice, for example, in the case of morphological transformations of the active surface during induction or steady-state operation [81]. The technical obstacles emerging in such cases can be overwhelming and are in existing practical applications unfortunately still too often "solved" by resorting to unsystematic and unjustified approximations.

Obviously, one way out of all of these problems would be that the kMC simulation determines by itself which processes need to be considered. Imagine there is such an automatized procedure that tells which processes lead out of any given system state. Apart from solving the problem of overlooking a possibly relevant process, this also addresses the limitations imposed by an exceeding total number of processes, as one would be able to concentrate on only those processes out of states that are actually visited by the system. In principle, this also supersedes the lattice approximation, unless one wants to keep it for state recognition when constructing dynamically evolving look-up tables where the information of all pathways of already visited states is kept [82]. In view of these merits, it is not surprising that such a procedure is hunted for like the holy grail. In fact, it gets reinvented over the years and features under names such as adaptive kMC [83], on-the-fly kMC [84], self-learning kMC [85], or kinetic activation-relaxation technique [86]. While the technical realizations vary, a core feature of all of these approaches is to resort to a transition-state search algorithm that only requires information about the initial state. Minimum mode following techniques like the dimer method [87, 88] would be a typical example. With such an algorithm at hand, the idea is to initiate a sufficiently large number of such transition-state searches to identify all low-lying saddle points around the current state of the system (potentially accelerated MD techniques [89] could be used for this, too). The rate constant for each of these pathways is then supplied to the kMC algorithm that propagates the system in a dynamically correct way to the next state, and the procedure starts again. In principle, this is a beautiful concept that gives access to the exact time evolution if indeed *all* bounding saddle points are found. Even if "only" the lowest-barrier escape pathways are determined, the dynamics may still be approximately ok as the omission of high-barrier pathways and their corresponding small rate constants in Eq. (3.5) does not introduce an exceeding error. In practice, however, it is essentially impossible to demonstrate that all (or at least all low-barrier) pathways have been found through a finite number of transition-state searches, especially for the high-dimensional PESs of systems that are of interest in catalysis research. In this situation, it is mostly a matter of taste whether one thinks that the holy grail has already been found or not. Due to the prohibitive number of first-principles energy and force evaluations required even when employing only the smallest justifiable number of searches from each state, such adaptive or on-the-fly approaches are at present in any case practically not really feasible for anything but the most trivial model systems – and on a philosophical note it is precisely those crooked PES alleys that are likely *not* to be found by a few transition-state searches that often correspond just to those nonobvious pathways that are overlooked by chemical intuition.

In examples of the type of the $RuO_2(110)$ system, the computational constraints do not extend to the level of the actual kMC simulations. Compared to the costs of setting up the catalog of first-principles rate constants, running these simulations up to the required extended time spans and even for a manifold of environmental conditions represents normally an insignificant add-on, in particular as these calculations can be performed on cheaper capacity compute infrastructures. However, this must not necessarily always be the case, and the mentioned low-barrier problem, that is, a process operating on a much faster timescale, provides already a prototypical example where the statistical kMC simulations themselves can quickly become a bottleneck, too. This holds equally for problems with a substantially increased number of processes as often automatically implied by larger simulation cells. On a bookkeeping level, significant speedups to overcome corresponding limitations are still realized by ever more efficient algorithms both with respect to the searches involved in the process selection step and with respect to updating the list of possible processes for a new system configuration [19, 82, 90–92]. Beyond this, developing concepts resort to either spatial or temporal coarse-graining, for example, through grouping of lattice sites into coarse cells or by executing several processes at a time with approximate postcorrections of ensuing correlation errors [19].

While the focus of such work is to improve the efficiency within the kMC framework, a more problematic issue at the general statistical mechanics level is the Markov assumption underlying the master equation (3.1). In parts, this concerns already the possibility of locally adiabatic correlated events, for example, a sliding motion over several PES minima that are separated only by low barriers. However, if only the existence of such events is known, this kind of problem is quickly patched up by an appropriate augmentation of the list of possible kMC processes. Particularly in view of the often largely exothermic surface catalytic processes, a much more severe challenge arises instead from non-Markovian behavior resulting from a finite heat dissipation. A classic example of this would be the often debated hot adatom motion, in which after a largely exothermic dissociative adsorption process, the reaction intermediate travels larger distances before it has equilibrated with the underlying substrate [93–95]. Rather than assuming a constant global temperature T as in prevailing kMC work, a consideration of such effects requires a model for the local heat transfer including, for example, an assignment of a local time-dependent temperature to each surface species accounting for its past process history.

Apart from dissipation through electronic friction and substrate phonons, a heat transfer model for steady-state catalysis at ambient pressures would further need to include a heat release channel into the fluid environment. This then connects to a last frontier that actually leads outside the methodological first-principles kMC framework as sketched in this chapter. In the work of the latter type, the focus is at present almost exclusively on the surface reaction chemistry, with only a rudimentary treatment of the reactive flow field over the catalyst surface. Apart from the constant global temperature mimicking an infinitely efficient heat dissipation, this concerns foremost the complete neglect of mass transfer limitations in the fluid environment as, for example, expressed by equating the reactant partial pressures entering the surface impingement (cf. Eq. (3.13)) with those at the inlet and disregarding the built-up of a finite product concentration over the active surface as done in the described $RuO_2(110)$ work. In order to improve on this, the kMC simulations need to be adequately interfaced with a computational fluid dynamics modeling of the macro-scale flow structures in a given reactor setup [96]. While the latter are naturally described at the continuum level, the crucial difficulty is that their effects intertwine with all lower scales, with local concentration changes (or fluctuations) in the gas-phase directly and nonlinearly connecting to inhomogeneities in the spatial distribution of the chemicals at the surface (e.g., reaction fronts), or the local heat dissipation at the catalyst surface intimately coupling to the kinetics and statistics of the elementary processes. The correspondingly required robust and self-consistent link of surface reaction chemistry modeling as provided by first-principles kMC with an account of the effects of heat and mass transfer as provided, for example, by time-dependent Navier–Stokes equations is only just emerging [97]. In order to overcome the empirical parameters and (unscrutinized) simplifying approximations entering seminal engineering-style work tackling this coupling [98, 99] will necessitate quite some further methodological developments at all involved length- and timescales and has already been identified as a core critical need in a report [100] on chemical industrial technology vision for the year 2020.

3.5 Conclusions

The purpose of this chapter was both to introduce the first-principles kinetic Monte Carlo approach in its specific flavor geared toward heterogeneous catalytic applications and to provide an impression of its present capabilities and open frontiers. As a hierarchical approach, first-principles kMC jointly addresses two complementary and crucial aspects of reactive surface chemistry, namely, an accurate description of the elementary processes together with a proper evaluation of their statistical interplay. For the former, it resorts to first-principles electronic structure theories and for the latter to a master equation-based description tackling the rare event evolution of surface catalytic systems. The result is a powerful technique that gives access to the surface chemical kinetics with an unprecedented accuracy, in particular by including a sound quantum mechanical description of the ubiquitous bond making and breaking and by including a full account of the atomic-scale correlations, fluctuations, and spatial distributions of the chemicals at the active surface sites. As a technique based on elementary processes with a clear-cut microscopic meaning, first-principles kMC is naturally transferable and can provide comprehensive insight into the ongoing surface chemistry over a wide range of temperature and pressure conditions.

In its philosophy, it is clearly a bottom-up approach that aims to propagate the predictive power of first-principles techniques up to increasing length- and timescales. For this it is important to realize that at both the electronic structure and the statistical mechanics levels, the various applied approximations introduce uncertainty, and the bridging from one scale to the other gives rise to additional uncertainty. With such multiple sources of uncertainty affecting the final result, only a stringent error control will ever allow to assign the desired predictive quality to the simulation. Most pressing issues for such a robust link at the transition between the molecular and the mesoscopic scale are presently the identification of and ensuing coarse-graining to the relevant elementary processes, as well as the propagation of error in the first-principles rate constants to the ensemble properties resulting from the statistical interplay.

With the stated claim of predictive quality, first-principles kMC simulations inherently require a level of detail and effort that is uncommon, if not far beyond standard effective modeling based on empirical parameters. Critical issues here are the necessity of a well-established microscopic characterization of the active surface sites and the computational cost to determine accurate first-principles rate constants for a number of elementary processes that virtually explodes with increasing system complexity. As a result, existing practical applications are still restricted to single-crystal model catalysts and reaction schemes of modest complexity, but this will undoubtedly change with further algorithmic developments and ever increasing computational power. The demanding characteristics of the approach must also be borne in mind when attempting the now due interfacing of first-principles kMC with their focus on the surface chemistry with macroscopic theories that model the heat and mass transport in the system, that is, the flow of chemicals over the catalyst surface and the propagation of the heat released during the ongoing chemical reactions.

Corresponding developments will necessitate incisive methodological advances at all involved length- and timescales, and have to overcome the present standard of uncontrolled semiempiricism and effective treatments. Yet, they are fired by the imagination of an ensuing error-controlled multiscale modeling approach that starting from the molecular-level properties will yield a quantitative account of the catalytic activity over the entire relevant range of reactive environments. Most important, this also encompasses a full "reverse mapping" capability to the mechanistic aspects, that is, the power to analyze in detail how the electronic structure (bond breaking and bond making) actuates the resulting macroscopic catalytic properties, function, and performance. I am convinced that it is ultimately only the concomitant new level of understanding that will pave the way for a rational design of tailored material-, energy-, and cost-efficient catalysts, and it is this vision that makes working on these problems so exciting and worthwhile.

Acknowledgments

I began using kinetic Monte Carlo simulations to model surface catalytic problems some 8–9 years ago. During almost all of this time I worked at the Fritz-Haber Institut, and much of the progress and understanding reviewed in this chapter would have been impossible without the enlightening and stimulating environment offered by this unique institution. In this respect, it is a pleasure to gratefully acknowledge the countless discussions with all members of the Theory Department that ultimately led to the specific vision of the first-principles kMC approach described here. This holds prominently and foremost for Matthias Scheffler, with whom most of the work on the $RuO_2(110)$ system was carried out.

References

1 Yip, S. (ed.) (2005) *Handbook of Materials Modeling*, Springer, Berlin.
2 Reuter, K., Stampfl, C., and Scheffler, M. (2005) Ab initio atomistic thermodynamics and statistical mechanics of surface properties and functions, in *Handbook of Materials Modeling* (ed. S. Yip), vol. 1, Springer, Berlin, pp. 149–194.
3 Kramer, C.J. (2004) *Essentials of Computational Chemistry: Theories and Models*, 2nd edn., John Wiley & Sons, Inc., Chichester.
4 Martin, R.M. (2004) *Electronic-Structure: Basic Theory and Practical Methods*, Cambridge University Press, Cambridge.
5 Carter, E.A. (2008) *Science*, **321**, 800.

6 van Kampen, N.G. (1980) *Stochastic Processes in Physics and Chemistry*, North-Holland, Amsterdam.
7 Chandler, D. (1987) *Introduction to Modern Statistical Mechanics*, Oxford University Press, Oxford.
8 Voter, A.F. (2007) Introduction to the kinetic Monte Carlo method, in *Radiation Effects in Solids* (eds K.E. Sickafus, E.A. Kotomin, and B.P. Uberuaga), Springer, Berlin.
9 Masel, R.I. (1996) *Principles of Adsorption and Reaction on Solid Surfaces*, John Wiley & Sons, Inc., New York.
10 Masel, R.I. (2001) *Chemical Kinetics and Catalysis*, John Wiley & Sons, Inc., New York.
11 Chorkendorff, I. and Niemantsverdriet, J.W. (2003) *Concepts of Modern Catalysis and Kinetics*, Wiley-VCH Verlag GmbH, Weinheim.
12 Wales, D.J. (2003) *Energy Landscapes*, Cambridge University Press, Cambridge.
13 Frenkel, D. and Smit, B. (2002) *Understanding Molecular Simulation*, 2nd edn., Academic Press, San Diego.
14 Landau, D.P. and Binder, K. (2002) *A Guide to Monte Carlo Simulations in Statistical Physics*, Cambridge University Press, Cambridge.
15 Gillespie, D.T. (1976) *J. Comput. Phys.*, **22**, 403.
16 Fichthorn, K.A. and Weinberg, W.H. (1991) *J. Chem. Phys.*, **95**, 1090.
17 Gillespie, D.T. (1977) *J. Phys. Chem.*, **81**, 2340.
18 Amar, J.G. (2006) *Comput. Sci. Eng.*, **8**, 9.
19 Chatterjee, A. and Vlachos, D.G. (2007) *J. Comput.-Aided Mater. Des.*, **14**, 253.
20 Bortz, A.B., Kalos, M.H., and Lebowitz, J.L. (1975) *J. Comput. Phys.*, **17**, 10.
21 Kang, H.C. and Weinberg, W.H. (1995) *Chem. Rev.*, **95**, 667.
22 Metropolis, N., Rosenbluth, A.W., Rosenbluth, M.N., Teller, A.H., and Teller, E. (1953) *J. Chem. Phys.*, **21**, 1087.
23 Kawasaki, K. (1972) *Phase Transitions and Critical Phenomena* (eds C. Domb and M. Green), Academic Press, London.
24 Wigner, E. (1932) *Z. Phys. Chem. B*, **19**, 203.
25 Eyring, H. (1935) *J. Chem. Phys.*, **3**, 107.
26 Haenggi, P., Talkner, P., and Borkovec, M. (1990) *Rev. Mod. Phys.*, **62**, 251.
27 Vineyard, G.H. (1957) *J. Phys. Chem. Solids*, **3**, 121.
28 Reuter, K., Frenkel, D., and Scheffler, M. (2004) *Phys. Rev. Lett.*, **93**, 116105; Reuter, K. and Scheffler, M. (2006) *Phys. Rev. B*, **73**, 045433.
29 Dellago, C., Bolhuis, P.G., and Geissler, P.L. (2002) *Adv. Chem. Phys.*, **123**, 1.
30 Garrett, B.C. and Truhlar, D.G. (2005) Variational transition state theory, in *Theory and Applications in Computational Chemistry: The First Forty Years* (eds C. Dykstra, G. Frenking, K. Kim, and G. Scuseria), Elsevier, Amsterdam, pp. 67–87.
31 Parr, R.G. and Yang, W. (1989) *Density Functional Theory of Atoms and Molecules*, Oxford University Press, Oxford.
32 Dreizler, R.M. and Gross, E.K.U. (1990) *Density Functional Theory*, Springer, Berlin.
33 Koch, W. and Holthausen, M.C. (2001) *A Chemist's Guide to Density Functional Theory*, 2nd edn., Wiley-VCH Verlag GmbH, Weinheim.
34 Perdew, J.P. and Schmidt, K. (2001) *Density Functional Theory and Its Application to Materials* (eds V. Van Doren, C. Van Alsenoy, and P. Geerlings), American Institute of Physics, Melville.
35 Cohen, A.J., Mori-Sanchez, P., and Yang, W. (2008) *Nature*, **321**, 792.
36 Scheffler, M. and Stampfl, C. (2000) Theory of adsorption on metal substrates, in *Handbook of Surface Science, Vol. 2: Electronic Structure* (eds K. Horn and M. Scheffler), Elsevier, Amsterdam.
37 Groß, A. (2002) *Theoretical Surface Science–A Microscopic Perspective*, Springer, Berlin.
38 Payne, M.C., Teter, M.P., Allan, D.C., Arias, T.A., and Joannopoulos, J.D. (1992) *Rev. Mod. Phys.*, **64**, 1045.
39 Henkelman, G., Johannesson, G., and Jonsson, H. (2000) Methods for finding saddle points and minimum energy paths, in *Progress on Theoretical Chemistry*

and *Physics* (ed. S.D. Schwarz), Kluwer Academic Publishers, New York.

40 Hratchian, H.P. and Schlegel, H.B. (2005) Finding minima, transition states and following reaction pathways on ab initio potential energy surfaces, in *Theory and Applications in Computational Chemistry: The First Forty Years* (eds C. Dykstra, G. Frenking, K. Kim, and G. Scuseria), Elsevier, Amsterdam, pp. 195–250.

41 Jonsson, H., Mills, G., and Jacobson, K.W. (1998) Nudged elastic band method for finding minimum energy paths of transitions, in *Classical and Quantum Dynamics in Condensed Phase Simulations* (eds B.J. Berne, G. Cicotti, and D.F. Coker), World Scientific, New Jersey.

42 Henkelman, G., Uberuaga, B.P., and Jonsson, H. (2000) *J. Chem. Phys.*, **113**, 9901.

43 Voter, A.F. (1986) *Phys. Rev. B*, **34**, 6819.

44 Fichthorn, K.A. and Scheffler, M. (2000) *Phys. Rev. Lett.*, **84**, 5371; Fichthorn, K.A., Merrick, M.L., and Scheffler, M. (2002) *Appl. Phys. A*, **75**, 17.

45 Sanchez, J.M., Ducastelle, F., and Gratias, D. (1984) *Phys. A*, **128**, 334.

46 De Fontaine, D. (1994) *Statics and Dynamics of Alloy Phase Transformations, NATO ASI Series* (eds P.E.A. Turchy and A. Gonis), Plenum Press, New York.

47 Zunger, A. (1994) First-principles statistical mechanics of semiconductor alloys and intermetallic compounds, in *Statics and Dynamics of Alloy Phase Transformations, NATO ASI Series* (eds P.E.A. Turchy and A. Gonis), Plenum Press, New York.

48 Stampfl, C., Kreuzer, H.J., Payne, S.H., Pfnür, H., and Scheffler, M. (1999) *Phys. Rev. Lett.*, **83**, 2993.

49 Drautz, R., Singer, R., and Fähnle, M. (2003) *Phys. Rev. B*, **67**, 035418.

50 Tang, H., van der Ven, A., and Trout, B.L. (2004) *Phys. Rev. B*, **70**, 045420; Tang, H., van der Ven, A., and Trout, B.L. (2004) *Mol. Phys.*, **102**, 273.

51 Fähnle, M., Drautz, R., Lechermann, F., Singer, R., Diaz-Ortiz, A., and Dosch, H. (2005) *Phys. Status Solidi (b)*, **242**, 1159.

52 Zhang, Y., Blum, V., and Reuter, K. (2007) *Phys. Rev. B*, **75**, 235406.

53 Hansen, E.W. and Neurock, M. (1999) *Chem. Eng. Sci.*, **54**, 3411.

54 Shustorovich, E. and Sellers, H. (1998) *Surf. Sci. Rep.*, **31**, 5.

55 Ruggerone, P., Ratsch, C., and Scheffler, M. (1997) Density-functional theory of epitaxial growth of metals, in *Growth and Properties of Ultrathin Epitaxial Layers, The Chemical Physics of Solid Surfaces*, vol. 8 (eds D.A. King and D.P. Woodruff), Elsevier, Amsterdam.

56 Kratzer, P., Penev, E., and Scheffler, M. (2002) *Appl. Phys. A*, **75**, 79.

57 Reuter, K. (2006) *Oil Gas Sci. Technol.*, **61**, 471.

58 Over, H. and Muhler, M. (2003) *Prog. Surf. Sci.*, **72**, 3.

59 Reuter, K. and Scheffler, M. (2003) *Phys. Rev. B*, **68**, 045407.

60 Brønsted, N. (1928) *Chem. Rev.*, **5**, 231; Evans, M.G. and Polanyi, N.P. (1936) *Trans. Faraday Soc.*, **32**, 1333.

61 Temel, B., Meskine, H., Reuter, K., Scheffler, M., and Metiu, H. (2007) *J. Chem. Phys.*, **126**, 204711.

62 Zhdanov, V.P. and Kasemo, B. (1994) *Surf. Sci. Rep.*, **20**, 111.

63 Ziff, R.M., Gulari, E., and Barshad, Y. (1986) *Phys. Rev. Lett.*, **56**, 2553; Brosilow, B.J. and Ziff, R.M. (1992) *Phys. Rev. A*, **46**, 4534.

64 Stoltze, P.I. (2000) *Prog. Surf. Sci.*, **65**, 65.

65 Reuter, K. and Scheffler, M. (2003) *Phys. Rev. Lett.*, **90**, 046103.

66 Reuter, K. and Scheffler, M. (2004) *Appl. Phys. A*, **78**, 793.

67 Wang, J., Fan, C.Y., Jacobi, K., and Ertl, G. (2002) *J. Phys. Chem. B*, **106**, 3422.

68 Matera, S., Meskine, H., and Reuter, K. (2011) *J. Chem. Phys.*, **134**, 064713.

69 Woodruff, D.P. and Delchar, T.A. (1994) *Modern Techniques of Surface Science*, Cambridge University Press, Cambridge.

70 Jansen, A.P.J. (1995) *Comput. Phys. Commun.*, **86**, 1.

71 Wendt, S., Knapp, M., and Over, H. (2004) *J. Am. Chem. Soc.*, **126**, 1537.

72 Rieger, M., Rogal, J., and Reuter, K. (2008) *Phys. Rev. Lett.*, **100**, 016105.

73 Stroppa, A. and Kresse, G. (2008) *New J. Phys.*, **10**, 063020.

74 Kroes, G.-J. (2008) *Science*, **321**, 794.
75 Boudard, M. and Tamaru, K. (1991) *Catal. Lett.*, **9**, 15.
76 Campbell, C.T. (1994) *Top. Catal.*, **1**, 353.
77 Dumesic, J.A. (1999) *J. Catal.*, **185**, 496.
78 Baranski, A. (1999) *Solid State Ionics*, **117**, 123.
79 Meskine, H., Matera, S., Scheffler, M., Reuter, K., and Metiu, H. (2009) *Surf. Sci.*, **603**, 1724.
80 Ala-Nissila, T., Ferrando, R., and Ying, S.C. (2002) *Adv. Phys.*, **51**, 949.
81 See, for example, the discussion in Rogal, J., Reuter, K., and Scheffler, M. (2007) *Phys. Rev. Lett.*, **98**, 046101; Rogal, J., Reuter, K., and Scheffler, M. (2008) *Phys. Rev. B*, **77**, 155410.
82 Mason, D.R., Hudson, T.S., and Sutton, A.P. (2005) *Comput. Phys. Commun.*, **165**, 37.
83 Henkelman, G. and Jonsson, H. (2001) *J. Chem. Phys.*, **115**, 9657.
84 Bocquet, J.L. (2002) *Defect Diffus. Forum*, **203**, 81.
85 Trushin, O., Karim, A., Kara, A., and Rahman, T.S. (2005) *Phys. Rev. B*, **72**, 115401.
86 El-Mellouhi, F., Mousseau, N., and Lewis, L.J. (2008) *Phys. Rev. B*, **78**, 153202.
87 Henkelman, G. and Jonsson, H. (1999) *J. Chem. Phys.*, **111**, 7010.
88 Heyden, A., Bell, A.T., and Keil, F.J. (2005) *J. Chem. Phys.*, **123**, 224101.
89 Voter, A.F., Montalenti, F., and Germann, T.C. (2002) *Annu. Rev. Mater. Res.*, **32**, 321.
90 Blue, J.L., Beichl, I., and Sullivan, F. (1994) *Phys. Rev. E*, **51**, R867.
91 Gibson, M.A. and Bruck, J. (2000) *J. Phys. Chem. A*, **104**, 1876.
92 Schulze, T.P. (2002) *Phys. Rev. E*, **65**, 036704.
93 Brune, H., Wintterlin, J., Behm, R.J., and Ertl, G. (1992) *Phys. Rev. Lett.*, **68**, 624.
94 Wintterlin, J., Schuster, R., and Ertl, G. (1996) *Phys. Rev. Lett.*, **77**, 123.
95 Schintke, S., Messerli, S., Morgenstern, K., Nieminen, J., and Schneider, W.-D. (2001) *J. Chem. Phys.*, **114**, 4206.
96 Deutschmann, O. (2008) Computational fluid dynamics simulation of catalytic reactors, in *Handbook of Heterogeneous Catalysis*, 2nd edn (eds G. Ertl, H. Knözinger, F. Schüth, and J. Weitkamp,), Wiley-VCH Verlag GmbH, Weinheim, pp. 1811–1828.
97 Matera, S. and Reuter, K. (2009) *Catal. Lett.*, **133**, 156; (2010) *Phys. Rev. B*, **82**, 085446.
98 Kissel-Osterrieder, R., Behrendt, F., Warnatz, J., Metka, U., Volpp, H.R., and Wolfrum, J. (2000) *Proc. Combust. Inst.*, **28**, 1341.
99 Vlachos, D.G., Mhadeshwar, A.B., and Kaisare, N.S. (2006) *Comput. Chem. Eng.*, **30**, 1712.
100 The U.S. Chemical Industry, American Chemical Society, American Institute of Chemical Engineers, The Chemical Manufactures Association, The Council for Chemical Research, and The Synthetic Organic Chemical Manufactures (1996) *Technology Vision 2020*, http://www.ccrhq.org/vision/welcome.html.

4
Modeling the Rate of Heterogeneous Reactions

Lothar Kunz, Lubow Maier, Steffen Tischer, and Olaf Deutschmann

4.1
Introduction

This chapter discusses links and existing gaps between modeling surface reaction rates on a fundamental, molecular approach on the one hand and on a practical reaction engineering approach on the other hand.

The mechanism of heterogeneously catalyzed gas-phase reactions can in principle be described by the sequence of elementary reaction steps of the cycle, including adsorption, surface diffusion, chemical transformations of adsorbed species, and desorption, and it is the basis for deriving the kinetics of the reaction. In the *macroscopic* regime, the rate of a catalytic reaction is modeled by fitting empirical equations, such as power laws, to experimental data to describe its dependence on concentration and pressure and to determine rate constants that depend exponentially on temperature. This macroscopic approach was used in chemical engineering for reactor and process design for many years [1].

Assumptions on reaction schemes (kinetic models) provide correlations between the surface coverages of intermediates and the external variables. Improved kinetic models could be developed when atomic processes on surfaces and the identification and characterization of surface species became available. Here, the progress of a catalytic reaction is described by a *microkinetics* approach by modeling the macroscopic kinetics by means of correlations of the atomic processes with macroscopic parameters within the framework of a suitable continuum model [2, 3]. Continuum variables for the partial surface coverages are, to a first approximation, correlated with external parameters (partial pressures and temperature) by the mean field (MF) approximation of a surface consisting of identical noninteracting adsorption sites. Because of this idealization of the catalytic process, the continuum model can describe the reaction kinetics only to a first approximation neglecting interactions between adsorbed species and nonidentical adsorption sites. Apart from the heterogeneity of adsorption sites, the surfaces may exhibit structural transformations.

The Langmuir–Hinshelwood–Hougen–Watson (LHHW) model has been a popular simplified approach of the mean field approximation for modeling

Modeling and Simulation of Heterogeneous Catalytic Reactions: From the Molecular Process to the Technical System,
First Edition. Edited by Olaf Deutschmann.
© 2012 Wiley-VCH Verlag GmbH & Co. KGaA. Published 2012 by Wiley-VCH Verlag GmbH & Co. KGaA.

Table 4.1 Approaches for modeling rates of heterogeneous catalytic reactions.

Method of modeling	Simplification	Application
Ab initio calculation Density functional theory	Most fundamental approach Replacement of the N electron wave function by the electron density	Simple chemical systems Dynamics of reactions, activation barriers, adsorbed structures, frequencies
(Kinetic) Monte Carlo	Details of dynamics neglected	Adsorbate–adsorbate interactions on catalytic surfaces and nanoparticles
Mean field approximation	Details on adsorbate structure neglected	Microkinetic modeling of catalytic reactions in technical systems
Langmuir–Hinshelwood–Hougen–Watson	Rate-determining step needed	Modeling of catalytic reactions in technical systems
Power law kinetics	All mechanistic aspects neglected	Scale-up and reactor design for black box systems

technical catalytic reactors for many years. It is based on a continuum model, in which the surface of the catalyst is described as an array of equivalent sites that do interact neither before nor after chemisorption. Furthermore, the derivation of rate equations assumes that both reactants and products are equilibrated with surface species and that a rate-determining step can be identified. Surface coverages are correlated with partial pressures in the fluid phase by means of Langmuir adsorption isotherms. Despite these oversimplifications, the LHHW kinetics model has been used for reactor and process design in industry until today. The kinetic parameters determined by fitting the rate equations to experimental data, however, do not have physical meaning in general. Sometimes, even less complicated simple power law kinetics for straightforward reactions (e.g., A + B) are used.

On the most fundamental level, density functional theory (DFT), molecular dynamics (MD), and Monte Carlo (MC) simulations are used to elucidate the molecular aspects of heterogeneous catalysis as discussed in the previous three chapters. Table 4.1 lists methods for modeling the chemical reaction rate of heterogeneous catalytic reactions in a hierarchical order.

A major objective of present research in catalysis is the development of methods that allow the incorporation of the molecular understanding of catalysis into the modeling of technical reactors. In principle, *ab initio* and DFT calculations can provide information that is fed into MC simulations of catalytic processes on individual nanoparticles, which then can compute surface reaction rates as function of the local (fluid phase) partial pressures, temperature, and adsorbate structure. These rates have then to be applied in models tractable for the simulation of technical systems. Hence, the gap still to be bridged in modeling technical systems is between MC simulations and reactor simulation. In the past two decades, the MF approximation has been used as workaround in order to overcome the much simpler Langmuir–Hinshelwood or even power law approaches and to include some of the elementary aspects of catalysis into models suitable for numerical simulation of catalytic reactors.

This chapter focuses on two major items: the MC simulation as a potential tool for the derivation of surface reaction rates and the MF approach as state-of-the-art modeling of reaction rates in technical systems. Eventually, a local chemical source term, R_i^{het}, is needed to provide the specific net rate of the production of species i due to heterogeneous chemical reactions at a certain macroscopic position of a catalytic surface in the technical reactor. This source term as function of the local conditions can then be implemented into fluid dynamics and heat transport simulations of the technical system, which will be discussed in the next chapters.

Since elementary-step reaction mechanisms were first introduced in modeling homogeneous reaction systems and since homogeneous reactions in the fluid phase do also play a significant role in many technical catalytic reactors, this chapter will start with a short introduction on the well-established approach of modeling the rates of chemical reactions in the gas phase.

4.2
Modeling the Rates of Chemical Reactions in the Gas Phase

In many catalytic reactors, the reactions do not only occur on the catalyst surface but also in the fluid flow. In some reactors, even the desired products are mainly produced in the gas phase, for instance, in the oxidative dehydrogenation of paraffins to olefins over noble metals at short-contact times and high temperatures [4–11]. Such cases are dominated by the interaction between gas-phase and surface kinetics and transport. Therefore, reactor simulations often need to include an appropriate model for the homogeneous kinetics along with the heterogeneous reaction models. The species governing equations in fluid flow simulations usually contain a source term such as R_i^{hom} denoting the specific net rate of production of species i due to homogeneous chemical reactions. Considering a set of K_g chemical reactions among N_g species A_i

$$\sum_{i=1}^{N_g} v'_{ik} A_i \rightarrow \sum_{i=1}^{N_g} v''_{ik} A_i, \tag{4.1}$$

with v'_{ik}, v''_{ik} being the stoichiometric coefficients, and an Arrhenius-like rate expression, $AT^\beta \exp(-E_a R^{-1} T^{-1})$, this source term can be expressed by

$$R_i^{hom} = M_i \sum_{k=1}^{K_g} (v''_{ik} - v'_{ik}) A_k T^{\beta_k} \exp\left[\frac{-E_{a_k}}{RT}\right] \prod_{j=1}^{N_g} \left(\frac{Y_j \varrho}{M_j}\right)^{a_{jk}}. \tag{4.2}$$

Here, A_k is the preexponential factor, β_k is the temperature exponent, E_{a_k} is the activation energy, and a_{jk} is the order of reaction k related to the concentration of species j. The advantage of the application of elementary reactions is that the reaction orders a_{jk} in Eq. (4.2) equal the stoichiometric coefficients ν'_{jk}.

Various sets of elementary reactions are available for modeling homogeneous gas phase reactions, for instance, for total [12] and partial oxidation, and pyrolysis [13, 14] of hydrocarbons. Table 4.2 lists a selection (far from being complete) of gas-phase reaction mechanisms, which may also be considered in the simulation of heterogeneous chemical systems.

Even though implementation of Eq. (4.2) into CFD codes for the simulation of chemical reactors is straightforward, an additional highly nonlinear coupling is introduced into the governing equations leading to considerable computational efforts. The nonlinearity, the large number of chemical species, and the fact that chemical reactions exhibit a large range of timescales render the solution of these equation systems challenging. In particular for turbulent flows, and sometimes even for laminar flows, the solution of the system is too CPU time consuming with present numerical algorithms and computer capacities. This calls for the application of reduction algorithms for large reaction mechanisms, for instance, through the extraction of the intrinsic low dimensional manifolds of trajectories in chemical space [15]. Another approach is to use "as little chemistry as necessary." In these so-called adaptive chemistry methods, the construction of the reaction mechanism only includes steps relevant for the application studied [16].

4.3
Computation of Surface Reaction Rates on a Molecular Basis

In this section, we discuss the derivation of rates of heterogeneous reactions from a molecular point of view. Since the previous chapters have already discussed in detail DFT, MD, and kMC simulations, we focus on ways to make such molecular methods – in particular MC simulations – a useful tool for simulation of technical systems.

4.3.1
Kinetic Monte Carlo Simulations

From a macroscopic point of view, effects on overall reaction rates resulting from lateral interactions of adsorbates are inherently difficult to treat. In the mean field approximation, as discussed below, they are either neglected or incorporated by mean rate coefficients. If the specific adsorbate–adsorbate interactions are understood quantitatively, then (kinetic) Monte Carlo (kMC) simulations can be carried out.

From a microscopic point of view, the coarse graining of rates for chemical reactions starting from elementary processes requires averaging over many reaction events. As rare events, chemical reactions require a long simulation time, which limits the use of MD simulations and requires methods such as kinetic Monte Carlo simulations.

Table 4.2 Selection of homogeneous reaction mechanisms relevant for modeling catalytic reactors.

	Number species/ irreversible reactions	Mechanism type/ species considered	Range of conditions
Warnatz [104, 105]	26/189	C_1–C_2	Methane and natural gas combustion
Miller/Bowman [106]	.../141	C_1–C_4	Methane and natural gas combustion
GRI 1.2 (Frenklach) [107]	25/168	C_1–C_2	Natural gas flames and ignition, flame propagation
GRI 3.0 (Smith/ Golden/ Frenklach) [108]	53/325	C_1–C_3	1000–2500 K, 1.0–1000 kPa, equivalence ratios: $\Phi = 0.1$–5 for premixed systems
Qin [109]	.../258	C_1–C_3	Propane flame speed and ignition
Konnov [110]	127/1200	C_1–C_3, OH, NOx, and NH$_3$ kinetics	Species profiles in flow reactor, ignition delay times in shock waves, laminar flame speed, and species profiles
Hidaka/Tanaka [111, 112]	48/157	C_1–C_2	Reflected shock waves 1350–2400 K, 162–446 kPa.
Leeds 1.4 (Hughes) [113]	37/351	C_1–C_2	Species profiles in laminar flames, flame speeds, ignition delay
GRI extented (Eiteneer/Frenklach) [114]	71/486	C_1–C_3 (GRI 3.0) + C_2H_x, C_3H_x, C_4H_x	Acetylene ignition in shock tubes
GDF-Kin (El Bakali/ Dagaut) [115]	99/671	C_1–C_6	Laminar flame speeds, jet stirred reactor at 1 atm, $\Phi = 0.75 : 1.0 : 1.5$
San Diego [116]	86/362	C_1–C_3 (excluding low-temperature fuel–peroxide kinetics)	Below about 100 atm, above about 1000 K, $\Phi < 3$, deflagration velocities and shock tube ignition.
UBC 2.0 [117]	55/278	C_1–C_2 (GRI 1.2) + C_3 + CH_3O_2, $C_2H_5O_2$, $C_3H_7O_2$	Ignition delay in reflected shock, 900–1400 K, 16–40 bar
Ranzi [118]	79/1377	C_1–C_3	High-temperature pyrolysis, partial oxidation and combustion

(Continued)

Table 4.2 (Continued)

	Number species/ irreversible reactions	Mechanism type/ species considered	Range of conditions
Healy/Curran [119, 120]	289/3128	C_1–C_3	740–1550 K, 10–30 atm, $\Phi = 0.3$–3.0 in a high-pressure shock tube and in a rapid compression machine
Gupta [121]	190/1150	C_1–C_6 + Frenklach/Warnatz 1987 soot model [122]	Low-temperature pyrolysis in fuel cells (900–1200 K)
Younessi-Sinaki [123]	75/242	C_1–C_2 (GRI 1.2) + C > 2 (excluding the oxygenates) + Appel 2000 soot model [124]	Homogeneous thermal (oxygen free) decomposition
Norinaga [125]	227/827	C_1–C_4 + Frenklach/Warnatz 1987 soot model [122]	Pyrolysis, flow reactor at 900 °C, pressures of 2–15 kPa, and residence times of up to 1.6 s.
LLNL [126]	857/3606	C_1–C_8	550–1700 K; 1–45 bar; 70–99% N_2(Ar)
Glaude [127]	367/1832	C_1–C_8	500–1100 K; 1–20 bar
Golovitchev [128]	130/690	C_1–C_8	640–1760 K; 1–55 bar; 76–95% Ar
Battin-Leclerc [129]	.../7920	C_1–C_{10}	Jet-stirred reactor, premixed laminar flame, 550–1600 K
Bikas/Peters [130]	67/600	C_1–C_{10}	Premixed flame at 100 kPa, flat flames, shock waves, jet-stirred reactor
Ristori/Dagaut [28]	242/1801	C_1–C_{16}	Jet-stirred reactor, 1000–1250 K, 100 kPa, residence time of 70 ms, $\Phi = 0.5$, 1, and 1.5.
Nancy [131]	265/1787	C_1–C_{16}	Jet-stirred reactor, 1000–1250 K

MC simulations today appear as the model of choice to bridge the gap between molecular modeling and reactor modeling. Therefore, this section is specifically devoted to MC simulations that try to include more and more effects of real catalytic particles such as the three-dimensional structure of the catalyst nanoparticles,

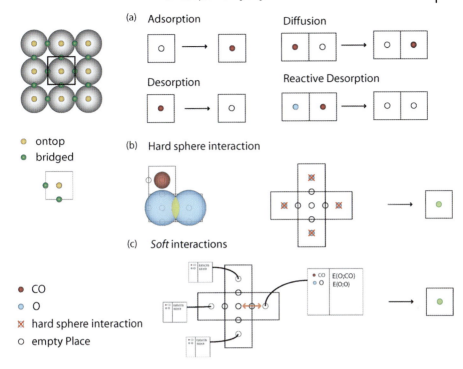

Figure 4.1 Surface description by reactant and product patterns. (a) Examples for the main processes occurring in catalysis. (b) Incorporation of a hard sphere interaction specific to oxygen. (c) Modeling of pairwise soft interactions with interaction energy tables. Adapted from Ref. [32, 33].

support, and spillover effects. Since MC simulation is the specific topic of the previous chapter, only the fundamentals need to be summarized here.

In MC simulations for surface processes (Figure 4.1), the evolution of the surface coverage configuration is calculated as a correct sequence in time. The foundation of this method is the master equation:

$$\frac{dP_\alpha}{dt} = \sum_\beta [k_{\alpha\beta} P_\beta - k_{\beta\alpha} P_\alpha]. \tag{4.3}$$

This equation describes the evolution of probability P_α for the system being in the surface configuration state α. Here, $k_{\alpha\beta}$ defines the transition probability from state α to state β. The transition in the sense of surface simulation can, for instance, be a diffusion step or a reaction with rate $k_{\alpha\beta}$. Analytical solutions to the master equation can be derived only for simple cases. In general, a numerical solution is required. A MC simulation starts from a state α and repeatedly picks a random possible process and advances in time. Averaging over several trajectories leads to a numerical solution of the master equation.

The time step and process have to be chosen in accordance with the likelihood that state α is left and that a specific process has occurred. Two main algorithms have been developed, the variable step size method (VSSM) and the first reaction method

(FRM). In the VSSM, the time increment Δt is derived from a uniform random number r on $(0,1)$ by

$$\Delta t = -\frac{\ln r}{\sum_\beta k_{\beta\alpha}}. \qquad (4.4)$$

The process is picked independent of the time step at a likelihood k/k_{total}, where k is the rate of the picked process and k_{total} is the sum of all processes that can start from the current state. In the FRM for every possible process with rate k starting from state α, a time increment

$$\Delta t = -\frac{\ln r}{k} \qquad (4.5)$$

is drawn and finally the process with the smallest Δt is chosen as the next step. A variant of the VSSM method is called random site method (RSM). For this a site, a process, and a time increment are picked independently. With a probability of $k/\max_p k_p$, the process is performed. VSSM is the most widely used method, also known as n-fold way or BKL method after Bortz, Kalos, and Libowitz [17]. In general, the choice of algorithm depends on the specific problem. In some cases, a combination of the variants leads to a good performance [18].

4.3.2
Extension of MC Simulations to Nanoparticles

The description of the surface configuration requires only a mapping from adsorption sites to species. The layout of the sites can be not only a lattice, like crystal surfaces, but also an off-lattice. The widely used lattices originate from the periodic surfaces resulting in a low number of processes since they are independent of their global location.

But catalytic surfaces are certainly nonuniform; site heterogeneity exists because the surface of practical catalyst particles is characterized by terraces of different crystal structures, steps, edges, additives, impurities, and defects. Therefore, it is required to enable MC simulations for such systems in order to derive technical meaningful rates and give insight into geometric and communication effects.

Prior attempts have been made to perform MC simulations on nanoparticles. One approach is to regard a single lattice without periodic boundary conditions as particle and describe the facets as different regions [19–21]. Another simulation approach uses three-dimensional particles, which can vary their height to mimic shape transformation. These models use a single lattice with additional information about the particle height for each adsorption place [22]. Both models neglect the nature of different facets regarding their neighborhood because they are limited to one lattice type and cannot represent the different neighborhoods of combinations like fcc(111) and fcc(100) faces. A hybrid approach between a lattice and an off-lattice method can overcome these limitations. The facets of the catalyst particle and the support are each described by a lattice, which are linked along their edges (Figure 4.2). Since such models lead to a high number of different processes, it is favorable to have a general implementation, which is not restricted to a specific mechanism and allows different particle shapes.

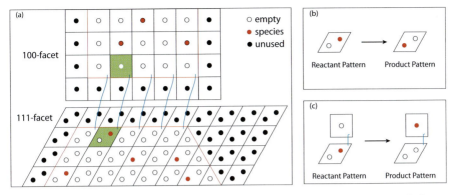

Figure 4.2 (a) An extract from a surface model for a particle with a (100) and a (111) facet, linked along their common edge. (b) A diffusion process for the (111) facet. (c) A diffusion process across the facets. One matching position is marked green. Adapted from Ref. [32, 33].

For this, an abstract view on kinetic Monte Carlo simulations has been developed [23]: Surface processes that occur in catalytic processes as adsorption, desorption, diffusion, and reaction can be seen as an exchange of patterns. One pattern describes the positions of reactants on the surface before the process – the reactant pattern – and another one the position of products after the process – the product pattern (Figure 4.1a).

Lateral interactions can be incorporated into this abstract view as well. From the modeling point of view, one distinguishes between hard sphere and soft interactions. Hard sphere interactions are very strong lateral interactions, in which the adsorbed species behave as hard spheres and exclude neighboring places from being occupied. This can be incorporated into the reactant pattern (Figure 4.1b). Soft interactions are of lower energy and cannot be treated as hard sphere interactions. They occur as pairwise interactions and can also be of many species interactions if detailed cluster expansions are used. One interaction describes how specific neighboring adsorbates influence the activation energy barrier. These interactions can be derived from lattice gas Hamiltonians for the initial state and the transition state. Interactions between the species, which the reactant pattern requires in order to be applied at a specific position, can be incorporated into the fixed activation energy barrier. Interactions between the optional neighboring species are kept in interaction energy tables. Therefore, the energy difference resulting from this place occupied by a specific species is tabulated. Before the process is applied, the neighbor-dependent interaction energy contributions are added to the fixed activation energy. This works only for pairwise interactions; many species interactions are treated separately.

The surface model assigns each place an occupation and allows determination of neighboring places. In general, places are not regularly distributed and the surface has to be modeled as a graph, with places as nodes and edges to neighboring places. Places on crystal surfaces obey a translational symmetry, which can be used to systematically enumerate the unit cells and places in each cell. In single-lattice codes, the surface can be represented by a three-dimensional array of integers, that is, two spatial dimensions and one dimension to account for different lattice places in each unit cell. The extension to structured surfaces uses the same model for each facet. Each place is then described by a lattice id, two spatial coordinates, and a place id. Within each facet, the neighboring places can be determined via the neighborhood of unit cells. Across the facets, neighboring unit cells are connected by links (Figure 4.2a).

This surface model of connected lattices allows an extension of the concept of pattern exchanges to a nanoparticle surface model. A process description has a reactant pattern and a product pattern, which are specific parts of the surface model. Within each facet, the processes are described as before (see Figure 4.2b). Processes spanning several facets exhibit more complex reactant and product pattern parts since they include the links as subsets of the surface model, as well (Figure 4.2c).

Despite this complexity, all described MC algorithms stay applicable because the following basic procedures are executable for complex processes: Check if a reactant pattern matches the occupation of the surface at a specific position, perform a process at a specific position, that is, copying of the product pattern to the surface, and calculation of the interaction energy.

Single-lattice codes apply the pattern at a specific position of the lattice and iterate over the array in order to compare the pattern with the surface. Every pattern has an origin, for which a specific position of the lattice, the application point, indicates where the pattern is applied to the surface. For processes spanning different facets, the complete reactant pattern has only one global origin and each pattern part has a local origin. The global application point defines the local application point of only one pattern part directly. The other local application points have to be determined via a depth search over the graph setup by the surface links (Figure 4.3).

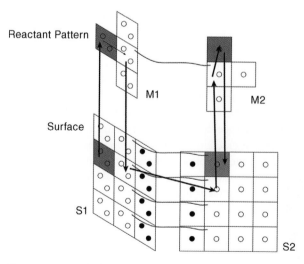

Figure 4.3 Example for the matching of reactant pattern consisting of two parts M_1 and M_2: The surface consists of the two parts S_1 and S_2. The global application point for the reactant pattern is given by the position of M_1 on S_1 (red). From there, the local application point for M_2 on S_2 (green) can be derived as it is indicated by the arrows. Adapted from Ref. [32, 33].

After the performance of a product pattern, the algorithm VSSM requires an update of the list of possible reactions. An update step has to match the reactant pattern at and around the application point. Since many processes are not even on the same facet or since the product pattern of the step before is contradictory to possible reactant patterns, a list of candidate reactions for each product pattern can be generated to limit the number of process candidates that have to be tested.

If duplicate facets occur within the model, which vary in their shape, but describe the same processes, further optimization is possible. Duplicate facets can be modeled by one single-crystal lattice with unused places separating them from each other. This procedure saves the effort to describe the processes for each facet again. The gap of unused places between the mapped facets is determined by the maximal size of reactant pattern for these facets.

Since no complete set of rates is available for a supported nanoparticle, rates for the facets for single-crystal surfaces have been taken from literature. The aim of these examples is not to show a well-elaborated set of processes, but rather to show the capabilities and limits of the presented methods.

The shape of Pd particles has been resolved with STM for some support materials [24–26], even though the (100) facets have not been resolved in atomic resolution. Despite this, CO oxidation on Pd particles marks a well-examined system. However, the form of the particles has not been determined under reaction conditions. Therefore, it is assumed in this example that the form is in accordance with the STM measurements and keeps its shape under reaction conditions. It was transferred to a MC model of a single particle without a support description and to two supported particles (Figure 4.4). The system has been described by a MC simulation before [27]. A MC model for the processes on Pd(100) crystal surfaces was derived from TPD experiments [28], while for Pd(111) crystal surfaces process rate values have been taken from simulations [29–31].

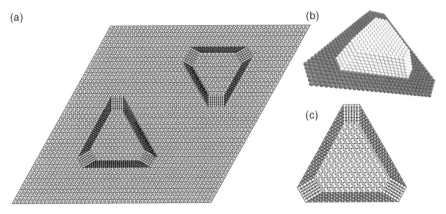

Figure 4.4 (a) Two supported nanoparticles surface model and (b) view of the atoms for the (c) lattice of the single-particle model. Adapted from Ref. [32, 33].

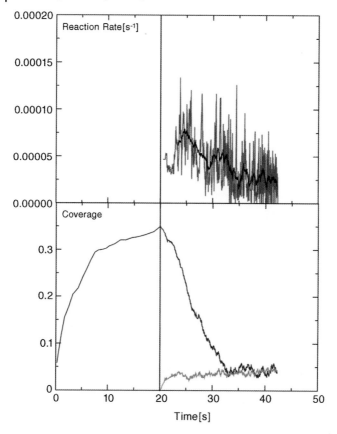

Figure 4.5 Transition behavior of CO oxidation on the single-particle Pd model. In the first 20 s the particle was saturated with partial oxygen pressure of $p_{O2} = 2 \times 10^{-7}$ mbar. After that additionally a CO partial pressure of $p_{CO} = 8 \times 10^{-7}$ mbar was applied. *Top*: reaction rate per unit cell per second (gray) and its time average (black). *Bottom*: oxygen (dark gray) and CO (light gray) coverage. Adapted from Ref. [32, 33].

4.3.3
Reaction Rates Derived from MC Simulations

As for single-crystal surfaces, rate models for mean field calculations can be derived from MC simulations. For a given set of gas-phase concentrations and temperature, the surface reaction rate can be derived by averaging over a system in equilibrium or periodic fluctuations (Figure 4.5) yielding coefficients for Eq. (4.7) [32, 33]. The surface reaction rate can now be directly related to the local conditions in the technical system. Since nanoparticles in technical catalyst exhibit strong variations in size and shape, these variations have to be taken into account in the derivation of rates by kMC simulations leading to numerous simulations of individual particles. Knowing these distributions, averaged rates may be determined for a given distribution of nanostructures of the catalyst and external conditions. Today, this approach can work only for exemplary cases.

Furthermore, since the implementation of MC simulations into reactor simulations are too computer time consuming in general, a workaround has to be developed to use the results of MC simulations in CFD simulations of technical systems. Potential methods are the establishment of table lookup strategies or the derivation of rate expressions from the reaction rates computed as function of gas-phase concentration and temperature by MC simulations. In both methods, the individual state of the catalyst still has to be incorporated because the state depends on the history of the particle that means the transients of the reactor behavior and associated local conditions.

4.3.4
Particle–Support Interaction and Spillover

The catalyst particles cannot be considered as isolated particles. They in general communicate with their neighborhood not only by gas-phase processes (desorption, diffusion, and readsorption) but also by diffusive transport on the solid surface. The detailed description of support and nanocatalysts presented above leads, in contrast to single-crystal surfaces, to heterogeneous surfaces. Therefore, insight into the influence of geometrical and communication effects on the rate as they occur in real catalyst systems can be gained. The analysis of a MC simulation identifies locations of reactions and diffusion intensities. This provides information on the communication effects both between the facets and between the support and the particle in the form of spillover effects and reverse spillover effects [34].

The expansion of the reverse spillover effect, that is, the capture zone of each particle, can be derived by a reverse MC analysis or a net diffusion analysis (Figure 4.6). A reverse MC analysis rewinds a simulation run and results in the location, where a reactant has adsorbed depending on the particle, where it reactively desorbed. A strong overlapping of capture zone areas leads to a lowering of the reaction rate per particle (Figure 4.7).

The net diffusion analysis sums up the diffusion directions for a specific species at each place. The resulting direction points to the particle where this location mainly *delivers* its adsorbates to. Mean field models taking into account nanocatalyst densities on the support may incorporate the capture zone effect.

4.3.5
Potentials and Limitations of MC Simulations for Derivation of Overall Reaction Rates

The described kinetic Monte Carlo method with multiple lattices shows the feasibility of a detailed description of the surface and the communication with the support despite a large number of processes. The presented analysis by reversing the simulation and following the diffusion path of each species results in a general

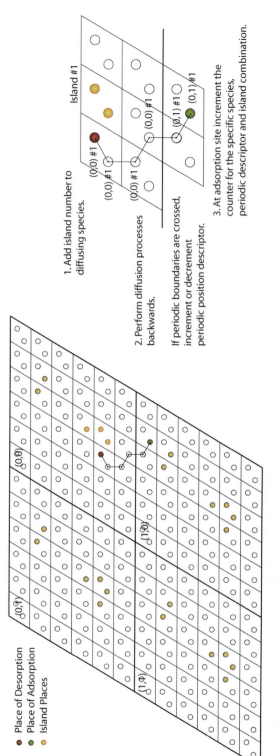

Figure 4.6 Reverse MC analysis: each species is traced backward from the particle, where it reactively desorbed to its original place, where it adsorbed. For periodic models, the periodic boundary conditions have to be unfolded for species traveling across the boundary. This can be achieved by a two-dimensional descriptor that is incremented or decremented during boundary passage. Reproduced from Ref. [33].

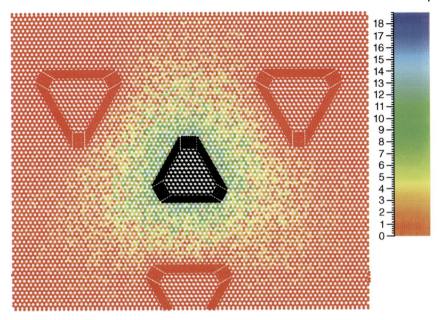

Figure 4.7 Capture zone of the black particle shown in the center. The number of adsorbed species, which later reactively desorb from the central particle, is counted for each individual adsorption site. Reproduced from Ref. [33].

usable analysis of communication and spillover effects. The hybrid manner between on-lattice and off-lattice approaches provides the possibility to locally use an off-lattice description, which may be useful for clusters consisting of a small number of atoms and lacking any local periodicity.

Of course, the main problems of MC simulations still remain, that is, the identification of a list with a complete set of processes and the presence of fast processes lowering the simulation advance in time. In contrast to crystal surfaces, an empirical derivation of process rates has a much higher effort since the experimental results are averaged over statistical distributions of particle size and shape. In the presented examples, no rates for processes at edges were available and therefore only diffusion was considered. If reactions at edges are preferred, this will result in a significant error. Therefore, the presented example can be considered only as a starting point for an exact and geometrical complete model. A detailed analysis of the elementary processes and lateral interactions is still necessary.

In order to fully bridge the geometric gap between *ab initio* crystal surface simulations and real catalytic surfaces, further development is needed. Several problems have to be solved, which is the connection to the support, the exact description of the impingement rate on and near particles, the account for surface oxidations, and the reshaping of particles during catalysis.

In the presented method, the connection to the support requires a commensurable lattice between support and particle, otherwise the model results in a rather large number of different processes across the links. The latter might be avoided by using an interatomic potential [35] for the interactions rather than discrete interactions as in LGH. Furthermore, the support can induce strain into the particle that destroys the local periodicity of the facet that is exploited in the given model [36].

Reshaping of supported nanoparticles has been observed for systems such as Rh [37] and Cu [38] and poses MC a similar problem to surface oxidations. Simulation of these processes can be made available by stacking several two-dimensional lattices, linking three-dimensional lattices, or switching to a completely three-dimensional particle model.

So far, the impingement rate for adsorption from kinetic gas theory used in MC simulations is based on the assumption of a free half space above the surface. This assumption is not given for facets and supports because the reflection–adsorption ratio and desorption from other surface parts have to be accounted for and are coverage dependent. A possible way to further investigate this is coupling of the described MC simulation to molecular dynamics simulations [39].

As mentioned above, the direct coupling of MC simulations with continuum mechanics simulation is in general not tractable due to computational efforts, at least for most of the technically relevant systems. However, there are a few valuable exceptions. In catalytic combustion, Kissel-Osterrieder *et al.* [40] coupeld a real time-dependent MC simulation with the surrounding stagnation-point flow field to study the ignition of CO oxidation on Pt and compared it with simulations using the mean field approximation. The same flow configuration was recently used by Matera and Reuter [41] to study transport limitations and bistability of CO oxidation at $RuO_2(110)$ single-crystal surface. They revealed that the coupling of gas-phase transport and surface kinetics leads to an additional complexity, which needs to be accounted for in the interpretation of dedicated *in situ* experiments. Hence, the coupling of MC and CFD seems to be of great interest for an atomic-scale understanding of the function of heterogeneous catalysts used in surface science experiments at technologically relevant gas-phase conditions. While surface science was used to operate the lab reactors at conditions (low pressure and controlled temperature), where mass and heat transfer do not matter, the efforts for bridging the pressure gap call for more sophisticated computational tools for interpretation of the measured data.

It is worth to mention that Kolobov *et al.* [42] recently coupled atomistic and continuum models for multiscale simulations of gas flows both by combinations of MC and CFD and by combinations of direct Boltzmann solvers with kinetic CFD schemes; the codes were applied to study catalytic growth of vertically aligned carbon nanotubes.

4.4
Models Applicable for Numerical Simulation of Technical Catalytic Reactors

Since the rate of catalytic reactions is very specific to the catalyst formulation, global rate expressions have been used for many years [1, 43, 44]. The reaction rate has often been based on catalyst mass, catalyst volume, reactor volume, or catalyst external surface area. The implementation of this *macrokinetics* approach is straightforward; the reaction rate can be easily expressed by any arbitrary function of gas-phase concentrations and temperature at the catalyst surface calculated at any computational cell containing catalytically active particles or walls. It is evident that this approach cannot account for the complex variety of phenomena of catalysis and that the rate parameters must be evaluated experimentally for each new catalyst and various external conditions.

The direct computation of surface reaction rates from the molecular situation as discussed above and, in more detail throughout Chapters 1–3 of this book, leads more and more to a comprehensive description, at least for idealized systems. However, DFT, MD, and MC simulations cannot be implemented in complex flow field simulations of technically relevant systems due to missing algorithms and, more importantly, due to the immense amount of computational time needed. Treating the catalytic system as a *black box* is not the alternative; rather, the knowledge gained from experimental and theoretical surface science studies should be implemented in chemical models used in reaction engineering simulation. A tractable approach was proposed for the treatment of surface chemistry in reactive flows in the early 1990s [45–47], when the rate equations known from modeling homogeneous reactions systems were adapted to model heterogeneous reactions. This approach was then first computationally realized in the SURFACE CHEMKIN [48] module of the CHEMKIN [49, 50] software package; later on, other packages such as CANTERA [51] and DETCHEM [52] as well as many of the commercially available multipurpose CFD codes adopted this methodology or directly implemented those modules. In the remainder of this section, this concept at present being state of the art in modeling of catalytic reactors will be discussed.

4.4.1
Mean Field Approximation and Reaction Kinetics

The mean field approximation is related to the size of the computational cell in the flow field simulation, assuming that the local state of the active surface can be represented by mean values for this cell. Hence, this model assumes randomly distributed adsorbates on the surface, which is viewed as being uniform (Figure 4.8). The state of the catalytic surface is described by temperature T and a set of surface coverages θ_i, which is the fraction of the surface covered with surface species i. The surface temperature and the coverage depend on time and the macroscopic position in the reactor, but are averaged over microscopic local fluctuations. Under these assumptions, a chemical reaction can be defined in analogy to Eq. (4.1) by

$$\sum_{i=1}^{N_g + N_s + N_b} v'_{ik} A_i \rightarrow \sum_{i=1}^{N_g + N_s + N_b} v''_{ik} A_i. \tag{4.6}$$

The difference is that now A_i denote not only gas-phase species (e.g., H_2) but also surface species (e.g., H(s)) and bulk species (e.g., H(b)). The N_s surface species are

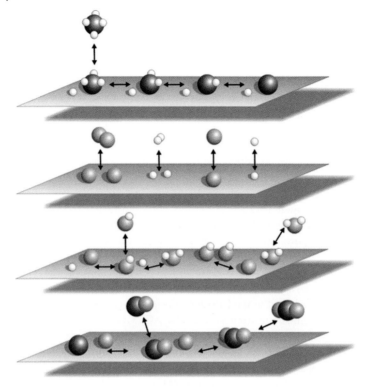

Figure 4.8 Sketch of the major reaction steps in catalytic partial oxidation of methane over Rh applying the mean field approximation [87].

those that are adsorbed on the top monoatomic layer of the catalytic particle, while the N_b bulk species are those found in the inner solid catalyst.

Steric effects of adsorbed species and various configurations, for example, the type of chemical bonds between adsorbate and solid, can be taken into account using the following concept: The surface structure is associated with a surface site density Γ that describes the maximum number of species that can adsorb on a unit surface area, given, for example, in mol/m². Then each surface species is associated with a coordination number σ_i describing the number of surface sites that are covered by this species. Under the assumptions made, a multistep (quasielementary) reaction mechanism can be set up. The local chemical source term, R_i^{het}, needed in the CFD simulation of the technical reactor is then derived from the molar net production rate $\dot{s}_i$ by

$$R_i^{het} = \dot{s}_i M_i = M_i \sum_{k=1}^{K_s} \nu_{ik} k_{f_k} \prod_{j=1}^{N_g + N_s + N_b} c_j^{\nu'_{jk}}. \tag{4.7}$$

Here, K_s is the number of surface reactions and c_i are the species concentrations, which are given, for example, in mol/m² for the N_s adsorbed species and in, for example, mol/m³ for the N_g and N_b gaseous and bulk species. According to Eq. (4.7) and the relation $\Theta_i = c_i \sigma_i \Gamma^{-1}$, the variations in surface coverages follow

$$\frac{\partial \Theta_i}{\partial t} = \frac{\dot{s}_i \sigma_i}{\Gamma}. \tag{4.8}$$

Since the reactor temperature and concentrations of gaseous species depend on the local position in the reactor, the set of surface coverages also varies with position. However, no lateral interaction of the surface species between different locations on the catalytic surface is modeled. This assumption is justified by the fact that the computational cells in reactor simulations are usually much larger than the range of lateral interactions of the surface processes. In each of these cells, the state of the surface is characterized by mean values (mean field approximation). The set of differential Eqs. (4.8) has to be solved simultaneously with flow field equations for every computational cell containing catalytic material. At steady state, the left side of Eq. (4.8) becomes zero, and a set of algebraic equations has to be solved. The timescales to reach the steady state of Eq. (4.8) are commonly much shorter than the timescales of significant variations in species concentrations and temperature in the fluid. Therefore, a quasisteady-state assumption with vanishing left side of Eq. (4.8) is frequently justified even for transient reactor operation. For fast transients, usually on the order of $<10^{-4}$ s, this assumption breaks down and the straight implementation of this procedure into CFD simulations becomes more sophisticated [53, 54].

The binding states of adsorption of all species vary with the surface coverage. This additional coverage dependence is modeled in the expression for the rate coefficient k_{f_k} in Eq. (4.7) by two additional parameters, μ_{i_k} and ε_{i_k} [47, 48, 55]:

$$k_{f_k} = A_k T^{\beta_k} \exp\left[\frac{-E_{a_k}}{RT}\right] \prod_{i=1}^{N_s} \Theta_i^{\mu_{i_k}} \exp\left[\frac{\varepsilon_{i_k} \Theta_i}{RT}\right]. \tag{4.9}$$

For adsorption reactions, sticking coefficients are commonly used, which can be converted to conventional rate coefficients [47] by

$$k_{f_k}^{ads} = \frac{S_i^0}{\Gamma^\tau} \sqrt{\frac{RT}{2\pi M_i}}, \quad \text{where } \tau = \sum_{j=1}^{N_s} \nu'_{jk}. \tag{4.10}$$

S_i^0 is the initial (uncovered surface) sticking coefficient. At high sticking coefficients, a correction needs to be applied [47].

4.4.2
Thermodynamic Consistency

A crucial issue with many of the mechanisms published is thermodynamic consistency. Even though most of the mechanisms lead to consistent enthalpy diagrams, many are not consistent regarding the entropy change in the overall reaction due to lack of knowledge on the transition states of the individual reactions and therefore on the preexponentials in the rate equations. Lately, optimization procedures enforcing overall thermodynamic consistency have been applied to overcome this problem [56, 57]; here, we present one [57] of these approaches.

The equilibrium of a reversible chemical reaction,

$$\sum_i v'_{ik} A_i \underset{k_{r_k}}{\overset{k_{f_k}}{\rightleftarrows}} \sum_i v''_{ik} A_i, \qquad (4.11)$$

is completely defined by the thermodynamic properties of the participating species. Expressed in terms of the equilibrium constant, K_{pk}, the equilibrium activities, a_j^{eq}, obey the equation

$$K_{pk} = \prod_i \left(a_i^{eq}\right)^{v_{ik}} = \exp\left(-\frac{\Delta_k G^0}{RT}\right). \qquad (4.12)$$

The change in free enthalpy at normal pressure p^0 is

$$\Delta_k G^0 = \sum_i v_{ik} G_i^0(T). \qquad (4.13)$$

In case of gases, the activities can be approximated by their partial pressures $a_i = p_i/p^0$ and in case of surface species by their coverages $a_i = \Theta_i$. When the dependence of the heat capacities on temperature is given by a forth-order polynomial [45] and standard enthalpies and entropies of formation, the standard free enthalpies can be expressed in terms of seven coefficients, $a_{0,i}, \ldots, a_{6,i}$:

$$G_i^0(T) = a_{0,i} + a_{1,i} T + a_{2,i} T^2 + a_{3,i} T^3 + a_{4,i} T^4 + a_{5,i} T^5 + a_{6,i} T \ln T. \qquad (4.14)$$

In order to predict the correct equilibrium, the rate coefficients for the forward and the reverse reaction must obey the equation

$$\frac{k_{f_k}}{k_{r_k}} = K_{pk} \prod_i (c_i^0)^{v_{ik}}. \qquad (4.15)$$

The c_i^0 are reference concentrations at normal pressure, that is, $c_i^0 = p^0/RT$ for gas-phase species and $c_i^0 = \Gamma/\sigma_i$ for surface species.

However, one problem in setting up a reaction mechanism is the difficulty to define the thermodynamic data for intermediate surface species. Therefore, Eq. (4.15) cannot be used to calculate the rate coefficient of the reverse reaction. The forward and the reverse reaction are then defined separately with their own rate laws. Nevertheless, these rates cannot be chosen independently.

Therefore, the following method is proposed [57]: Assume an initial guess for a surface reaction mechanism. The rate coefficients for forward and reverse reactions may have been adjusted separately. Suppose that the thermodynamic data for species 1...N_u are unknown. For each pair of reversible reactions, we can calculate an equilibrium constant using Eq. (4.15) and, according to Eq. (4.12), its logarithm yields the change in free enthalpy. (*Note*: coverage-dependent activation energies are not considered at this stage.) Separation of the known and the unknown variables in Eq. (4.13) leads to

$$\Delta_k G^0 = \sum_{i=1}^{N_u} v_{ik} \tilde{G}_i^0(T) + \sum_{i=N_u+1}^{N} v_{ik} G_i^0(T), \qquad (4.16)$$

that is a linear equation system for the unknown free enthalpies $\tilde{G}_i^0$. Since most species are involved in more than one reaction, this system is usually overdetermined. Inserting Eq. (4.14) for several temperatures, T_j gives a system of linear equations in the unknown coefficients $\tilde{a}_{l,i}$

$$\sum_{i=1}^{N_u} \sum_{l=0}^{6} v_{ik} t_{lj} \tilde{a}_{l,i} = g_{kj} \quad \text{for all } k \text{ and } j, \tag{4.17}$$

where the following abbreviations have been used:

$$g_{kj} = \Delta_k G^0(T_j) - \sum_{i=N_u+1}^{N} v_{ik} G_i^0(T_j) \text{ and} \tag{4.18}$$

$$t_{lj} = \begin{cases} T_j^l, & \text{if } l < 6, \\ T_j \ln T_j, & \text{if } l = 6. \end{cases} \tag{4.19}$$

An "optimal" set of parameters $\tilde{a}_{l,i}$ is determined by a weighted least square approximation. The weights can be chosen individually for each pair of reactions according to a sensitivity analysis of the reaction mechanism. This guarantees that the equilibrium of crucial reaction steps will be shifted less than others after adjustment.

The newly adjusted thermodynamic coefficients are then used to calculate the change in free enthalpy for each reaction (4.16), the equilibrium constant (4.12), and the rate coefficient of the reverse reaction (4.15). In case the reverse reaction shall be expressed in terms of Arrhenius coefficients, another least square approximation using the rate constants at the discrete temperatures, T_j, is performed.

Since writing of surface reaction mechanisms as pairs of irreversible reactions is often more convenient, this procedure has to be repeated during mechanism development after modification of rate coefficients belonging to any of these pairs. Figure 4.9 illustrates the algorithm. The difference between this method

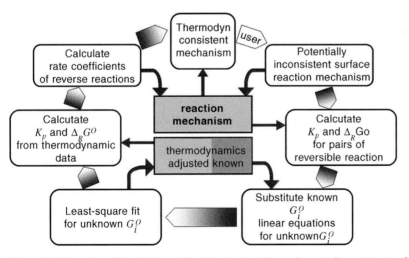

Figure 4.9 Illustration of the adjustment algorithm to set up thermodynamically consistent surface reaction mechanisms [57].

and the scheme proposed by Mhadeshwar et al. [56] is the fact that in the procedure described there is no need to select a linearly independent set of reactions. Instead of distinguishing reactions between linear base and linear combinations, all reactions are treated equally by solving the same linear problem using a least square fit.

Exemplarily, Table 4.3 lists a surface reaction mechanism for steam reforming over Ni-based catalysts, which has been adjusted to be thermodynamically consistent according to this procedure. The mechanism has successfully been used to model steam reforming of methane over Ni/Al_2O_3 catalysts [57] and thermocatalytic conversion of partially preformed methane in solid–oxide fuel cells (SOFC) [58, 59].

4.4.3
Practicable Method for Development of Multistep Surface Reaction Mechanisms

The development of a reliable surface reaction mechanism is a complex process following the scheme given in Figure 4.10. A tentative reaction mechanism can be proposed based on experimental surface science studies, on analogy to gas-phase kinetics and organometallic compounds, and on theoretical studies, increasingly including DFT calculations [60, 61] and in the future also MD and MC simulations as discussed above [32, 34]. This mechanism should include all possible paths for the formation of the chemical species under consideration in order to be "elementary-like" and thus applicable over a wide range of conditions. The mechanism idea then needs to be evaluated by numerous experimentally derived data, which are compared with theoretical predictions based on the mechanism. Here, the simulations of the laboratory reactors require appropriate models for all significant processes in order to evaluate the intrinsic kinetics.

A key step in improving detailed reaction mechanisms is the application of sensitivity analysis, which leads to the crucial steps in the mechanism for which refined kinetic experiments and data may be needed [62]. This technique has been applied in understanding homogeneous combustion processes for decades [12]. A similar approach called *degree of rate control* has been introduced by Campbell [63]. Sensitivity analysis allows for identification of the individual reaction steps that are most influential to the system. Several methods can be used in order to calculate the sensitivity of the solution, that is, the time behavior of the different species profiles with respect to the rate coefficients. The most straightforward way is to simply run a calculation and then rerun it after changing the parameters successively. Although very simple, this approach has the disadvantage of requiring a large amount of computational power for the n_p parameters for which sensitivities have to be calculated. Furthermore, here the errors result from the fact that the parameters have to be changed by a finite amount, which is sufficiently large to observe an effect on the overall solution, which is not superimposed by discretization errors.

A more elegant way is the direct solution of the sensitivity equations together with the solution of the system equations [64]. In order to derive the governing equations

Table 4.3 Thermodynamically consistent surface reaction mechanism for steam reforming of methane over nickel-based catalysts [57].

	Reaction	A (cm, mol, s)	E_a (kJ/mol)	β (−)
R_1	$H_2 + 2\,Ni(s) \rightarrow 2\,H(s)$	1.000×10^{-02}	0.0	0.0
R_2	$2\,H(s) \rightarrow 2\,Ni(s) + H_2$	$2.545 \times 10^{+19}$	81.21	0.0
R_3	$O_2 + 2\,Ni(s) \rightarrow 2\,O(s)$	1.000×10^{-02}	0.0	0.0
R_4	$2\,O(s) \rightarrow 2\,Ni(s) + O_2$	$4.283 \times 10^{+23}$	474.95	0.0
R_5	$CH_4 + Ni(s) \rightarrow CH_4(s)$	8.000×10^{-03}	0.0	0.0
R_6	$CH_4(s) \rightarrow CH_4 + Ni(s)$	$8.705 \times 10^{+15}$	37.55	0.0
R_7	$H_2O + Ni(s) \rightarrow H_2O(s)$	1.000×10^{-01}	0.0	0.0
R_8	$H_2O(s) \rightarrow H_2O + Ni(s)$	$3.732 \times 10^{+12}$	60.79	0.0
R_9	$CO_2 + Ni(s) \rightarrow CO_2(s)$	1.000×10^{-05}	0.0	0.0
R_{10}	$CO_2(s) \rightarrow CO_2 + Ni(s)$	$6.447 \times 10^{+07}$	25.98	0.0
R_{11}	$CO + Ni(s) \rightarrow CO(s)$	5.000×10^{-01}	0.0	0.0
R_{12}	$CO(s) \rightarrow CO + Ni(s)$	$3.563 \times 10^{+11}$	$111.27 - 50\theta_{CO(s)}$	0.0
R_{13}	$CH_4(s) + Ni(s) \rightarrow CH_3(s) + H(s)$	$3.700 \times 10^{+21}$	57.7	0.0
R_{14}	$CH_3(s) + H(s) \rightarrow CH_4(s) + Ni(s)$	$6.034 \times 10^{+21}$	61.58	0.0
R_{15}	$CH_3(s) + Ni(s) \rightarrow CH_2(s) + H(s)$	$3.700 \times 10^{+24}$	100.0	0.0
R_{16}	$CH_2(s) + H(s) \rightarrow CH_3(s) + Ni(s)$	$1.293 \times 10^{+23}$	55.33	0.0
R_{17}	$CH_2(s) + Ni(s) \rightarrow CH(s) + H(s)$	$3.700 \times 10^{+24}$	97.10	0.0
R_{18}	$CH(s) + H(s) \rightarrow CH_2(s) + Ni(s)$	$4.089 \times 10^{+24}$	79.18	0.0
R_{19}	$CH(s) + Ni(s) \rightarrow C(s) + H(s)$	$3.700 \times 10^{+21}$	18.8	0.0
R_{20}	$C(s) + H(s) \rightarrow CH(s) + Ni(s)$	$4.562 \times 10^{+22}$	161.11	0.0
R_{21}	$CH_4(s) + O(s) \rightarrow CH_3(s) + OH(s)$	$1.700 \times 10^{+24}$	88.3	0.0
R_{22}	$CH_3(s) + OH(s) \rightarrow CH_4(s) + O(s)$	$9.876 \times 10^{+22}$	30.37	0.0
R_{23}	$CH_3(s) + O(s) \rightarrow CH_2(s) + OH(s)$	$3.700 \times 10^{+24}$	130.1	0.0
R_{24}	$CH_2(s) + OH(s) \rightarrow CH_3(s) + O(s)$	$4.607 \times 10^{+21}$	23.62	0.0
R_{25}	$CH_2(s) + O(s) \rightarrow CH(s) + OH(s)$	$3.700 \times 10^{+24}$	126.8	0.0
R_{26}	$CH(s) + OH(s) \rightarrow CH_2(s) + O(s)$	$1.457 \times 10^{+23}$	47.07	0.0
R_{27}	$CH(s) + O(s) \rightarrow C(s) + OH(s)$	$3.700 \times 10^{+21}$	48.1	0.0
R_{28}	$C(s) + OH(s) \rightarrow CH(s) + O(s)$	$1.625 \times 10^{+21}$	128.61	0.0
R_{29}	$H(s) + O(s) \rightarrow OH(s) + Ni(s)$	$5.000 \times 10^{+22}$	97.9	0.0
R_{30}	$OH(s) + Ni(s) \rightarrow H(s) + O(s)$	$1.781 \times 10^{+21}$	36.09	0.0
R_{31}	$H(s) + OH(s) \rightarrow H_2O(s) + Ni(s)$	$3.000 \times 10^{+20}$	42.7	0.0
R_{32}	$H_2O(s) + Ni(s) \rightarrow H(s) + OH(s)$	$2.271 \times 10^{+21}$	91.76	0.0
R_{33}	$OH(s) + OH(s) \rightarrow H_2O(s) + O(s)$	$3.000 \times 10^{+21}$	100.0	0.0
R_{34}	$H_2O(s) + O(s) \rightarrow OH(s) + OH(s)$	$6.373 \times 10^{+23}$	210.86	0.0
R_{35}	$C(s) + O(s) \rightarrow CO(s) + Ni(s)$	$5.200 \times 10^{+23}$	148.1	0.0
R_{36}	$CO(s) + Ni(s) \rightarrow C(s) + O(s)$	$1.354 \times 10^{+22}$	$116.12 - 50\theta_{CO(s)}$	−3.0
R_{37}	$CO(s) + O(s) \rightarrow CO_2(s) + Ni(s)$	$2.000 \times 10^{+19}$	$123.6 - 50\theta_{CO(s)}$	0.0
R_{38}	$CO_2(s) + Ni(s) \rightarrow CO(s) + O(s)$	$4.653 \times 10^{+23}$	89.32	−1.0
R_{39}	$CO(s) + H(s) \rightarrow HCO(s) + Ni(s)$	$4.019 \times 10^{+20}$	132.23	−1.0
R_{40}	$HCO(s) + Ni(s) \rightarrow CO(s) + H(s)$	$3.700 \times 10^{+21}$	$0.0 + 50\theta_{CO(s)}$	0.0
R_{41}	$HCO(s) + Ni(s) \rightarrow CH(s) + O(s)$	$3.700 \times 10^{+24}$	95.8	−3.0
R_{42}	$CH(s) + O(s) \rightarrow HCO(s) + Ni(s)$	$4.604 \times 10^{+20}$	109.97	0.0

Parameter definition according to Eq. (4.9). Parameter A in Reactions 1, 3, 5, 7, 9, and 11 is the intial sticking coefficient according to Eq. (4.10). The mechanism can be downloaded from www.detchem.com.

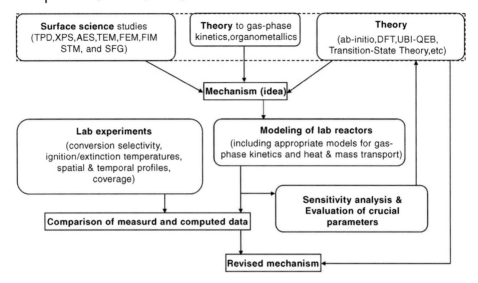

Figure 4.10 Survey of the methodology of the development of a surface reaction mechanism [55].

for the sensitivity coefficients, we consider the differential algebraic equation systems describing the chemical reactor under consideration:

$$\overline{\overline{B}} \frac{\partial \vec{y}}{\partial t} = \vec{F}(\vec{y}, t; \vec{p}), \tag{4.20}$$

where $\vec{y}$ is the n-dimensional vector of the dependent variables (species partial pressures, temperature, coverages, etc.), $\vec{p}$ is the vector of the n_p system variables (rate parameters, transport coefficients, etc.), and $\vec{F}$ is the vector function of these variables. The sensitivity coefficients are defined as

$$s_{ij} = \frac{\partial y_i}{\partial p_j}. \tag{4.21}$$

Partial differentiation of Eq. (4.20) with respect to n_p parameters p_j leads to

$$\overline{\overline{B}} \frac{\partial \overline{\overline{S}}(t)}{\partial t} = \overline{\overline{F_y}}\,\overline{\overline{S}}(t) + \overline{\overline{F_p}}(t), \tag{4.22}$$

where $\overline{\overline{S}}$ is the matrix of the time-dependent sensitivity coefficients, $\overline{\overline{F_y}}$ is the Jacobian, $\partial F_i / \partial y_j$, and $\overline{\overline{F_p}}$ is the matrix of the derivation of $\vec{F}$ with respect to the system parameters $\vec{p}$, that is, $\partial F_i / \partial p_j$. In case of rate coefficients k_j as system parameters, we get

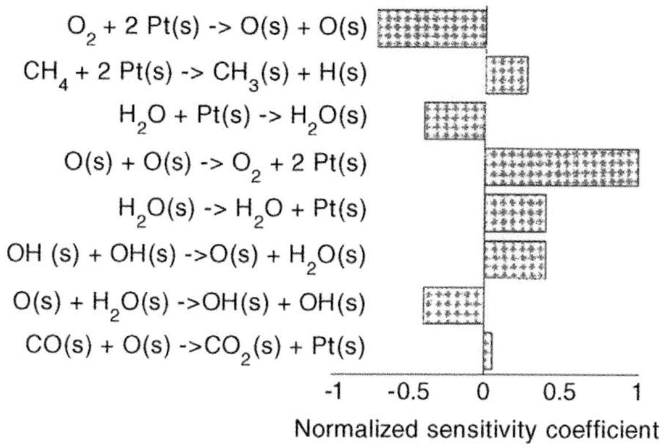

4.4 Models Applicable for Numerical Simulation of Technical Catalytic Reactors | 137

$O_2 + 2\, Pt(s) \rightarrow O(s) + O(s)$
$CH_4 + 2\, Pt(s) \rightarrow CH_3(s) + H(s)$
$H_2O + Pt(s) \rightarrow H_2O(s)$
$O(s) + O(s) \rightarrow O_2 + 2\, Pt(s)$
$H_2O(s) \rightarrow H_2O + Pt(s)$
$OH(s) + OH(s) \rightarrow O(s) + H_2O(s)$
$O(s) + H_2O(s) \rightarrow OH(s) + OH(s)$
$CO(s) + O(s) \rightarrow CO_2(s) + Pt(s)$

-1   -0.5   0   0.5   1
Normalized sensitivity coefficient

Figure 4.11 Sensitivity of the uncovered surface area on rate coefficients of surface reactions immediately before catalytic ignition of CH_4 oxidation on a platinum foil [53].

$$s_{ij} = \frac{\partial y_i}{\partial \ln k_j}. \qquad (4.23)$$

Consequently, the macroscopic variable y_i will be changed by 1% times s_{ij} for a change in k_j by 1% [64].

Exemplarily, Figure 4.11 shows the normalized sensitivity coefficients concerning the sensitivity of the vacancies (uncovered adsorption sites) on rate coefficients of surface reactions immediately before catalytic ignition of methane oxidation in a stagnation-point flow onto a platinum foil [53]. The decisive parameters for ignition are the rates of adsorption and desorption of oxygen on the surface, that is, the adsorption–desorption equilibrium of oxygen on Pt, which becomes clear after studying the surface coverages as shown in Figure 4.12. Oxygen coverage inhibits methane adsorption before ignition.

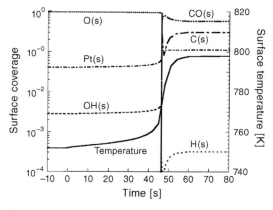

Figure 4.12 Calculated time-dependent surface coverage and surface temperature during heterogeneous ignition of a $CH_4/O_2/N_2$ mixtures flowing in a stagnation-point flow onto a platinum foil [53].

The application of sensitivity analysis is also a powerful tool in identifying the rate-limiting steps in a detailed chemical reaction mechanism, which then can be used to set up even simpler rate expressions such as those used in the LHHW model.

The coupling of several complex models introduces a large number of parameters into the simulations. For instance, detailed reaction schemes may have hundreds of kinetic parameters, each value associated with a certain inaccuracy. Hence, agreement between predicted and experimentally observed overall conversion and selectivity alone is not sufficient to evaluate individual submodels. Consequently, once established mechanisms require continuous evaluation. Eventually, reaction mechanisms are rather falsified by applying them; they can never be verified! One can never be sure not to have missed certain reaction pathways. Hence, the practical development and testing shall include large sets of experimental data at a wide variety of external conditions (temperature, chemical composition, and contact time). The use of experimental data for the same chemical system but in different flow configurations limits the risk of having transport effects, primarily internal diffusion, included in the intrinsic kinetics. The use of experimental data of different groups helps in a similar way.

Recently, a new platform – available on the Web site www.detchem.com – has been established to assist in the development of surface reaction mechanisms. On this platform, experimental data and the corresponding computational tools (flow models) for their numerical simulation are given. The user can choose a set of *real* and *numerical* experiments for a certain chemical system, for example, NO oxidation over Pt, and a corresponding reaction mechanism, which can be modified, and let the underlying computers do the simulation. As results, comparisons of experimental and numerical predicted conversion, selectivities, profiles, or any other relevant variables are presented. Using this computational tool, one can easily test a new or modified kinetic scheme versus a large set of experimental data points. This procedure will not only boost the development of reaction mechanisms but also help identify experimental situations (conditions, support effects, and false measurements) that need further considerations.

Time and locally resolved profiles provide a more stringent test for model evaluation. Useful data arise from spatial and temporal species profiles by *in situ*, noninvasive methods such as Raman and laser-induced fluorescence (LIF) spectroscopy. For instance, an optically accessible catalytic channel reactor can be used to evaluate models for heterogeneous and homogeneous chemistry as well as transport by simultaneous detection of stable species by Raman measurements and OH radicals by planar laser-induced fluorescence (PLIF) [65, 66]. For instance, the reliability of different heterogeneous and homogeneous reaction schemes proposed in literature was investigated by comparison of the experimentally derived ignition distances with numerical elliptic two-dimensional simulations of the flow field using combinations of a variety of schemes [67–70]. While some models perform well, others lead to quite inaccurate predictions. Model evaluation is discussed in more details in Chapter 7.

Since the early 1990s, many groups have developed surface reaction mechanisms following the mean field approximation. Oxidation reactions over noble metals in particular have been modeled extensively such as of hydrogen [46, 71–76], CO [77–79], methane [53, 80–84], and ethane [7, 11, 85, 86] over Pt and formation of synthesis gas over Rh [84, 87]. Lately, mechanisms have been established for more complex reaction systems, for instance, three-way catalysts [88] or chemical vapor deposition (CVD) reactors for the formation of diamond [89, 90], silica [91], and nanotubes [92]. Table 4.4 lists a selection of heterogeneous reaction mechanisms following the concept of the mean field approximation; the list is far from complete.

4.4.4
Potentials and Limitations of the Mean Field Approximation

Catalytic surfaces are certainly nonuniform; site heterogeneity exists because the surface of practical catalyst particles is characterized by terraces of different crystal structures, steps, edges, additives, impurities, and defects. In the mean field approximation, the site heterogeneity is averaged by mean rate coefficients. If the distribution of the different types of adsorption sites and the reaction kinetics on these sites are known, this concept can be easily used to set up a reaction mechanism, which consists of several submechanisms for the different surface structures [48]. This concept was applied in the framework of a two-adsorption site model for the simulation of CO combustion on polycrystalline Pt [93]. The site heterogeneity can be described by the probability that an arbitrary site is characterized by the associated reaction kinetics. In the models discussed so far, this probability function is a sum over a finite number of surface structures; however, continuous functions are used in literature, as well [2]. Here, the problem is the limited knowledge of distribution of the different types of surface patches and the kinetics of these patches, again.

Effects resulting from lateral interactions of the adsorbates are inherently more difficult to treat. In the mean field approximation, they are either neglected or incorporated by mean rate coefficients. If the specific adsorbate–adsorbate interactions are understood quantitatively, a Monte Carlo simulation of the surface chemistry can be carried out as discussed above.

The amount of catalyst in a technical system is generally determined rather by the active catalytic surface area than the total amount of catalysts used, which means the dispersion of the catalyst matters. The active catalytic surface area can be experimentally determined by chemisorption studies using CO or H_2. Since the molar surface reaction rate, Eq. (4.7), is given per catalyst surface area $(mol/(m^2 s))$, the active catalytic surface area serves as scaling parameter to compute the chemical source term due to heterogeneous reactions per volumetric or surface element of the reactor. Recently, it was shown that the active catalytic surface area can also be used to describe the change in reaction rates with varying catalyst loading and due to hydrothermal aging [94, 95].

Another crucial issue of MF kinetics is the origin of the kinetic data, in particular the activation energies. Since data from DFT simulation become more and more available and the commercial codes are easy to use even for nonexperts, DFT-derived data have found their way into reaction engineering modeling. Two concerns have to

Table 4.4 Selection of heterogeneous reaction mechanisms using the mean field approach.

Catalyst	Reactants and products	Systems/reactor configurations primarily[a]
Pt	$H_2/O_2 + H_2O$	Pt(111) [132–136], Pt foil [71, 72, 137], wire [73], plate [68, 70], Pt [138, 139]
Pt	$CO/O_2 + CO_2$	Pt(100) [140], Pt(111) [141], Pt(110) [142], Pt [73, 143], Pt wire [144, 145]
Pt	$CH_4/O_2 + H_2, CO, CO_2, H_2O$	Pt foil [53, 146, 147], Pt [71, 83, 148], Pt/Al_2O_3 monolith [146, 149, 150], Pt gauze [151]
Pt	$C_2H_4/H_2 + C_2H_6$	Single crystals [2, 152, 153]
Pt	$C_2H_6/O_2 + C_2H_4$	Pt-coated monolith [7, 86, 154, 155]
Pt	$N_2/H_2 + NH_3$	Single-crystal studies [2, 156–159]
Pt	$NH_3/O_2 + NO, H_2O$	Pt wire [160]
Pt	$C_3H_6/CH_4/CO/NOx + N_2, CO_2, H_2O$	Three-way automotive catalyst, NO_x storage/reduction automotive catalyst [161]
Rh	$H_2/O_2 + H_2O$	Rh foil [148]
Rh	$CO/O_2 + CO_2$	Rh(111) [162], Rh/Al_2O_3 [163, 164]
Rh	$CH_4/O_2 + H_2, CO, H_2O, CO_2$	Rh/Al_2O_3-coated monoliths [87, 146, 150, 165, 166]
Rh	$CH_4/H_2O/CO_2/ + CO, H_2$	Flow reactor [167, 168]
Rh	$CH_4/H_2O + CO, H_2$	Rh/Al_2O_3-coated microreactor [169, 170], channel [170]
Rh	$CO/NO + CO_2, N_2$	Rh(111) [171], Rh/Al_2O_3 in automotive converters [163]
Rh	$CO/N_2O + CO_2, N_2$	Rh(111) [172]
Rh	$C_3H_8/O_2/H_2O + CO, H_2$	Rh/Al_2O_3 monolith [173]
Pt/Rh	$CO/NO + CO_2, N_2$	Flow reactor [174]
Pt/Rh/Al_2O_3	$C_3H_6/CO/NO + N_2, CO_2, H_2O$	Three-way automotive catalyst [88]
Pt/BaO/Al_2O_3	$NO/NH_3/H_2 + N_2$	NO_x storage and reduction automotive monolith catalysts [175]
Pd	$H_2/O_2 + H_2O$	Pd wire [53], coated microreactor [176, 177]
Ag	$C_2H_4/O_2 + C_2H_4O$	Kinetics of epoxidation, ethylene combustion, TPD, TPR [178]
Cu	$CO/H_2O + H_2, CO_2$	Cu(111) [179–182], Cu(110) [183], Cu/Al_2O_3; SiO_2 [179]
Cu	$CO/H_2 + CH_3OH$	Cu(100) [184, 185]
Ni	$CH_4/H_2O/CO_2 + H_2, CO$	Ni/$MgAl_2O_3$-MgO [186], Ni/YSC [58], Ni/Al_2O_3 [57]
Ni	$CH_4/H_2O + H_2, CO$	Ni(111) [187]

[a] Catalysts without crystal face denote polycrystalline material.

be mentioned. First, not all DFT data published are trustworthy; the assumptions used in the simulation, such as size of the unit cell, effect of coadsorbates, or choice of the functional, are often not justified. Furthermore, the uncertainties of the data computed may be too large for a meaningful implementation into a complex reaction mechanism. The second point here lies in the intrinsic assumptions of the mean field approximation, which is that the individual morphology and size of the catalyst particle is not taken into account. In particular, the individual kinetics of the general crystal phase is usually not considered, even though it is in principle possible. As discussed above and in previous chapters, the molecular processes are rather too complex for an adequate description within the concept of the mean field approximation. If only one single-crystal phase or a certain defect structure determines the overall reaction rate of the catalyst particle, then DFT simulations of these phases and structures may be implemented directly into a mean field model, but often this is not the case. Reuter and coworkers recently studied this phenomenon in detail [96] highlighting the problems that arise when MF simulations are based on first-principles computations only. The averaging over individual microscopic events implies that – in general – the direct use of DFT data in MF simulations is at least questionable. We often found that MF-based mechanisms are more reliable when rather based on semiempirical methods such as UBI-QEP [97] than on pure DFT data. The right way seems to be to use DFT data in MC simulations and feed the MC-computed rates into a kind of rate expression instead of using DFT directly in MF. Unfortunately, the approach from DFT to MC to MF is quite tricky. Therefore, the model of choice in reaction engineering simulations is likely going to remain the MF approach, combining knowledge gained from surface science and experiments at technical relevant conditions in a pragmatic way. This approach, which definitely has its deficiencies, has also shown some robustness and reliability and has quite often proven not only to help understand the complex interactions of heterogeneous chemical kinetics with mass and heat transport in laboratory reactors but also to support design and optimization of technical catalytic reactors.

4.5
Simplifying Complex Kinetic Schemes

Even though the previous section discusses a variety of simplifications used in the established reaction mechanisms, these kinetic schemes are sometimes still too complex for direct use in CFD simulations. They introduce an additional highly nonlinear coupling into the governing equations of CFD simulations, which leads to considerable computational efforts, in particular when used in simulations of turbulent flows (e.g., in flow reactors) and in simulations with continuously varying inlet and boundary conditions (e.g., in automobile catalytic converters). The nonlinearity, the large number of chemical species in different phases, and the fact that surface reactions in particular exhibit a large range of timescales make the solution of those equation systems challenging. In particular for transient simulations such as simulations of the behavior of catalytic converters during entire driving cycles [98], the solution of the system is too CPU time consuming with present numerical algorithms and computer capacities.

Therefore, many groups use global reaction schemes with kinetic data derived from experimental data such as conversion and selectivity as function of temperature through a pure fitting process using well-established optimization procedures [99]. Due to the complexity in the chemical processes, many additional parameters, for example, through the so-called inhibition terms, have to be introduced to let the model match a large set of experimental data [94, 100]. Sometimes, the number of parameters needed to describe all effects in such global reaction schemes can even exceed the number of parameters used in molecular multistep reaction mechanisms (MF approximation).

As a consequence, the increase in computing time with an increasing number of species and reactions implemented in the CFD simulation calls for the application of reduction algorithms for large reaction mechanisms. One approach is the extraction of the intrinsic low dimensional manifolds of trajectories in chemical space [15], once developed for homogeneous reaction mechanisms but also applied for heterogeneous reactions [101].

Votsmeier *et al.* recently developed several alternative strategies to use data-based models in the form of splines, neural networks, and lookup tables in CFD simulations of complex technical reactors [102, 103]. They demonstrated, for instance, that precomputed rate data can be used to enable the numerically efficient implementation of mechanistic kinetics in a reactor model for an automotive ammonia slip catalyst [103]. The source terms of the gas species are mapped (80 000 data points) as a function of gas composition and temperature to construct a spline interpolation function resulting in a speedup of about two orders of magnitude at an error of less than 1% [103].

4.6
Summary and Outlook

Today, most of the models for description of the heterogeneous reaction rates used in reaction engineering computations are based on the mean field approximation, in which the local state of the surface is described by its coverage with adsorbed species averaged on a microscopic scale. The averaging procedure questions the direct use of kinetic data derived from molecular surface science experiments and simulations such as density functional theory. In general, an intermediate step needs to be introduced. The most promising flow of information in this case is from DFT to MC to MF: DFT-derived activation energies for reactions and diffusion processes on nanoparticle catalysts are used in MC simulations, which directly provide the local heterogeneous reaction rates as function of gas-phase concentrations, temperature, and state of the catalyst particle (coverage, oxidation state, size, and morphology). These reaction rates can then be implemented into MF models. MD simulations may complete this pathway, coming in between DFT and MC. Although tremendous progress has recently been made in the efficient and reliable provision of DFT data and their use in MC simulations as well as the implementation of "real catalyst" effects such as spillover and catalyst–support interaction into the MC techniques, this approach can at present be used for simple and exemplary cases only. The computational efforts are simply too large.

On the other hand, global reaction kinetics with dozens of fitted parameters with little or even black box models without any mechanistic insight are still in use in scale-up and design of technical catalytic reactors. Hence, one of the challenges in modeling heterogeneous reaction rates still remaining is the transfer of the knowledge gained in experimental and theoretical surface science into models tractable for numerical simulation of technical catalytic reactors. There is still a large gap to be bridged. At present, the only way seems to be the rather tedious one, collecting all the information from surface science into the mechanistic models that need to be evaluated by experimental studies under technical conditions to come up with (intrinsic) kinetic rate expressions usable in CFD simulations. Alternatively, sophisticated (e.g., neuronal networks) generic, data-based models treating the reactor as black box but covering all potential ranges of reactor operation can be applied. However, these models will not allow exploration of the full optimization potential of the reactor, in particular concerning the optimized interplay between chemical reactions in different phases and mass and heat transfer.

Some of the present major challenges in modeling heterogeneous reactions in technical systems are catalyst–support interactions, spillover, multifunctional catalysts, operando modifications of the catalyst (oxidation/reduction cycles, storage/regeneration, partial melting, etc.), nonthermally equilibrated nanoparticles, structure sensitivity, aging (agglomeration, coking, and oxidation), and implementation of detailed kinetic models into numerical simulation of catalytic reactors, as discussed in the last chapters of the book.

Acknowledgment

The authors would like to thank Y. Dedecek (Karlsruhe Institute of Technology) for editorial corrections of the manuscript. Financial support of our work on modeling heterogeneous catalysis by the German Research Foundation (DFG) and the Helmholtz Association through the Helmholtz Research School Energy-Related Catalysis is gratefully acknowledged.

References

1 Deutschmann, O., Knözinger, H., Kochloefl, K., and Turek, T. (2009) *Heterogeneous Catalysis and Solid Catalysts*, 7th edn, Wiley-VCH Verlag GmbH, Weinheim.

2 Dumesic, J.A., Rudd, D.F., Aparicio, L.M., Rekoske, J.E., and Trevino, A.A. (1993) *The Microkinetics of Heterogeneous Catalysis*, American Chemical Society, Washington, DC.

3 Boudart, M. (2000) *Catal. Lett.*, **65**, 1.

4 Beretta, A., Forzatti, P., and Ranzi, E. (1999) *J. Catal.*, **184**, 469.

5 Beretta, A. and Forzatti, P. (2001) *J. Catal.*, **200**, 45.
6 Beretta, A., Ranzi, E., and Forzatti, P. (2001) *Chem. Eng. Sci.*, **56**, 779.
7 Zerkle, D.K., Allendorf, M.D., Wolf, M., and Deutschmann, O. (2000) *J. Catal.*, **196**, 18.
8 Subramanian, R. and Schmidt, L.D. (2005) *Angew. Chem. Int. Edit.*, **44**, 302.
9 Krummenacher, J.J. and Schmidt, L.D. (2004) *J. Catal.*, **222**, 429.
10 Schmidt, L.D., Siddall, J., and Bearden, M. (2000) *Am. Inst. Chem. Eng. J.*, **46**, 1492.
11 Donsi, F., Williams, K.A., and Schmidt, L.D. (2005) *Ind. Eng. Chem. Res.*, **44**, 3453.
12 Warnatz, J., Dibble, R.W., and Maas, U. (1996) *Combustion, Physical and Chemical Fundamentals, Modeling and Simulation, Experiments, Pollutant Formation*, Springer, New York.
13 Dean, A.M. (1990) *J. Phys. Chem.*, **94**, 1432.
14 Dente, M., Ranzi, E., and Goossens, A.G. (1979) *Comput. Chem. Eng.*, **3**, 61.
15 Maas, U. and Pope, S. (1992) *Combust. Flame*, **88**, 239.
16 Susnow, R.G., Dean, A.M., Green, W.H., Peczak, P., and Broadbelt, L. (1997) *J. Phys. Chem. A*, **101**, 3731.
17 Bortz, A.B., Kalos, M.H., and Lebowitz, J.L. (1975) *J. Comput. Phys.*, **17**, 10.
18 Lukkien, J.J., Segers, J.P.L., Hilbers, P.A.J., Gelten, R.J., and Jansen, A.P.J. (1998) *Phys. Rev. E*, **58**, 2598.
19 Maillard, F., Eikerling, M., Cherstiouk, O.V., Schreier, S., Savinova, E., and Stimming, U. (2004) *Faraday Discuss.*, **125**, 357.
20 Zhdanov, V.P. and Kasemo, B. (1998) *Surf. Sci.*, **405**, 27.
21 Zhdanov, V.P. and Kasemo, B. (1998) *Phys. Rev. Lett.*, **81**, 2482.
22 Kovalyov, E.V., Resnyanskii, E.D., Elokhin, V.I., Bal'zhinimaev, B.S., and Myshlyavtsev, A.V. (2003) *Phys. Chem. Chem. Phys.*, **5**, 784.
23 Segers, J.P.L. (2000) PhD thesis, Technische Universiteit Eindhoven (Eindhoven).
24 Silly, F. and Castell, M.R. (2005) *Phys. Rev. Lett.*, **94**, 046103.
25 Hansen, K.H., Worren, T., Stempel, S., Laegsgaard, E., Baumer, M., Freund, H.J., Besenbacher, F., and Stensgaard, I. (1999) *Phys. Rev. Lett.*, **83**, 4120.
26 Shaikhutdinov, S.K., Meyer, R., Lahav, D., Baumer, M., Klüner, T., and Freund, H.J. (2003) *Phys. Rev. Lett*, **91** 076102.
27 Hoffmann, J. (2003) PhD Thesis, Freie Universität Berlin (Berlin).
28 Liu, D.-J. and Evans, J.W. (2006) *J. Chem. Phys.*, **124**, 154705.
29 Honkala, K., Pirilä, P., and Laasonen, K. (2001) *Phys. Rev. Lett*, **86** 5942.
30 Mitsui, T., Rose, M.K., Fomin, E., Ogletree, D.F., and Salmeron, M. (2005) *Phys. Rev. Lett.*, **94**, 036101.
31 Lynch, M. and Hu, P. (2000) *Surf. Sci.*, **458**, 1.
32 Kunz, L.W.H. (2006) Diploma thesis, Universität Karlsruhe (TH) (Karlsruhe).
33 Kunz, L.W.H. and Deutschmann, O. (2011) *Phys. Rev. B*, to be submitted.
34 Kunz, L.W.H. and Deutschmann, O. (2011) *J. Catal.*, to be submitted.
35 Bos, C. (2006) PhD thesis, Universität Stuttgart (Stuttgart).
36 Meier, J., Schiøtz, J., Liu, P., Nørskov, J.K., and Stimming, U. (2004) *Chem. Phys. Lett.*, **390**, 440.
37 Nolte, P., Stierle, A., Jin-Phillipp, N.Y., Kasper, N., Schulli, T.U., and Dosch, H. (2008) *Science*, **321**, 1654.
38 Hansen, P.L., Wagner, J.B., Helveg, S., Rostrup-Nielsen, J.R., Clausen, B.S., and Topsøe, H. (2002) *Science*, **295**, 2053.
39 Pomeroy, J.M., Jacobsen, J., Hill, C.C., Cooper, B.H., and Sethna, J.P. (2002) *Phys. Rev. B*, **66**, 235412.
40 Kissel-Osterrieder, R., Behrendt, F., and Warnatz, J. (1998) *Proc. Combust. Inst.*, **27**, 2267.
41 Matera, S. and Reuter, K. (2010) *Phys. Rev. B*, **82**, 085446.
42 Kolobov, V., Arslanbekov, R., and Vasenkov, A. (2007) Computational Science: ICCS 2007. Pt 1, Proceedings, vol. 4487 (eds Y. Shi, G.D. VanAlbada, J. Dongarra, and P.M.A. Sloot), p. 858.
43 Baerns, M., Hofmann, H., and Renken, A. (1992) *Chemische*

Reaktionstechnik, Georg Thieme, Stuttgart.
44 Hayes, R.E. and Kolaczkowski, S.T. (1997) *Introduction to Catalytic Combustion*, Gordon and Breach Science Publ., Amsterdam.
45 Coltrin, M.E., Kee, R.J., and Rupley, F.M. (1991) *Int. J. Chem. Kinet.*, **23**, 1111.
46 Warnatz, J. (1992) *Proc. Combust. Inst.*, **24**, 553.
47 Kee, R.J., Coltrin, M.E., and Glarborg, P. (2003) *Chemically Reacting Flow*, Wiley-Interscience.
48 Coltrin, M.E., Kee, R.J., and Rupley, F.M. (1991) SURFACE CHEMKIN (Version 4.0): A Fortran Package for Analyzing Heterogeneous Chemical Kinetics at a Solid-Surface–Gas-Phase Interface, SAND91-8003B, Sandia National Laboratories.
49 Kee, R.J., Rupley, F.M., and Miller, J.A. (1998) CHEMKIN-II: A Fortran Chemical Kinetics Package for the Analysis of Gas-Phase Chemical Kinetics, SAND89-8009, Sandia National Laboratories.
50 Kee, R.J., Rupley, F.M., Miller, J.A., Coltrin, M.E., Grcar, J.F., Meeks, E., Moffat, H.K., Lutz, A.E., Dixon-Lewis, G., Smooke, M.D., Warnatz, J., Evans, G.H., Larson, R.S., Mitchell, R.E., Petzold, L.R., Reynolds, W.C., Caracotsios, M., Stewart, W.E., Glarborg, P., Wang, C., and Adigun, O. (2000) Chemkin, 3.6 edn, Reaction Design, Inc., San Diego, www.chemkin.com.
51 Goodwin, D.G. (2003) CANTERA. An open-source, extensible software suite for CVD process simulation, www.cantera.org.
52 Deutschmann, O., Tischer, S., Correa, C., Chatterjee, D., Kleditzsch, S., and Janardhanan, V.M. (2004) DETCHEM Software Package, 2.0 edn, Karlsruhe, www.detchem.com.
53 Deutschmann, O., Schmidt, R., Behrendt, F., and Warnatz, J. (1996) *Proc. Combust. Inst.*, **26**, 1747.
54 Raja, L.L., Kee, R.J., and Petzold, L.R. (1998) *Proc. Combust. Inst.*, **27**, 2249.
55 Deutschmann, O. (2007) Chapter 6.6 in *Handbook of Heterogeneous Catalysis*, 2nd edn (eds H.K.G. Ertl, F. Schüth, and J. Weitkamp), Wiley-VCH Verlag GmbH, Weinheim.
56 Mhadeshwar, A.B., Wang, H., and Vlachos, D.G. (2003) *J. Phys. Chem. B*, **107**, 12721.
57 Maier, L., Schädel, B., Herrera Delgado, K., Tischer, S., and Deutschmann, O. (2011) *Top. Catal.*, DOI: 10.1007/s11244-011-9702-1.
58 Hecht, E.S., Gupta, G.K., Zhu, H.Y., Dean, A.M., Kee, R.J., Maier, L., and Deutschmann, O. (2005) *Appl. Catal. A Gen.*, **295**, 40.
59 Janardhanan, V.M. and Deutschmann, O. (2006) *J. Power Sources*, **162**, 1192.
60 Inderwildi, O.R., Lebiedz, D., Deutschmann, O., and Warnatz, J. (2005) *J. Chem. Phys.*, **122**, 154702.
61 Heyden, A., Peters, B., Bell, A.T., and Keil, F.J. (2005) *J. Phys. Chem. B*, **109**, 4801.
62 Behrendt, F., Deutschmann, O., Maas, U., and Warnatz, J. (1995) *J. Vac. Sci. Technol. A*, **13**, 1373.
63 Campbell, C.T. (2001) *J. Catal.*, **204**, 520.
64 Deutschmann, O. (1996). Dissertation thesis, Universität Heidelberg (Heidelberg).
65 Appel, C., Mantzaras, J., Schaeren, R., Bombach, R., Kaeppeli, B., and Inauen, A. (2003) *Proc. Combust. Inst.*, **29**, 1031.
66 Reinke, M., Mantzaras, J., Schaeren, R., Bombach, R., Kreutner, W., and Inauen, A. (2002) *Proc. Combust. Inst.*, **29**, 1021.
67 Dogwiler, U., Benz, P., and Mantzaras, J. (1999) *Combust. Flame*, **116**, 243.
68 Mantzaras, J., Appel, C., and Benz, P. (2000) *Proc. Combust. Inst.*, **28**, 1349.
69 Mantzaras, J. and Appel, C. (2002) *Combust. Flame*, **130**, 336.
70 Reinke, M., Mantzaras, J., Schaeren, R., Bombach, R., Inauen, A., and Schenker, S. (2004) *Combust. Flame*, **136**, 217.
71 Williams, W.R., Marks, C.M., and Schmidt, L.D. (1992) *J. Phys. Chem.*, **96**, 5922.
72 Hellsing, B., Kasemo, B., and Zhdanov, V.P. (1991) *J. Catal.*, **132**, 210.

73. Rinnemo, M., Deutschmann, O., Behrendt, F., and Kasemo, B. (1997) *Combust. Flame*, **111**, 312.
74. Veser, G. (2001) *Chem. Eng. Sci.*, **56**, 1265.
75. Bui, P.-A., Vlachos, D.G., and Westmoreland, P.R. (1997) *Ind. Eng. Chem. Res.*, **36**, 2558.
76. Andrae, J.C.G. and Björnbom, P.H. (2000) *Am. Inst. Chem. Eng. J.*, **46**, 1454.
77. Mai, J., von Niessen, W., and Blumen, A. (1990) *J. Chem. Phys.*, **93**, 3685.
78. Zhdanov, V.P. and Kasemo, B. (1994) *Appl. Surf. Sci.*, **74**, 147.
79. Aghalayam, P., Park, Y.K., and Vlachos, D.G. (2000) *Proc. Combust. Inst.*, **28**, 1331.
80. Veser, G., Frauhammer, J., Schmidt, L.D., and Eigenberger, G. (1997) *Stud. Surf. Sci. Catal.*, **109**, 273.
81. Bui, P.-A., Vlachos, D.G., and Westmoreland, P.R. (1997) *Surf. Sci.*, **386**, L1029.
82. Dogwiler, U., Benz, P., and Mantzaras, J. (1999) *Combust. Flame*, **116**, 243.
83. Aghalayam, P., Park, Y.K., Fernandes, N., Papavassiliou, V., Mhadeshwar, A.B., and Vlachos, D.G. (2003) *J. Catal.*, **213**, 23.
84. Hickman, D.A. and Schmidt, L.D. (1993) *Am. Inst. Chem. Eng. J.*, **39**, 1164.
85. Huff, M. and Schmidt, L.D. (1993) *J. Phys. Chem.*, **97**, 11815.
86. Huff, M.C., Androulakis, I.P., Sinfelt, J.H., and Reyes, S.C. (2000) *J. Catal.*, **191**, 46.
87. Schwiedernoch, R., Tischer, S., Correa, C., and Deutschmann, O. (2003) *Chem. Eng. Sci.*, **58**, 633.
88. Chatterjee, D., Deutschmann, O., and Warnatz, J. (2001) *Faraday Discuss.*, **119**, 371.
89. Ruf, B., Behrendt, F., Deutschmann, O., and Warnatz, J. (1996) *Surf. Sci.*, **352**, 602.
90. Harris, S.J. and Goodwin, D.G. (1993) *J. Phys. Chem.*, **97**, 23.
91. Romet, S., Couturier, M.F., and Whidden, T.K. (2001) *J. Electrochem. Soc.*, **148**, G82.
92. Scott, C.D., Povitsky, A., Dateo, C., Gokcen, T., Willis, P.A., and Smalley, R.E. (2003) *J. Nanosci. Nanotechnol.*, **3**, 63.
93. Kissel-Osterrieder, R., Behrendt, F., Warnatz, J., Metka, U., Volpp, H.R., and Wolfrum, J. (2000) *Proc. Combust. Inst.*, **28**, 1341.
94. Hauff, K., Tuttlies, U., Eigenberger, G., and Nieken, U. (2010) *Appl. Catal. B Environ.*, **100**, 10.
95. Boll, W., Tischer, S., and Deutschmann, O. (2010) *Ind. Eng. Chem. Res.*, **49**, 10303.
96. Temel, B., Meskine, H., Reuter, K., Scheffler, M., and Metiu, H. (2007) *J. Chem. Phys.*, **126**, 12.
97. Shustorovich, E. and Sellers, H. (1998) *Surf. Sci. Rep.*, **31**, 1.
98. Braun, J., Hauber, T., Többen, H., Windmann, J., Zacke, P., Chatterjee, D., Correa, C., Deutschmann, O., Maier, L., Tischer, S., and Warnatz, J. (2002) SAE Technical Paper, 2002-01-0065.
99. Tuttlies, U., Schmeisser, V., and Eigenberger, G. (2004) *Chem. Eng. Sci.*, **59**, 4731.
100. Schmeisser, V., Eigenberger, G., and Nieken, U. (2009) *Top. Catal*, **52**, 1934.
101. Yan, X. and Maas, U. (2000) *Proc. Combust. Inst.*, **28**, 1615.
102. Votsmeier, M. (2009) *Chem. Eng. Sci.*, **64**, 1384.
103. Votsmeier, M., Scheuer, A., Drochner, A., Vogel, H., and Gieshoff, J. (2009) *Catal. Today*, **151**, 271.
104. Warnatz, J. (1981) *Proc. Combust. Inst.*, **18**, 369.
105. Warnatz, J., Dibble, R.W., and Maas, U. (1996) *Combustion* Springer, New York.
106. Miller, C.T.B.J.A. (1989) *Prog. Energy Combust. Sci.*, **15**, 287.
107. Frenklach, M., Wang, H., Yu, C.L., Goldenberg, M., Bowman, C.T., Hanson, R.K. et al. (1995) http://web.galcit.caltech.edu/EDL/mechanisms/library;<http://diesel.me.berkeley.edu/wgri_mech/new21/>.
108. Smith, G.P., Golden, D.M., Frenklach, M., Moriarty, N.W., Eiteneer, B., Goldenberg, M., Bowman, C.T., Hanson, R.K., Song, S., Gardiner, W.C., Jr., Lissianski, V.V., and Qin, W. (1999) <http://www.me.berkeley.edu/gri_mech/>.
109. Qin, Z., Yang, H., Gardiner, W.C., Davis, S.G., and Wang, H. (2000) *Proc. Combust. Inst.*, **28**, 1663.

110 Konnov, A (2000) http://homepages.vub.ac.be/~akonnov.
111 Hidaka, Y., Sato, K., Henmi, Y., Tanaka, H., and Inami, K. (1999) *Combust. Flame*, **118**, 340.
112 Hidaka, Y., Sato, K., Hoshikawa, H., Nihimori, T., Takahashi, R., Tanaka, H. et al. (2000) *Combust. Flame*, **120**, 245.
113 Hughes, K.J., Clague, A., and Pilling, M.J. (2001) *Int. J. Chem. Kinet.*, **33**, 513.
114 Eiteneer, B. and Frenklach, M. (2003) *Int. J. Chem. Kinet.*, **35**, 391.
115 El Bakali, A., Dagaut, P., Pillier, L., Desgroux, P., Pauwels, J.F., Rida, A. et al. (2004) *Combust. Flame*, **137**, 109.
116 Petrova, M.V. and Williams, F.A. (2006) *Combust. Flame*, **144**, 526.
117 Huang, J., and Bushe, W.K. (2006) *Combust. Flame*, **144**, 74.
118 Ranzi, E. et al. (2007) http://www.chem.polimi.it/CRECKModeling/kinetic.html.
119 Healy, D., Curran, H.J., Dooley, S., Simmie, J.M., Kalitan, D.M., Petersen, E.L., and Bourque, G. (2008) *Combust. Flame*, **155**, 451.
120 Healy, D., Curran, H.J., Simmie, J.M., Kalitan, D.M., Zinner, C.M., Barrett, A.B., Petersen, E.L., and Bourque, G. (2008) *Combust. Flame*, **155**, 441.
121 Gupta, G.K., Hecht, E.S., Zhu, H., Dean, A.M., and Kee, R.J. (2006) *J. Power Sources*, **156**, 434.
122 Frenklach, J.W.M. (1987) *Combust. Sci. Tech.*, **51**, 265.
123 Maryam, Y.-S. Matida, E.A., and Hamdullahpur, F. (2009) *Int. J. Hydrogen Energy*, **34**, 3710.
124 Bockhron, H., Appel, J., and Frenklach, M. (2000) *Combust. Flame*, **121**, 122.
125 Norinaga, K. and Deutschmann, O. (2007) *Ind. Eng. Chem. Res.*, **46**, 3547.
126 Curran, H.J., Gaffuri, P., Pitz, W.J., and Westbrook, C.K. (2002) *Combust. Flame*, **129**, 253.
127 Glaude, P.A., Conraud, V., Fournet, R., Battin-Leclerc, F., Come, G.M., Scacchi, G., Dagaut, P., and Cathonnet, M. (2002) *Energy Fuel*, **16**, 1186.
128 Golovitchev, V.I. and Chomial, L. (1999) SAE paper 1999-01-3552; http://www.tfd.chalmers.se/~valeri/MECH.html.
129 Battin-Leclerc, F., Fournet, R., Glaude, P.A., Judenherc, B., Warth, V., Côme, G.M., and Scacchi, G. (2000) *Proc. Combust. Inst.*, **28**, 1597.
130 Bikas, G. and Peters, N. (2001) *Combust. Flame*, **126**, 1456.
131 http://www.enisc.u-nancy.fr/ENSIC/DCPR/global/accueil.html.
132 Zhdanov, V.P. and Kasemo, B. (1994) *Surf. Sci. Rep.*, **20**, 111.
133 Norton, P.R. (1982) *The Chemical Physics of Solid Surfaces and Heterogeneous Catalysis*, vol. **4** (eds D.A. King and D.P. Woodroft), Elsevier, Amsterdam.
134 Gland, J.L., Fisher, G.B., and Kollin, E.B. (1982) *J. Catal.*, **77**, 263.
135 Engel, T. and Kuipers, H. (1979) *Surf. Sci.*, **90**, 181.
136 Ogle, K.M. and White, J.M. (1984) *Surf. Sci.*, **139**, 43.
137 Warnatz, J., Allendorf, M.D., Kee, R.J., and Coltrin, M.E. (1994) *Combust. Flame*, **96**, 393.
138 Park, Y.K., Aghalayam, P., and Vlachos, D.G. (1999) *J. Phys. Chem. A*, **103**, 8101.
139 Bui, P.A., Vlachos, D.G., and Westmoreland, P.R. (1997) *Surf. Sci.*, **385**, L1029.
140 van Santen, R.A., van der Runstraat, A., and Gelten, R.J. (1997) *Stud. Surf. Sci. Catal.*, **109**, 61.
141 Rinnemo, M., Kulginov, D., Johansson, S., Wong, K.L., Zhdanov, V.P., and Kasemo, B. (1997) *Surf. Sci.*, **276**, 297.
142 Harold, M.P. and Garske, M.E. (1991) *J. Catal.*, **127**, 553.
143 Aghalayam, P., Park, Y.K., and Vlachos, D.G. (2000) Proceedings of Combustion Institute, vol. 28, Edinburgh, Scottland.
144 Harold, M.P. and Garske, M.E. (1991) *J. Catal.*, **127**, 524.
145 Garske, M.E. and Harold, M.P. (1992) *Chem. Eng. Sci.*, **47**, 623.
146 Hickman, D.A. and Schmidt, L.D. (1993) *AICHE J.*, **39**, 1164.
147 Deutschmann, O., Behrendt, F., and Warnatz, J. (1994) *Catal. Today*, **21**, 461.
148 Mallen, M.P., Williams, W.R., and Schmidt, L.D. (1993) *J. Phys. Chem.*, **97**, 625.
149 Deutschmann, O., Maier, L., Riedel, U., Stroemann, A.H., and Dibble, R.W. (2000) *Catal. Today*, **59**, 141.

150 Hickman, D.A., Haupfear, E.A., and Schmidt, L.D. (1993) *Catal. Lett.*, **17**, 223.
151 Quiceno, R., Perez-Ramirez, J., Warnatz, J., and Deutschmann, O. (2006) *Appl. Catal. A Gen.*, **303**, 166.
152 Cortright, R.D., Goddard, S.A., Rekoske, J.E., and Dumesic, J.A. (1991) *J. Catal.*, **127**, 342.
153 Goddard, S.A., Cortright, R.D., and Dumesic, J.A. (1992) *J. Catal.*, **137**, 186.
154 Huff, M.C. and Schmidt, L.D. (1996) *AICHE J.*, **42**, 3484.
155 Vincent, R.S., Lindstedt, R.P., Malik, N.A., Reid, I.A.B., and Messenger, B.E. (2008) *J. Catal.*, **260**, 37.
156 Dumesic, J.A. and Trevino, A.A. (1989) *J. Catal.*, **116**, 119.
157 Stoltze, P. and Nørskov, J.K. (1985) *Phys. Rev. Lett.*, **55**, 2502.
158 Bowker, M., Parker, I.B., and Waugh, K.C. (1985) *Appl. Catal.*, **14**, 101.
159 Bowker, M., Parker, I., and Waugh, K.C. (1988) *Surf. Sci.*, **197**, L223.
160 Pignet, T. and Schmidt, L.D. (1974) *Chem. Eng. Sci.*, **29**, 1123.
161 Koop, J. and Deutschmann, O. (2009) *Appl. Catal. B Environ.*, **91**, 47.
162 Schwartz, S.B., Schmidt, L.D., and Fisher, G.B. (1986) *J. Phys. Chem.*, **90**, 6194.
163 Oh, S.H., Fisher, G.B., Carpenter, J.E., and Goodman, D.W. (1986) *J. Catal.*, **100**, 360.
164 Fisher, G.B., Oh, S.H., Carpenter, J.E., DiMaggio, C.L., and Schmieg, S.J. (1987) in *Catalysis and Automotive Pollution Control* (eds Crucq, A., Frenet, A.), Elsevier, Amsterdam.
165 Deutschmann, O., Schwiedernoch, R., Maier, L.I., and Chatterjee, D. (2001) *Stud. Surf. Sci. Catal.*, **136**, 251.
166 Maestri, M., Vlachos, D.G., Beretta, A., Forzatti, P., Groppi, G., and Tronconi, E. (2009) *Top. Catal.*, **52**, 1983.
167 Mhadeshwar, A.B. and Vlachos, D.G. (2005) *J. Phys. Chem. B*, **109**, 16819.
168 Maestri, M., Vlachos, D.G., Beretta, A., Groppi, G., and Tronconi, E. (2008) *J. Catal.*, **259**, 211.
169 Thormann, J., Maier, L., Pfeifer, P., Kunz, U., Deutschmann, O., and Schubert, K. (2009) *Int. J. Hydrogen Energy*, **34**, 5108.
170 Schadel, B.T., Duisberg, M., and Deutschmann, O. (2009) *Catal. Today*, **142**, 42.
171 Zhdanov, V.P. and Kasemo, B. (1996) *Catal. Lett.*, **40**, 197.
172 Belton, D.N. and Schmieg, S.J. (1992) *J. Catal.*, **138**, 70.
173 Hartmann, M., Maier, L., Minh, H.D., and Deutschmann, O. (2010) *Combust. Flame*, **157**, 1771.
174 Ng, K.Y.S., Belton, D.N., Schmieg, S.J., and Fisher, G.B. (1994) *J. Catal.*, **146**, 394.
175 Xu, J., Harold, M.P., and Balakotaiah, V. (2009) *Appl. Catal. B: Environ.*, **89**, 73.
176 Kramer, J.F., Reihani, S.A.S., and Jackson, G.S. (2003) *Proc. Combust. Inst.*, **29**, 989.
177 Seyed-Reihani, S.A. and Jackson, G.S. (2004) *Chem. Eng. Sci.*, **59**, 5937.
178 Stegelmann, C., Schiødt, N.C., Campbell, C.T., and Stoltze, P. (2004) *J. Catal.*, **221**, 630.
179 Ovesen, C.V., Clausen, B.S., Hammershøi, B.S., Steffensen, G., Askgaard, T., Chorkendorff, I., Nørskov, J.K., Rasmussen, P.B., Stoltze, P., and Taylor, P. (1996) *J. Catal.*, **158**, 170.
180 Waugh, K.C. (1996) *Chem. Eng. Sci.*, **51**, 1533.
181 Tserpe, E. (1997) *Stud. Surf. Sci. Catal.*, **109**, 401.
182 Ovesen, C.V., Stoltze, P., Nørskov, J.K., and Campbell, C.T. (1992) *J. Catal.*, **134**, 445.
183 Ernst, K.-H., Campbell, C.T., and Moretti, G. (1992) *J. Catal.*, **134**, 66.
184 Rasmussen, P.B., Holmblad, P.M., Askgaard, T., Ovesen, C.V., Stoltze, P., Nørskov, J.K., and Chorkendorff, I. (1994) *Catal. Lett.*, **26**, 373.
185 Askgaard, T.S., Norskov, J.K., Ovesen, C.V., and Stoltze, P. (1995) *J. Catal.*, **156**, 229.
186 Aparicio, L.M. (1997) *J. Catal.*, **165**, 262.
187 Blaylock, D.W., Ogura, T., Green, W.H., and Beran, G.J.O. (2009) *J. Phys. Chem. C*, **113**, 4898.

5
Modeling Reactions in Porous Media
Frerich J. Keil

5.1
Introduction

Catalysis is the phenomenon in which one or more substances isothermally augment the rate of a chemical reaction and enables new reaction paths without appearing in the stoichiometric equation of the reaction. Up to now, heterogeneous catalysis is dominated by experimental research. Many thousands of experiments have led to rather few commercially successful catalysts. These catalysts are then modified and optimized in many subsequent experimental investigations. Recently, Metiu [1] has stated that finding new catalysts has been akin to mining for diamonds: most workers dig tunnels and only very few find diamonds. Although some books on "catalyst design" [2–4] have already been published (see also Ref. [5]), there is only a modest progress in catalyst design based on well-founded principles. The molecular view of heterogeneous catalysis, which could lead to a rational catalyst design, has been hampered by difficulties to bridge the "pressure gap" and the "material gap." The "pressure gap" refers to the difference in operating conditions of a real catalyst, mostly between 1 and 300 bars, and the ultrahigh vacuum conditions employed in surface science investigations [6]. This is particularly important for oxide catalysts [7, 8]. The term "material gap" refers to the difference between model single-crystal surfaces and the nanosized crystals distributed in porous catalyst particles. Over the last few years, a combination of first-principles calculations and statistical thermodynamics [9, 10] and various spectroscopic methods [11–14, 15] has led to insight into the molecular level. Examples are given by Thomas [16], Hellman *et al.* [17], Norskov *et al.* [18], van Santen *et al.* [19], Heyden *et al.* [20, 21], and Brüggemann *et al.* [22–24], among others.

When modeling phenomena within porous catalyst particles, one is confronted with several problems. First, one has to model the porous structure of the support (see Figure 5.1b). Many suggestions have been made how to describe the shape of pores and their surfaces. This item will be discussed in detail below. The next problem is the multicomponent diffusion of reactants and products in the porous network (see Figure 5.1a). This problem is still under debate.

Modeling and Simulation of Heterogeneous Catalytic Reactions: From the Molecular Process to the Technical System,
First Edition. Edited by Olaf Deutschmann.
© 2012 Wiley-VCH Verlag GmbH & Co. KGaA. Published 2012 by Wiley-VCH Verlag GmbH & Co. KGaA.

(a)

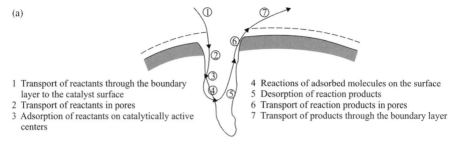

1 Transport of reactants through the boundary layer to the catalyst surface
2 Transport of reactants in pores
3 Adsorption of reactants on catalytically active centers
4 Reactions of adsorbed molecules on the surface
5 Desorption of reaction products
6 Transport of reaction products in pores
7 Transport of products through the boundary layer

(b)

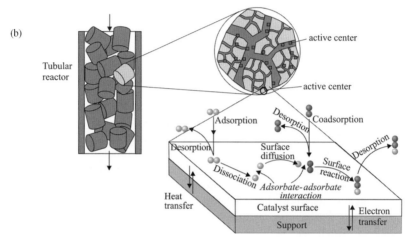

Figure 5.1 Phenomena in porous catalyst supports (a) transport, adsorption, and reaction and (b) pore structure and phenomena on catalytically active crystallites.

Although molecular approaches give reliable results of the diffusion process, many efforts are still made to develop simplified models, mostly based on approximations of Boltzmann's equation, where the model parameters may be taken from molecular simulations or experiments. Here problems occur. Often the detailed structure of the pore walls and the pore space are unknown on an atomic level, in particular, for amorphous mesoporous materials. Exceptions are some well-defined crystalline materials, like zeolites, metal–organic frameworks (MOFs), and some others. The same holds for the potential parameters for molecular simulations of the fluids and solids of the catalytic system under investigation. Therefore, simplified models are still justified. A further reason is that in iterative process engineering calculations, molecular simulations would be too time-consuming. The porous particle size is determined by a low-pressure drop along a packed bed and, in contrast, short diffusion paths. Between these opposing requirements, one has to find a compromise. When the rate of reaction is small compared with the rate of diffusion, the size of the pellet presents no problem for the concentration of reactants in the pores. At the center it is little different from that at the surface. In contrast, the concentration of reactants is depleted by reaction before it has a chance to diffuse within the pellet if the rate of reaction is large in comparison with the rate of diffusion. The catalyst in the interior of the pellet is not being used to any extent.

The next steps are adsorption, reaction, and desorption of the products [19, 25–29]. The active site and its nanoscopic environment are decisive for a good catalyst. Heterogeneous catalysis involves the adsorption of at least one of the reactants. The forces that hold the reactants on the surface are either physical (physisorption), such as van der Waal's attraction, or chemical (chemisorption), as when the binding is that of a chemical bond. If there is an appreciable interaction between adsorbed molecules, there may be phase changes in the adsorbed state. Also, the surface may not be uniform, such that the energy of binding at different sites may be variable. This can be caused by lateral interactions between the molecules too. It is the concept of adsorption that leads to the simple forms of reaction rate expressions in catalytic kinetics. In chemical engineering practice, the Langmuir–Hinshelwood (LH) kinetics has been the workhorse in kinetic modeling, although its assumptions like no lateral interaction between adsorbed atoms or molecules (mean field approximation), equivalence of all adsorption sites, heat of adsorption independent of surface coverage, and each adsorbed molecule occupies only one site do not apply in most cases. Furthermore, the reaction mechanism assumed is in general extremely simplified. The model parameters are fitted to experimental data. Owing to the assumptions and limitations of the Langmuir–Hinshelwood approach, the kinetics fitted to experiments cannot be extended to operating conditions beyond the values included into the fit. Nevertheless, besides simple power laws, which one obtains from the Freundlich isotherm, the Langmuir–Hinshelwood approach turned out to be quite useful in practice [30–32]. This approach does not consider the details of the active sites, which have to be studied by quantum chemical [19, 28, 33, 34] and experimental spectroscopic tools [6, 13, 14, 35, 36] combined with transition state theory (TST) [37–43], and statistical thermodynamics [7, 9, 10, 44, 45]. These approaches give deep insight into the details of reaction mechanisms. Density functional theory (DFT) is mostly used for the quantum chemical calculations [46–51]. Unfortunately, up to now, functionals used in DFT (e.g., LDA, GGA) do not give results of sufficient accuracy for van der Waal's interactions, although there is considerable progress in this direction [52, 53]. Therefore, sometimes one has to refer to *ab initio* methods [54–57]. Molecular simulations of the statistical thermodynamics part are executed by various Monte Carlo and molecular dynamics (MD) approaches [58–64].

Modeling of reactions in porous media is a multidiscipline and multiscale task that spans fields including physics, mathematics, chemical engineering, and material science. The timescales range from femtoseconds to hours and the length scales from nanometers to meters [65–70].

In the subsequent paragraphs, modeling of pore structures, pore surfaces, and diffusion/reaction phenomena will be reviewed. Some other subjects like density functional theory, modeling of surface reactions, and their dynamics will be mentioned, but are dealt with in more detail in other chapters of this book.

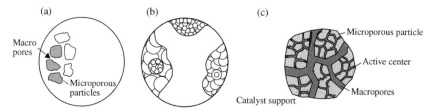

Figure 5.2 Catalyst support manufactured by compression of microparticles. (a) Two-dimensional cut. (b) Three-dimensional view. (c) Pore network, made of transport pores (macropores) and micropores.

5.2
Modeling Porous Structures and Surface Roughness

Porous solids are often used in many branches of science and technology. Most catalyst particles are formed by pressing a suitable powder into a mold. The particles of the powder are themselves porous but pressing them together will also create a system of macropores that are in fact the interstices of the powder particles and whose dimensions span the entire size of the particles. This leads to a bimodal pore structure (Figure 5.2).

Gas reactions catalyzed by solid materials occur partly on the exterior but mainly on interior surfaces of the porous catalyst support. The specific rate of reaction is also a function of the accessible surface area. The greater the amount of accessible internal surface area, the larger the amount of reactants converted per unit time and unit mass of catalyst. The accessibility depends on the void volume geometric structure and the relative size of the reactant molecules in relation to the surface roughness. The void volume may be regarded as a domain of Euclidean space. An approximation introduced in general is the concept of pores. Although this is a rather crude approximation, it turned out to be very useful in modeling. Pore size distribution, connectivity, and shape of pores influence the selectivity and effective rate of reactions. Owing to possible deposition of molecules, for example, carbon deposition, the pore structure may change as a function of time. Finally, the accessibility of the pores will drop to zero, such that the pellet will be catalytically inactive. Proper optimization of the pore structure may extend the operating time of a catalyst.

Wheeler [71] was one of the first who has considered pore models. He proposed that the mean radius $\bar{r}$ and the length $\bar{L}$ of pores in a pellet are determined in such a way that the sum of the surface areas of all the pores in a pellet is equal to the BET surface area [72] and that the sum of the pore volumes is equal to the experimental pore volume (Figure 5.3a).

For the average pore radius and the average pore length, one obtains the following equations:

$$\bar{r} = \frac{2 V_g}{S_g} \sigma (1-\psi), \tag{5.1}$$

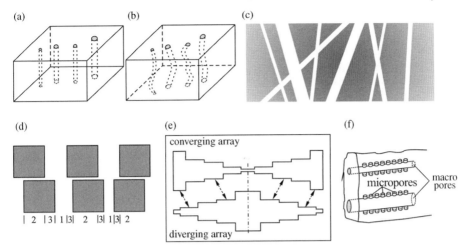

Figure 5.3 (a) Parallel pore model [71]. (b) Tortuous pore model [73]. (c) Cylindrical pore model [74]. (d) Model of Wakao and Smith [75]. (e) Model of Foster and Butt [76]. (f) Micro/macropore model [77].

$$\bar{L} = \sqrt{2} V_p / S_x, \tag{5.2}$$

where V_g is the specific pore volume, S_g is the BET specific surface area, and σ is a pore wall roughness factor, ψ is the pellet porosity, V_p is the total volume of the catalyst pellet and S_x is the external surface area of a catalyst particle. The model provides $\bar{r}$ and $\bar{L}$ in terms of measurable quantities. Figure 5.3b shows a model with tortuous pores that is slightly more realistic than the parallel pore model. The term "tortuosity" has been coined by Carman [73] (see also [78, 79]). Wakao and Smith [75] introduced a model (Figure 5.3d) for bidisperse pore systems resulting from a pellet made by pressing together particles that are themselves porous. The authors distinguish three types of passages through two layers of particles: (1) through the macropore structure of each; (2) through the micropore structure of each; and (3) through the macropores of one layer and the micropores of the next. The model has been used quite successfully for diffusion calculations of porous media. The drawback of this model is the absence of connections between pores. In fact, it is a refined version of the parallel pore model. A similar approach has been developed by Mann and Thomson [77]. Foster and Butt [76] introduced a computational model that represents the tortuosities and constrictions of actual pores by the notion of a major pore composed of minor pores. The void volume within the solid is considered to be composed of two major arrays of pores, centrally convergent and centrally divergent, respectively, interconnected at specified intervals within the arrays. The exact shape of these arrays is determined uniquely from the volume–area distribution of the porous structure. The model has been applied to computation of counterdiffusional flux through a porous solid as measured in a Wicke–Kallenbach [80] experiment. A comparison with experimental data reported for a particularly well-defined system has been given. Progress has been made by the model of Johnson and Stewart [74] that employs randomly oriented capillary axes (Figure 5.3c). Similar models were also used by Feng and Stewart [81], and Wang and Smith [82]. In these models, the capillary axes are randomly oriented and cross-linked. The effect of the local fluctuating concentration field on the macroscopic flux is neglected. This approximation is called smooth field approximation (SFA) [83]. The macroscopic concentration field is assumed to be a smooth function of position in the porous medium. The SFA holds rigorously only for networks of infinitely long, straight, nonoverlapping capillaries.

Beeckman and Froment [84] and Reyes and Jensen [85] have introduced Bethe lattices (Figure 5.4a) as models for catalyst supports.

In these models, no closed loops and a fixed connectivity are employed. The number of bonds connected to a site is called its coordination number or connectivity. Bethe lattices can be investigated by percolation theory [86]. The classical percolation theory centers around the bond and the site percolation problem. In the bond percolation problem, the bonds of the network are either occupied (i.e., they are open to flow, diffusion, and reaction) randomly and independent of each other with probability p or vacant (i.e., they are closed to flow or have been plugged) with probability $1 - p$. For a large network, this assignment is equivalent to removing a fraction of $1 - p$ of all bonds at random. Two sites are called connected if there exists at least one path between them consisting solely of occupied bonds. A set of connected sites bounded by vacant bonds is called a cluster. At some well-defined value of p, there must be a transition in the topological structure of the random network; this value is called the bond percolation threshold p_{cb}. This is the largest fraction of occupied bonds below which there is no sample-spanning cluster of occupied bonds. Similarly, a site percolation problem can also be defined. In this case, sites of network are occupied with probability p and vacant with probability $1 - p$. Two nearest-neighbor sites are called connected if they are both occupied, and connected clusters on the network are again defined in the obvious way. As before, there is a percolation threshold p_{cs} above which an infinite cluster of occupied sites spans the network. Most engineering problems of interest are discussed in the scheme of site percolation. Exact values of the percolation threshold could be obtained only for Bethe lattices and some two-dimensional lattices. For the Bethe lattice, one obtains

$$p_{cb} = p_{cs} = 1/(Z-1), \tag{5.3}$$

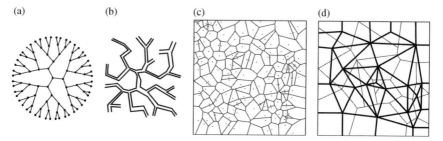

Figure 5.4 (a) Bethe lattice, coordination number $Z = 3$. (b) A possible corresponding pore structure. (c) Voronoi polyhedra. (d) Voronoi polyhedra micro/macro networks.

where Z is the coordination number. For all other networks, one has to use numerical techniques for finding percolation thresholds. A program is given by Stauffer and Aharony [86] that uses an efficient algorithm developed by Hoshen and Kopelman [87]. Detailed discussions of the percolation theory are, for example, given in Refs [88–90]. As will be discussed below, percolation theory is often employed for catalyst deactivation models. Percolation theory is valid strictly for systems that are infinitely large. Computer simulations, however, deal with finite systems. For this purpose, Fisher [91] has developed a finite-size scaling theory. The variation of any percolation property K_L of a system of size L can be written as

$$K \sim L^{-\gamma} f(u), \qquad (5.4)$$

where $f(u)$, which is a function of the variable $u = L^{1/\nu}(P-P_c)(L/\zeta)^{1/\nu}$, is a universal and nonsingular function in that its shape depends only on the dimensionality of the network and not on its type. The terms γ and ν are constants and ζ is the so-called correlation length. This is the typical radius of the connected clusters for $P < P_c$, that is, the properties of the system are in this case independent of its linear size. In any Monte Carlo simulations of percolation, one must have $L \gg \zeta$. Near the percolation threshold, most percolation parameters obey scaling laws. At the percolation threshold, the correlation length ζ is infinite, so the sample-spanning cluster is fractal at any length scale. An early paper on the application of percolation concepts on diffusion and catalysis has been presented by Mohanty et al. [92].

A totally random topology can be described by Voronoi polyhedra (Figure 5.4c and d) [93–95]. By employing Voronoi polyhedra, one can create isotropic networks. For this purpose, one subdivides the space by a so-called Voronoi tessellation by enclosing a set of points placed at random in space with the smallest polygon that can be formed by intersecting planes that bisect the lines connecting the point and its neighbor.

Reyes and Iglesia [96] simulated sintered porous solids by random–loose aggregates of spheres that are distributed in size and partially overlapped to achieve the required porosity. The resulting porous networks closely capture the morphological details of diffusing channels within granular materials commonly used as catalyst supports. Effective diffusivities in these model solids have been calculated by Monte Carlo techniques that allow the probing of representative regions of the void space throughout the Knudsen, transition, and molecular diffusion regimes. Simulated diffusivities and tortuosity factors have been in excellent agreement with experimental observations. These simulations also allow the calculation of accurate pore-size distributions and the transition region diffusivities, previously estimated by simple geometric arguments and by the Bosanquet approximation, respectively. The simulations have shown that the Bosanquet equation underestimates diffusivities (maximum error ~20%), suggesting that diffusing molecules are able to "detect" and use neighboring larger pore openings in parallel with the diffusing channels, instead of following the strictly segmental mechanism implied by the Bosanquet equation. This clearly demonstrates the superiority of molecular simulations. Drewry and Seaton [97] have developed an efficient method based on a first passage time (FPT) random walk approach for the simulation of diffusion and reaction in a supported catalyst. The model catalyst is composed of spheres

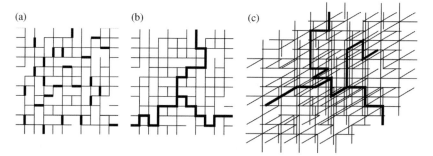

Figure 5.5 Cubic random pore network models. (a) Two-dimensional random micro/macropore network. (b) Macropores spanning the entire network. (c) Three-dimensional micro/macropore network, the macropores span the entire network.

representing the support and active sites. By varying the algorithm used to generate the model catalyst, a range of structures can be created. The effect of the structure and the size and distribution of active sites on the reaction rate were studied.

In the nineties, three-dimensional pore network models became fashionable. In Figure 5.5, two- and three-dimensional random cubic networks are presented.

Jerauld et al. [98] and Winterfeld et al. [94] have shown that as long as the average coordination number of a topologically disordered system is equal to the coordination number of a regular network, the transport properties of the two systems are essentially identical. In Figure 5.5a, the micro- and mesopores are randomly distributed, whereas in Figure 5.5b, the more realistic case is presented where the macropores cover the entire network. The transport properties of these two networks are quite different [99, 100]. The random structure (Figure 5.5a) might lead to unrealistic high values for the tortuosity. Bimodal catalyst supports are described realistically by the network presented in Figure 5.5b, where the macropores span the entire network. Random three-dimensional networks are advantageous in several respects:

1) Any type of network can be modeled (e.g., regular, irregular, and Voronoi).
2) The effect of pore connectivity can be taken into account.
3) Any pore size distribution and any pore geometry (e.g., cylindrical, slit-like) can be used.
4) Local heterogeneities, for example, spatial variation in mean pore size, can be modeled. This is an important point, as Hollewand and Gladden [101, 102] detected that pellet pore structures are quite heterogeneous.
5) Any distribution of catalytic active centers may be taken into account. Due to various impregnation and reduction methods, the distribution of active centers will be different. This causes different diffusional fluxes in the pore network.
6) Percolation phenomena can be described. This is particularly useful for modeling deactivation phenomena.
7) The pore walls can be smooth, irregular, or fractal.
8) Of particular importance is that parameters like tortuosity can be avoided. Tortuosity is a fitting parameter that, in most cases, contains all model deficiencies. As Aris [103] has stated: "When models are made of the configu-

ration of the pore structure the tortuosity can be related to some other geometrical parameter but in general it is a fudge factor of greater or less sophistication."

In lattice-based models, diffusion and reaction phenomena are limited to the lattice framework. Many more details about pore networks are presented in various reviews and books [88, 104–111] (see also Ref. [112]).

For a real catalyst one has to find out its textural parameters. Standard procedures are mercury porosimetry and nitrogen adsorption (see, for example, Ref. [113]), where the results depend on the data evaluation algorithms. Besides the BET approach and its variations [72], there are many more elaborated methods available (e.g., Refs [114–116]). Three effects are responsible for the general characteristics of porosimetry and nitrogen sorption data of porous materials: (i) the geometrical shape of individual pores, (ii) the relation between voids and throats (sites and bonds in percolation terminology), and (iii) the cooperative percolation effect of the porous network. Pore connectivity may be obtained by methods developed by Seaton [117] and Liu et al. [118]. Pore connectivity and architecture can also be traced by NMR methods [119] and SAXS [120]. Computer reconstruction approaches are also in use [104, 105, 121–124]. Manwart and Hilfer [125] have applied a simulating annealing algorithm to the reconstruction of two-dimensional porous media with prescribed correlation functions. Gelb and Gubbins [126, 127] and Sahimi and Tsotsis [89] have developed models that mimic the catalyst support formation process.

The textural parameters have a large influence on diffusion/reaction phenomena. This will be discussed below. Besides the shape of the Euclidian pore space, approximated as pores, the surfaces of the pores are of importance for diffusion and reaction. As catalyst pellets are often made of microscopic powder particles, the walls are not smooth, but amorphous. This structure influences the adsorption of reactants and products and the reaction kinetics. Reaction equilibriums are shifted owing to the interaction of reactants and products with the pore walls. The yardsticks are the diffusing and adsorbing molecules or atoms. As a consequence, small molecules will "see" a much larger surface area than the large ones (Figure 5.6c). This property has inspired the introduction of fractal concepts [128–138]. The accessible surface area depends on the molecular diameter δ, following a power law $\sim \delta^{2-D}$, where D is the fractal dimension of the surface. Details are reviewed by Coppens and Wang [135]. Examples of fractal objects are presented in Figure 5.6.

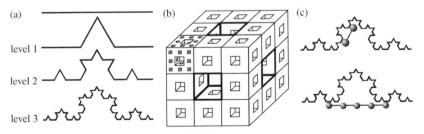

Figure 5.6 Self-similar objects. (a) Koch curve. (b) Menger sponge. (c) Molecules of different size on a Koch curve.

The fractal surface structure has a strong influence on the Knudsen diffusivity [139]. The Knudsen self-diffusivity decreases with increasing fractal dimension D and decreasing molecular diameter δ. This is an opposite effect to the better accessibility of the surface and, therefore, is of relevance in catalysis.

5.3
Diffusion

Four different diffusion mechanisms have been observed in porous catalyst particles: Knudsen diffusion, molecular diffusion, surface diffusion, and single-file diffusion. With Knudsen diffusion, the molecules collide more often with pore walls than with each other. This means the mean free path of the molecules is greater than the diameter of the catalyst pore. Molecular diffusion is significant for large pore sizes and high system pressures; here, molecule–molecule collisions dominate over molecule–wall collisions. In narrow pores, surface diffusion may be an important part not only of the transport processes but also in common catalyst supports, as has been detected experimentally by Feng et al. [140] and Rieckmann et al. [141], among others. In nanoporous materials, like carbon nanotubes (CNTs) or zeolites, single-file diffusion often occurs where the molecules very rarely bypass each other owing to the narrow pores. In common mesoporous catalyst supports, Knudsen, molecular and surface diffusion dominate.

Diffusion can be expressed in a variety of ways. In macroscopic experiments such as gravimetric measurements of the uptake, the diffusivity measured is the transport diffusion coefficient D_T. This coefficient is defined by Fick's first law:

$$J = -D_T(c)\nabla c, \tag{5.5}$$

where J is the sorbate flux when a concentration gradient ∇c is applied. In microscopic PFG-NMR experiments, the self-diffusivity D_S is measured. It is the diffusion of a tagged particle moving in a sea of other particles. In general, D_T and D_S are not the same. Self-diffusion is a measure of the mean square displacement:

$$D_S = \lim_{t\to\infty} \frac{1}{6Nt} \sum_{i=1}^{N} \Delta r_i(t)^2, \tag{5.6}$$

where Δr_i is the displacement of particle i at time t with respect to time zero, and N is the number of particles. According to thermodynamics, the driving force for the diffusion is the gradient in the chemical potential μ:

$$J(c) = -D_c(c)\nabla \mu = -\frac{L(c)}{k_B T}\nabla \mu, \tag{5.7}$$

where D^c is the collective, corrected, or Maxwell–Stefan (MS) diffusion coefficient, $L(c)$ is the single-component Onsager transport coefficient, k_B is the Boltzmann constant, and T is the absolute temperature. The collective diffusivity (also called the corrected diffusivity) is the collective diffusion of all adsorbate particles and can be interpreted as the movement of the center of mass of all particles together:

$$D_c = \lim_{t \to \infty} \frac{1}{6Nt} \left(\sum_{i=1}^{N} \Delta r_i(t) \right)^2. \tag{5.8}$$

In general, the collective diffusion is higher than the self-diffusion, because the collective diffusion contains interparticle correlations. For comparison of simulations with experiments, D_T is often converted to the corrected diffusivity D_c where it is frequently assumed that D_c is independent of loading, that is, $D_c(c) \approx D_c(0)$. With this approximation, one obtains

$$D_T(c) = D_S(0) \left(\frac{\partial \ln f}{\partial \ln c} \right)_T, \tag{5.9}$$

because the self-diffusivity and corrected diffusivity are identical when $c = 0$. This expression is often referred to as the Darken equation, although it is not historically correct [142]. Darken explicitly stated that D_c is composition dependent. Skoulidas and Sholl [143] and Snyder and Vlachos [144], among others, found substantial deviations from the Darken equation by molecular dynamics simulations and kinetic Monte Carlo (kMC), respectively. Skoulidas and Sholl [145] have computed D_C, D_T, and D_S for some light gases in four zeolites with different pore structures in order to find out the influences of pore shape and connectivity. Measurements of diffusivities are discussed in detail by Kärger and Ruthven [146]. The early theory of diffusion is compiled in books by Jackson [83], Dullien [104], Adler [121], Aris [103], and Cunningham and Williams [147]. As we have only limited space here, the two main theories in practical use, namely, the dusty gas model (DGM) and the mean pore transport model (MPTM) will be mentioned. The DGM was introduced by Evans et al. [148–150]. The MPTM has been developed by Rothwell [151], Schneider [152], Arnost and Schneider [153], among others. Both models result in almost the same equations, although they are based on different physical models. MPTM appears in various forms in the literature. Kerkhof and Geboers [154, 155] investigated the DGM and found a number of inconsistencies, for example, the "double accounting" of viscous contributions. A clear derivation providing new insight into the physical foundations of the diffusion in porous media has been presented by Young and Todd [156]. These authors follow Kerkhof's arguments and reject the DGM as a suitable foundation on which to construct a viable theory. A derivation of the governing equations for multicomponent convective–diffusive flow in capillaries and porous solids starting from a well-defined model and clear assumptions is presented. The solution for the continuum regime is discussed in detail, including a derivation of the diffusion slip boundary condition based on an improved momentum transfer theory. The Stefan–Maxwell species momentum equations have also been re-examined and important distinctions have been made between the local and tube-averaged equations. An equation for the pressure gradient is derived, and some examples of binary flows in capillaries have been discussed. A particular problem is the transition regime between Knudsen and molecular diffusion. Young and Todd have suggested an obvious method of interpolation for flow at arbitrary Knudsen number. The viscous term could also be included in a reasonable way. The extension to flow in porous media has been achieved by introducing a porosity–tortuosity factor, but unlike other treatments, this parameter is not absorbed into the gas diffusivities and flow permeability. It has been eliminated from all but one of the equations and, with appropriate boundary equations, the flux ratios could be obtained in terms of a mean pore radius only, which means they are independent of the porosity ε and tortuosity τ. It is only the absolute flux magnitudes that depend on ε/τ^2 and this parameter acquires a much clearer physical significance when it is viewed as a scaling factor for the length coordinate rather than a modifier of the gas diffusivities. In the cylindrical pore interpolation model (CPIM), the name given by Young and Todd to their model, the diffusivities depend only on the mean pore radius.

These developments have to be extended to multicomponent systems and reactions. Furthermore, adsorption and surface diffusion have to be included. If heat generation occurs, entropy production also has to be considered (Kjelstrup et al., 2005, [157]). In any case, models that are eventually based on approximations of the Boltzmann equation have to introduce some more or less justified heuristic assumptions. Within the framework of this approach, it is also difficult to include details of the intermolecular interactions. For example, as has been shown by Düren et al. [158], strong intermolecular interactions lead to deviations from Graham's law, owing to dragging effects.

The number of simulations of diffusion increased sharply over the last few years. Mostly nanoporous materials (zeolites, carbon nanotubes, and metal–organic frameworks) have intensively been investigated. The results have been reviewed in some papers [159–163]. The method of choice for diffusion simulations is molecular dynamics. In some cases, for example, benzene in MFI, molecules remain in the cages of zeolites for a rather long time. For such a situation, so-called rare event simulations are useful. There are various versions of this approach. Dubbeldam et al. [164] have proposed a rigorous extension of transition theory to finite loadings (dynamically corrected transition theory, dcTST). In this approach, the free energy barrier of a tagged particle is computed. The other particle–particle correlations can be taken into account by a proper definition of an effective hopping rate of a single particle. From this hopping rate of a molecule over a typical length scale given by the smallest repeating zeolite structure, that is, from the center of cage A to the center of cage B, the self-diffusivity can be computed. All other particles are regarded as a contribution to the external field exerted on this tagged particle. The diffusion coefficient corresponds exactly with the one that would be obtained from a molecular dynamics simulation if the assumptions underlying the rare event simulations hold, that is, once a particle hops over a free energy barrier, it remains sufficiently long in the free energy minimum such that it can fully equilibrate before jumping over the next barrier. Dubbeldam's approach has been employed in several applications where many surprising diffusion effects have not only been detected but could also be explained. For example, in Figure 5.7, results of molecular dynamics simulations of the diffusion coefficient of methane as a function of loading in various zeolites are presented.

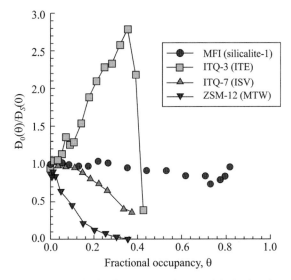

Figure 5.7 Molecular dynamics simulations of the loading dependence of the diffusion-corrected coefficient of methane in various zeolites [145].

An analysis by Beerdsen *et al.* [165, 166], employing the above method and some extensions, gave insight into the surprising diffusion behavior of different zeolites. For example, the increase of diffusion coefficients with increasing loading is amazing. Their analysis has led to a classification of zeolites correlated with their diffusional behavior. For a given molecule, Beerdsen *et al.* [165, 166] computed the free energy profile of a molecule hopping from one site to another. These profiles resulted in a classification (Figure 5.8):

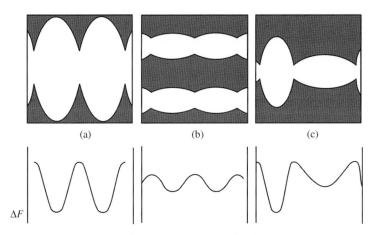

Figure 5.8 Free energy profiles of three categories of zeolites. (a) Cage-type (SAS). (b) Tube-type (AFI). (c) Intersecting-type (MFI) [165, 166]).

- **Cage-type structures**: These zeolites consist of large cages, connected by narrow windows forming large free energy barriers (e.g., SAS, LTA, ERI, and CHA). The molecules interaction with the zeolite wall is favorable. The preferential adsorption is in the cages. Hence, additional molecules will increase the bottom of the free energy profile but not the top. Therefore, additional molecules will lower the full energy barrier that leads to an increase of the diffusion coefficient.
- **Tube-type structures**: Again, the free energy in the interior of the cage rises when new molecules are added, but the effect on the free energy is even larger at the barriers. The result is a decreasing diffusivity with loading (e.g., AFI, TON, and MTW);
- **Intersecting two- or three-dimensional structures (e.g., MFI, BEC, ISV, and BOG)**: The barriers are formed by the horizontally aligned channels, creating entropic traps between consecutive vertical channels. The influence of loading on these systems is complex, as it involves effects as nonsimultaneous freezing in vertical and horizontal channels, owing to differences in channel diameter and length, causing a varying degree of commensurability of the particles with the structure (pore length) as a function of loading and direction. The self-diffusivity sharply decreases when the loading is increased, but the corrected diffusivity initially only slightly decreases with density until packing effects sharply decrease the corrected diffusivity as well.

Whether zeolite lattice flexibility has any significant influence on diffusivities has been under debate for a long time. Demontis et al. [167] and Fritzsche et al. [168] observed little influence of lattice flexibility on the diffusion of methane. Leroy et al. [169] have studied the effect of lattice flexibility for some n-alkanes in MFI. They found that the flexibility increases the diffusion coefficient, but only slightly for long hydrocarbons and at high loading. As has been shown by Leroy et al. [169] and Zimmermann et al. [170], the choice of the host model is important for the extent of the enhancement of the diffusion coefficient. For example, for the same system, different models for the zeolite interactions show either an increase or a decrease of the diffusion coefficient. At low loadings in nanoporous materials with smooth walls, one has to look out for flexibility effects on the diffusion. This has clearly been demonstrated by Jakobtorweihen et al. [171] for single-wall carbon nanotubes, where the diffusivity is reduced by an order of magnitude at zero loading owing to the wall flexibility.

Molecular simulations can be combined with classical approaches like the Maxwell–Stefan ansatz. In the MS formulation, the flux of any species in a mixture with n components is related to its chemical potential gradient [172–176] by

$$-\varrho \frac{\Theta_i}{k_B T} \nabla \mu_i = \sum_{\substack{j=1 \\ j \neq i}}^{n} \frac{\Theta_j N_i - \Theta_i N_j}{\Theta_{j,\text{sat}} \mathcal{D}_{ij}} + \frac{N_i}{\Theta_{i,\text{sat}} \mathcal{D}_i}, \quad i = 1, 2, \ldots, n, \quad (5.10)$$

where N_i is the flux of species i expressed in molecules per square meter per second, ϱ is the zeolite density expressed as the number of unit cells per cubic meter, Θ_i is the loading in molecules per unit cell, $\Theta_{i,\text{sat}}$ represents the saturation loading of species i, n is the total number of diffusion species, and k_B is the Boltzmann constant. Equation (5.10) defines two types of MS diffusivities: $\mathcal{D}_i$ and $\mathcal{D}_{ij}$. For a single sorbed component, only one $Đ_i$ is needed and in this case, $Đ_i$ is equivalent to the single component "corrected" diffusivity that can be obtained from molecular dynamics simulations. The binary exchange coefficients $Đ_{ij}$ reflect correlation effects in mixture diffusion. For mixture diffusion, the $Đ_{ij}$ tends to slow down the more mobile species and speed up the relatively sluggish ones. A lower value of the exchange coefficient $Đ_{ij}$ implies a stronger correlation effect. When $Đ_{ij} \to \infty$, correlation effects vanish. Krishna and van Baten [172] suggested an interpolation formula for estimating the binary exchange parameters $Đ_{ij}$ from pure component self-exchange coefficients $Đ_{ii}$:

$$\Theta_{j,\text{sat}} Đ_{ij} = \left[\theta_{j,\text{sat}} Đ_{ii}\right]^{\Theta_i/(\Theta_i+\Theta_j)} \left[\Theta_{i,\text{sat}} Đ_{jj}\right]^{\Theta_j/(\Theta_i+\Theta_j)}, \quad i,j = 1, 2, \ldots, n. \tag{5.11}$$

This interpolation formula has been tested for binary and ternary mixtures in various zeolites and turned out to be a reasonable estimation procedure for $Đ_{ij}$. The MS approach is a useful approximation to the challenging problem of quantitatively describing multicomponent transport of species in zeolite pores. The results can be compared with molecular dynamics calculations. The advantage of the MS method is that one can obtain multicomponent fluxes based on single component data by means of a rather simple formula. Merely for segregated adsorption [173] problems occur. For example, in mixtures of CO_2 and neon in LTA and DDR, CO_2 is preferentially lodged at the narrow window regions, and this hinders the diffusion of partner molecules between cages. Current phenomenological models do not include segregation effects on mixture diffusion.

5.4
Diffusion and Reaction

In chemical engineering, the Langmuir–Hinshelwood rate expression is very often employed to describe the kinetics of heterogeneously catalyzed reactions in mesoporous and microporous catalyst supports. Based on some experimental data, obtained from gaschromatographic measurements in differential or integral reactors under stationary conditions, several "reaction mechanisms" are proposed that could be in agreement with the experimental findings. Diffusional resistances are minimized by taking very small particles. These mechanisms are then transformed into the corresponding LH rate expressions whose parameters are fitted to the experimental data. The results are evaluated by various statistical methods to figure out which mechanism is the "best" corresponding to statistical criteria. Quite often more than one mechanism fits well the same data set. Sometimes one or more LH expressions can be excluded by doing additional experiments under instationary conditions by using, for example, mass spectrometry. Owing to severe limitations of this procedure, every chemical engineer knows that the eventually chosen mechanism and its corresponding LH rate expression or the set of differential equations for the instationary conditions are only descriptive. The rate expression can be used for reactor design within the range of measurements included in the data fit. Several approximations have been developed for simplifying the system of differential equations of the phenomenological model, like steady-state approximation, quasiequilibrium, irreversible step approximation, most abundant reaction intermediate (MARI), and the nearly empty surface approximation (see, for example, Ref. [32]).

As already mentioned in Section 5.1, the LH rate expressions are based on the Langmuir adsorption isotherm, and, therefore, includes its assumptions listed above. In practice, one employs the multicomponent version of Langmuir isotherm with the fractional occupancies θ_i given by

$$\theta_i = \Theta_i/\Theta_{i,\text{sat}} = b_i p_i \left/ \left(1 + \sum_{i=1}^{n} b_i p_i\right)\right. . \tag{5.12}$$

The fraction of vacant sites on the surface is given by

$$\theta_v = 1 - \sum_{i=1}^{n} \theta_i = 1 \left/ \left(1 + \sum_{i=1}^{n} b_i p_i\right)\right. , \tag{5.13}$$

where Θ_i are the molecular loadings, $\Theta_{i,\text{sat}}$ are the saturation capacities, p_i are the partial pressures in the gas phase, and b_i are the Langmuir adsorption constants. The LH rate equation for a reversible reaction would give

$$r = (k_f \Pi p_i - k_r \Pi p_j)\theta_v = \frac{k_f \Pi p_i - k_r \Pi p_j}{\left(1 + \sum_{k=1}^{n} b_k p_k\right)^q} . \tag{5.14}$$

Based on the Gibbs adsorption equation, Rao and Sircar [177] have developed thermodynamic consistency tests for binary adsorption equilibriums. For the Langmuir model, they found that it is only thermodynamically consistent if the saturation adsorption capacities of all components are equal. This fact has been used by Krishna and Baur [178] to demonstrate that configuration and size entropy effects in zeolites may lead to mixture adsorption isotherms and concentration profiles in fixed bed reactors that cannot be obtained qualitatively (maximum in the pure component isotherms) and quantitatively by means of the Langmuir isotherm. Furthermore, surface diffusion is not included in the Langmuir equation. Instead, a perfect mixing of the species on the surface is assumed that justifies the products of surface partial pressures (coverages) in the rate equation. As there are no lateral interactions in the model, clustering and phase transitions on the surface cannot occur. At higher loadings, which are common in commercial catalysis, one always has lateral interactions. Ternel et al. [179] have compared phenomenological kinetics with kinetic Monte Carlo calculations of CO oxidation on ruthenium (see Chapter 3). The authors have found that the phenomenological kinetics did not agree with the kMC results. Similar results have been obtained by Hansen et al. [70] for benzene ethylation in ZSM-5. Inhomogeneities of the surface coverage can be created by the interplay between the rates of adsorption, desorption, and vacancy creation by the chemical reactions. Corma et al. [180] have derived a generalized multicomponent adsorption isotherm that has been used to derive rate equations. These equations comprise the Langmuir conditions (all sites are identical), the Temkin model (linear decrease of adsorption energies with coverage), and the Freundlich isotherm (logarithmic decrease of adsorption energies with coverage). A comparison has then been made of the goodness of fit, for each of these rate equations, for 19 kinetic studies reported in the literature in which detailed kinetic data are available. In 10 out of 19 cases, the fitting has been comparably good for all models; in 6 cases, the Langmuir model gives a better fit, and in 3 cases, the fit by the Langmuir model has been much poorer than that by the Temkin or Freundlich models. The authors have given some possible reasons for this result (low-coverage Henry regime, variation of parameters over a too narrow range). A similar scenario can be deduced from the collection of published kinetic data by Mezaki and Inoue [30]. Ostrovskii [181] has executed a critical investigation of surface heterogeneity and has come to the conclusion that no surface heterogeneity reveals itself in catalysis and chemisorption at metals (see also Ref. [182]).

Inspired by the book written by Dumesic et al. [183], microkinetic modeling has been introduced by many groups. A review has been presented by Stoltze [184]. The starting point for microkinetic modeling is the detailed reaction mechanism, which explicitly includes the surface species and not only the apparent gas-phase reaction, that means one writes down the particular catalytic reaction mechanism of interest in terms of its most elementary steps. No rate-determining step is assumed. For each of the postulated elementary steps, rate parameters for all the forward and reverse reactions are needed to solve the equations representing the model. Application of various experimental techniques, for example, spectroscopic surface science methods, kinetic measurements under instationary condition [185, 186] combined with quantum chemical calculations, statistical thermodynamics, transition state theory, and molecular and mesoscopic simulation methods, leads to detailed microkinetic models. As has been demonstrated by Lynggaard et al. [187], measurement of reaction rates in a microreactor, even for simple catalytic reactions, in principle cannot disclose important mechanistic details. These authors have also shown that abrupt changes in slope in the Arrhenius plot, in general, cannot be taken as an evidence for, for example, a change in reaction path or diffusion limitations.

Inclusion of quantum chemical calculations and molecular modeling is important, because some species may not give a spectroscopic signal owing to a too short lifetime. Another reason is that the spectroscopic measurements are often incomplete and have to be interpreted by calculations. Besides quantum chemistry, mostly in the framework of density functional theory, kinetic Monte Carlo and transition state theory are in use. Norskov's group has shown that for some classes of chemical reaction, the same "universal" Brønsted–Evans–Polanyi relation is followed (see Ref. [188] and literature cited therein). The data obtained from these calculations can then be employed in rate equations that may be inserted, for example, into reactor mass balances. At present, most of the microkinetic models use a mean field approach even when known nonuniformity of the surface is present, but the number of papers taking adsorbate–adsorbate interactions into account is rapidly growing [9, 10, 189]. Mostly, DFT calculations on periodic systems, combined with transition state theory and kinetic Monte Carlo, are applied for this purpose. Up to now, these papers do not test the results with real reactor measurements. Sometimes it is possible to deduce simplified Langmuir–Hinshelwood kinetics from extensive sets of differential equations describing the detailed kinetics [21]. One of the ultimate goals in modeling heterogeneous catalytic reaction systems would be the development of a multiscale approach that could simulate the many transformations on an atomic scale that occur on a catalyst surface as they develop as a function of time, process conditions, catalyst surface and pore structure, and composition. This is beyond we can simulate in the near future. Chemisorption, dissociation/activation, recombination, and desorption can be described with reasonable accuracy if no multiconfigurational and dispersion effects play a dominant role. Typical accuracies for structural properties are 0.05 Å and 1–2°, overall adsorption and reaction energies are typically within 25–30 kJ/mol, and spectroscopic shifts that are within a few percent of experimental data. Van Santen and Neurock [19] compile many examples of DFT calculations for heterogeneous catalysis. An example of simulating a chemical reactor and various stationary and instationary experiments based on a microkinetic model from first principles, including more than 60 reactions and 40 species, has been presented by Heyden et al. [21]. Calculations over many timescales are still a problem. Molecular dynamics simulations have inherent timescale limitations, which can be partly overcome by methods like "hyper MD" [190], dissipative particle dynamics (DPDs) [191, 192], and rare event simulations [60].

Heterogeneous catalytic reactions occur in porous catalysts. Therefore, reactants and products interact with the porous walls and have to diffuse into and out of the pores. The reactants have to diffuse from the bulk phase through the laminar boundary layer surrounding the pellet to its external surface and pass their way through the porous structure to the catalytically active crystallites. Similarly, the reaction products must diffuse through the porous structure back to the external surface of the pellet and eventually into the bulk gas phase (see Figure 5.1). The transport of heat released or absorbed by the reaction follows the same path, but the pore walls itself are mainly involved as heat conducting material. The chemical reaction changes the reactant composition along the pores such that the active sites are surrounded by changing reactant compositions along the diffusion paths. The reactions on the surface create locally instationary conditions even though the outer catalyst pellet surface is circulated around under stationary conditions. The mass and heat transport processes influence not only the apparent activity of the catalysts but also the selectivity and, in some cases, may give rise to new phenomena like multiplicity of reaction rates, instabilities, and oscillations. Investigations of reaction and transport in catalyst particles have been executed by Damköhler [193], Thiele [194], and Zeldovich [195] in the 1930s and 1940s. In 1951, in a review paper, Wheeler brought up this matter again and expanded upon it. In the 1960s and 1970s, many papers on this subject were published that have been reviewed, for example, by Wicke [196], Schneider [197], Burghardt [198], Luss [199], among others. The multicomponent flow has been described by means of Maxwell–Stefan approach, augmented by terms of surface flux. A problem arises, as for mesoporous catalyst supports, the porous structure is not known in detail. The common characteristics used are usually the pore radii distribution, pore volume distribution, total porosity, and the bulk density. For the effect of internal mass and heat transfer, a pseudo-homogeneous model or a simple parallel pore bundle model has been used that has two adjustable parameters, namely, tortuosity and average pore radius. These data may be obtained from measurements in a Wicke–Kallenbach cell [80]. The parameters are then inserted into equations to obtain effective diffusivities. The Wicke–Kallenbach stationary method has been modified by Smith and Gibilaro and coworkers to dynamic experiments (see Ref. [198]). Effective diffusivities have also been measured by chromatographic methods. Sometimes the porosity is found by a fit to Wicke–Kallenbach experiments. For the transition region between Knudsen and molecular diffusion, various models mentioned in Section 5.3 have been employed. Over the past three decades, the dusty gas model [148, 149] or the model by Feng and Stewart [81] have been applied. The shortcomings of the DGM have been discussed above. The starting point for this model is a set of diffusion equations defining the transport of a gaseous species in the presence of temperature, pressure, and composition gradients and under the influence of external forces. These equations are sometimes called generalized Maxwell–Stefan equations. DGM assumes that the pore walls consist of heavy particles that are immobile and uniformly dispersed in the system. They represent the "dust" in the model. The dust particles are constrained by external forces, preventing them from moving despite the existing pressure gradient in the gas. There are no external forces acting on the gas particles. The porous structure of catalyst support is described by the porosity and tortuosity. In the Feng and Stewart model, the structure of the porous medium is represented by a statistically specified assembly of capillaries, closely interconnected, so that the smooth field assumption is valid. A far more realistic description of the pore structure may be achieved by employing three-dimensional network models. Rieckmann and Keil [200] have developed a three-dimensional network model where any kinetics can be included. Computations with this model have been executed for the selective hydrogenation of 1,2-dichloropropane to propane and hydrochloric acid. The results have then been compared with experiments in single-pellet diffusion reactor [141]. The importance of surface diffusion could clearly be demonstrated. The model has been extended to diffusion and reaction accompanied by capillary condensation [201]. Deepak and Bhatia [202] have compared the correlated random walk theory (CRWT) for cubic networks and the effective medium theory (EMT). Far away from the percolation threshold, EMT gives slightly more accurate results, but opposite to CRWT, the EMT cannot be used for inhomogeneous media. Sotirchos and Burganos [203] have developed diffusion flux models for isobaric diffusion of a multicomponent gas mixture in pore networks of distributed pore size and length constructed by arranging pore segments around the bonds of a lattice of constant coordination number.

Burghardt [198] has combined the dusty gas model and reaction kinetics to obtain a set of differential equations defining mass and energy transport in a symmetrical catalyst pellet, in which linearly independent reactions take place. Appropriate boundary conditions, which reflect the outer geometry of the pellet, have to be introduced. If one solves such a set of equations for a porous pellet, one obtains descending profiles from outside toward the pellet center for the reactant concentrations. These concentration profiles are caused by the diffusion resistance because reactants cannot diffuse in the bulk sufficiently rapidly. A small diffusion resistance gives a rather flat curve, whereas a steep curve is obtained for a large diffusion resistance. A consequence of this profile is a decreased averaged rate relative to that if the concentration were everywhere equal to the outer surface concentration. This loss of average reaction rate in a porous catalyst pellet is more than offset by the enormous increase in surface area of the pores. Thiele [194] and Zeldovich [195] defined the effectiveness factor:

$$\eta = \frac{\text{rate of reaction with pore resistance}}{\text{rate of reaction with surface conditions}}, \qquad (5.15)$$

$$\eta = \frac{(1/W_{cat}) \int r(C) dW_{cat}}{r(C_S)}, \qquad (5.16)$$

where W_{cat} is the catalyst mass, $r(C)$ is the rate at reactant concentration C at any position in the pellet, and $r(C_S)$ is the rate at surface conditions. Therefore, the actual rate that would be observed is

$$r_{obs} = \eta r(C_S). \qquad (5.17)$$

For reaction systems, each reaction has its own effectiveness factor. When the concentration profile found from the diffusion equation is substituted into the numerator of Eq. (5.16), this becomes for a first-order reaction

$$\eta = \frac{\tanh \Phi}{\Phi}, \qquad (5.18)$$

where Φ is the so-called Thiele modulus

$$\Phi = L\sqrt{k/D_{eff}}, \qquad (5.19)$$

where L is a length parameter, k is the rate constant, and D_{eff} is the effective diffusivity. Bischoff [204] has defined a general modulus that condenses the curves for a rather wide range of reaction types into a relatively narrow region. Effectiveness factors for the nth order reaction in bidisperse catalysts have been developed by Kulkarni et al. [205]. Effective diffusivities in pore networks have been calculated by Zhang and Seaton [206, 207]). Weisz and Hicks [208] have analyzed the reactivity behavior of porous catalyst particles subject to both internal mass concentration gradients and temperature gradients, in endothermic or exothermic reactions. Their results look typically as presented in Figure 5.9a and b. For large values of the Thiele modulus, the effectiveness factor is inversely proportional to this modulus. A general proof of this relation has been given by Murray (Murray, 1968).

Figure b shows multiplicities that are consequence of the strong nonlinearity of the heat generation term in the heat balance. Many investigations of these bifurcation phenomena have been reviewed by Luss [199] and Aris [103]. Kaza et al. [209] have developed a method for calculating diffusion effects in nonisothermal pellets for

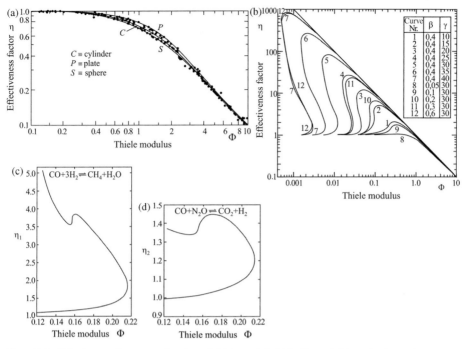

Figure 5.9 Effectiveness factors as a function of Thiele moduli: (a) isothermal, (b) nonisothermal, and (c and d) methanation reactions [209].

several components and reactions. The authors have applied their method to methanation reactions. The effectiveness factors as a function of the Thiele moduli are presented in Figure 5.9c and d. One should be aware that the curves presented in Figure 5.9b and d are rather simple. Far higher multiplicities are possible for more complex kinetics. In the 1990s, many investigations of the effect of volume changes in the pores owing to reactions on the selectivity or the effectiveness factor have been executed (see, for example, Ref. [210]). Dogu [211] has reviewed the pore models proposed in the literature and the experimental techniques for finding effective diffusivities. Meanwhile, there are many experimental findings that clearly indicate the importance of the pore size for optimal catalytic activity of mesoporous materials [212].

Molecular simulations that combine diffusion and reaction are still in their infancy. One of the early papers on this subject has been published by Keil et al. [210]. For the case of zeolites, the papers have been reviewed by Keil and Coppens [213]. A particular approach is the simulation of chemical reaction equilibriums by the reaction ensemble Monte Carlo method (RxMC) [214, 215]. The RxMC method was independently published by two different groups [216, 217]. The RxMC simulation method uses Monte Carlo sampling to directly simulate predefined forward and reverse reaction events in a simulation, yielding the equilibrium composition of the reacting mixture. In these simulations, completed forward and reverse reaction events are sampled, so the energetics of bond breaking and bond formation are unnecessary. Furthermore, since the activation barriers for the reactions are eliminated in these simulations, kinetic limitations are circumvented and equilibrium can be established very rapidly under various conditions. The RxMC approach can be performed in any number of basic ensembles, and its implementation is similar to a traditional grand canonical MC (GCMC) simulation for a multicomponent mixture, since it involves molecular insertions and deletions and changes of particle identities during the simulation. The primary utility of the RxMC approach is for efficiently predicting shifts of reaction equilibriums caused by nonideal environments. Hansen et al. [218] have combined the RxMC with the configurational-bias Monte Carlo (CBMC) approach, and then applied to the propene metathesis system. It has been found that the confined environment can increase the conversion significantly. A large change in selectivity between the bulk phase and the pore phase has been observed. A subsequent paper [219] derived a general acceptance rule for the RxMC approach in combination with the CBMC method. Using this derivation all other acceptance rules of any MC trials that are carried out in combination with the CBMC approach can be deduced from it, for example, regrowth, insertion/deletion, identity change, and Gibbs ensemble transfer/exchange. A review of the RxMC has been presented by Turner et al. [215].

Smit and Maesen [220] have reviewed their investigations on shape selectivity in zeolites. The authors have employed the "free energy landscape approach" to a molecular understanding of shape selectivity in zeolite catalysis. A central premise of this approach is that by ignoring the detailed chemical characteristics of a zeolite and simply quantifying instead how its topology affects the free energies of formation of the various reactants, intermediates, and products involved, one can identify the fundamental interactions and processes that control the shape selectivity of a particular transformation. The probability of a particular molecule forming during a zeolite-catalyzed process is directly proportional to its free energy of formation in the adsorbed phase. This approach is limited to simple reactions that occur at a single reaction site. Figure 5.10 presents how the free energy of formation of five intermediates involved in n-decane hydroconversion changes relative to that of n-decane as a function of zeolite structure (Figure 5.10a).

FAU is a large-pore zeolite that contributes little to the free energy differences. MEL and MFI have similar structures to each other but with pore widths intermediate between those of TON and FAU. The contributions of MEL and MFI are similar for most reaction intermediates, but there are marked differences in free energies for some specific intermediates. As can be seen from Figure 5.10b, MFI prefers to form 4,4-dimethyl octane because it is commensurate with the zigzag channel and hence forms a nice fit, whereas MEL prefers to form 2,4-dimethyl octane, which fits perfectly in the larger intersection. As a consequence, the reaction paths in MFI are dominated by the path $n\text{-}C_{10} \to \text{MeC}_9 \to 4,4\text{-MeC}_8$, whereas in MEL the dominant path is $n\text{-}C_{10} \to 5\text{-MeC}_9 \to 2,4\text{-MeC}_8$. This simulation-based methodology has predictive power as was demonstrated for the hydrodewaxing reaction. In hydrodewaxing, the zeolite catalyst thus needs to convert the longest hydrocarbons while leaving shorter hydrocarbons unchanged. A screening of zeolites for this reaction identified the zeolite GON as being suitable for this reaction, which will be

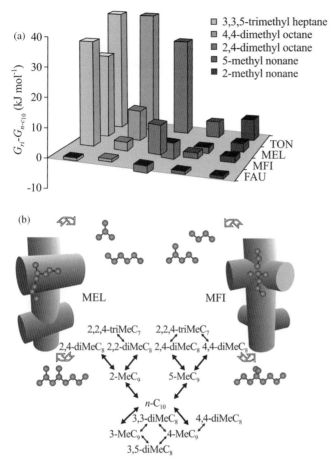

Figure 5.10 Free energy model applied to n-decane hydroconversion. (a) Free energy differences between n-decane and various products as a function of zeolites. (b) Comparison of product formation in MEL and MFI.

patented. Gounaris et al. [221] have developed a screening method for zeolites as selective separation or catalysis materials.

Hansen et al. [56, 70, 222, 223], for the first time, have investigated diffusion and reaction based on DFT, molecular simulations, and continuum modeling. The influence of adsorption thermodynamics and intraparticle diffusional transport on the overall kinetics of benzene alkylation with ethene over H-ZSM-5 has been analyzed, where the rate coefficients for the elementary steps of the alkylation have been taken from quantum chemical calculations and transition state theory. The intrinsic kinetics of two different reaction schemes have been analyzed, and their rate data have been combined with a continuum model based on the Maxwell–Stefan equations in combination with the ideal adsorbed solution theory (IAST). The parameters of the MS equations have been obtained from molecular dynamics

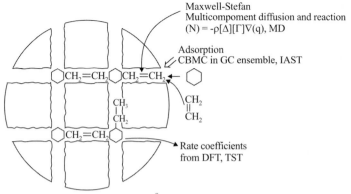

Radius 1 μm, corresponds to 7.9×10^8 unit cells

Figure 5.11 Approaches for a combined diffusion/reaction computation based on first principles and Maxwell–Stefan equations.

simulations, and pure component adsorption isotherms have been obtained from configurational-bias Monte Carlo simulations (Figure 5.11).

By taking diffusion limitation into account, experimentally determined reaction rates and the reaction orders in the partial pressures of the reactants could be reproduced. The results of this study have clearly shown that empirical power law or Langmuir–Hinshelwood rate expressions become inappropriate when used to correlate kinetic data over a broad range of conditions. In addition, it has been demonstrated that the usual approaches to determine effectiveness factors for reactions in porous media, which assume a constant effective diffusivity, may lead to substantial deviations from rigorous simulations, whereas the simulation model that has been developed in the paper can be employed to predict the effectiveness factor for zeolite particles for any set of reaction conditions (Figure 5.12a).

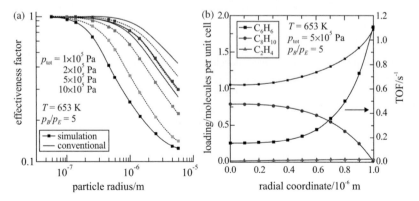

Figure 5.12 (a) Effectiveness factors as a function of the particle radius for different gas-phase pressures: $p_B \equiv$ partial pressure of benzene and $p_E \equiv$ partial pressure of ethylene. (b) Radial concentrations and turnover frequencies (TOFs).

Figure 5.12b shows the intraparticle concentration profiles of all species corresponding to the simulation for diffusion-limited conditions at 653 K along with the turnover frequency as a function of the radial coordinate. The figure shows that the benzene to ethene ratio on the exterior surface is higher by a factor of 6 than the ratio of partial pressures in the gas phase owing to the differences in adsorption strength. In the near future, the MS approach will be replaced by rare event simulations.

5.5
Pore Structure Optimization: Synthesis

It is obvious that the pore structure has a considerable influence on the diffusion and reaction phenomena, in particular, the relation between macroporous and microporous volume fractions in bimodal catalysts and the volume fraction of zeolites in mesoporous binders. In 1980, Hegedus [3, 224, 225] has optimized the structural variables of a Wakao/Smith model (1962) for an automobile emission control catalyst. In the 1990s, pore network models have been used for optimization of the catalytic hydrodemetallation of crude oil over a bimodal $CaO\text{-}MoO_3/Al_2O_3$ catalyst [226]. For this purpose, a bond-node pore network model has been used. For such a model, a diffusion–reaction equation is formulated for each pore in the network with the boundary conditions such that the concentrations at the two ends of each pore are equal to the concentration at the corresponding node where two or more pores intersect, which means at the inner nodes of the network, an equation similar to Kirchhoff's law holds. The fluxes of the species that enter a node must, therefore, be equal to the fluxes leaving the node. In order to cover the whole range of Thiele moduli, the outer dimension of the network has to be similar to the actual pore size used in operation. One has to introduce a performance index, for example, the mole flow of a certain product averaged over the whole pellet and a given period of time has to be maximized. A possible result for a demetallation reaction is presented in Figure 5.13 [226].

It could be demonstrated that the catalyst performance could be improved considerably by tuning the pore structure. It turned out that radius and the volume fraction of the micropores are of decisive importance for the performance of the catalyst. El-Nafaty *et al.* [227] have applied a two-dimensional stochastic network to find out the direct influence of pore structure assembly on diffusion and reaction in FCC catalyst particles. Coke burn-off in heavily coked particle has been used. Various pore architectural structures were tested, including random, positively spiraled, negatively spiraled, structured, and interspersed 2D networks. The interspersed network exhibited fastest burn-off kinetics relative to other structures, while the random configuration that probably characterized most current catalyst particles showed results better than the only negatively spiraled network. Prachayawarakorn and Mann [228] have used a two-dimensional bond-node pore network model for various pore architectures and calculated their effectiveness factors. A structure in which large pore channels are placed in the exterior shell of the catalyst, with nanopores in the interior, is the best of all tested structures. It turned out to be 800% better than random pore network at a

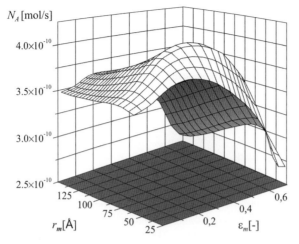

Figure 5.13 Optimization of catalyst support pore structures, specific demetallation rate as a function of microporous radius r_m, and micropore volume ε_m [218].

Thiele modulus of 1. Coppens and coworkers have investigated hierarchically structured porous catalysts [132, 133, 229–232]. The macropore, "distributor" network of a hierarchically structured porous catalyst has been optimized with the aim to maximize the effectiveness factor of a single reaction with general kinetics. The model was applied to a catalyst for power plant NO_x emission control, namely, the SCR reaction that is usually carried out in a honeycomb monolith reactor. It has been detected that the optimal, hierarchically structured porous catalyst uses 20–40% less catalytic material but is 80–180% more active than a purely mesoporous catalyst. Further results obtained by Coppens' group may be found in a review [135]. A compilation of the optimal distribution of active centers in the porous catalyst supports is presented in a book by Gavriilidis et al. [223].

Over the past two decades, impressing progress in the synthesis of defined porous structures has been made such that there is a good chance that the calculated optimal structures can be synthesized. Reviews of the experimental work on materials with a defined porous structure have been published [15, 234–252]. Many efforts are directed toward the circumvention of the diffusional limitation imposed by zeolitic structures. This is done by making zeolites with larger pores, making smaller zeolite particles, and inserting larger pores into the zeolite particles. Addition of mesopores to zeolites can increase diffusion coefficients more than two orders of magnitude. Templating techniques are mostly used to produce zeolite materials with hierarchical bi- or trimodal pore size distributions. Desilication is another technique [241]. Template-free "one-pot" synthetic methods include "oriented attachment," Ostwald ripening, and the Kirkendall effect for direct solid evacuation under mild reaction conditions. Ordered mesoporous materials have a high potential in catalysis [237]. Therefore, there are great perspectives for a pore structure engineering based on computationally optimized structures.

5.6
Conclusion

Phenomena of diffusion and reaction are inherent multiscale. Therefore, various approaches for modeling of these phenomena have been developed. Up to the end of the 1970s, classical continuum methods dominated, preferentially in the mathematical form of ordinary and partial differential equations. In the 1980s, molecular modeling (MD, MC) has gradually come up and has extensively been used from the 1990s on where quantum chemical approaches and microkinetic modeling also became a standard tool. These powerful methods gave far deeper insight into diffusion and reaction phenomena than what would have ever been possible by the more or less heuristic phenomenological approaches, although many of them are based on the well-founded Boltzmann equation. Nevertheless, details of bond cleavage and bond formation, chemisorption on surfaces, and dynamics of reactions are not comprehensible without quantum chemical approaches. Diffusion phenomena in confined media, in particular, in nanopores, turned out to be understandable by molecular simulations. A combination of quantum chemistry and molecular simulations results in a powerful tool that enables quantitative calculations of diffusion–reaction processes. Realistic pore network models have clearly demonstrated that the transport processes can be improved considerably. Owing to new synthesis methods, there is a good chance that these calculated optimum pore structures can be introduced into real chemical production processes.

Of course, the modeling methods need to be improved further to be able to tackle the problems over a wide range of time and length scales. Furthermore, even apparently simple reactions can turn out to be composed of many elementary steps that lead to a tremendous amount of transition state calculations and so on. The modeling of the full scenario of pore shape and surface modeling, multicomponent diffusion, dissociative adsorption combined with dynamic island formation and complicated phase changes of the surface, complex reaction networks, and desorption combined again with surface phase changes is still in its infancy. One can suppose that many new and unexpected phenomena will be discovered. Therefore, modeling of diffusion and reaction remains a fascinating field of research.

References

1 Metiu, H. (2008) Preface to special topic: a survey of some new developments in heterogeneous catalysis. *J. Chem. Phys.*, **128**, 182501.
2 Becker, E.R. and Pereira, C.J. (eds) (1993) *Computer-Aided Design of Catalysts*, Marcel Dekker, New York.
3 Hegedus, L.L. (ed.) (1987) *Catalyst Design*, John Wiley & Sons, Inc., New York.
4 Morbidelli, M., Gavriilidis, A., and Varma, A. (2001) *Catalyst Design: Optimal Distribution of Catalyst in Pellets, Reactors and Membranes*, Cambridge University Press, Cambridge, UK.
5 Bell, A.T. (1990) Impact of catalyst science on catalyst design and development. *Chem. Eng. Sci.*, **45**, 2013–2026.

6 Gates, B.C. and Knoezinger, H. (eds) (2000) *Impact of Surface Science on Catalysis*, Academic Press, New York.
7 Reuter, K. and Scheffler, M. (2006) First-principles kinetic Monte Carlo simulations for heterogeneous catalysis: application to the CO oxidation at $RuO_2(110)$. *Phys. Rev. B*, **73**, 045433.
8 Cinquini, F., Di Valentin, C., Finazzi, E., Giordano, L., and Pacchioni, G. (2007) Theory of oxides surfaces, interfaces and supported nano-clusters. *Theor. Chem. Acc.*, **117**, 827–845.
9 Stampfl, C. (2005) Surface processes and phase transitions from *ab initio* atomistic thermodynamics and statistical mechanics. *Catal. Today*, **105**, 17–35.
10 Stampfl, C. (2007) Predicting surface phase transitions from *ab initio* based statistical mechanics and thermodynamics. *Phase Transit.*, **80**, 311–332.
11 Somorjai, G.A. and Park, J.Y. (2008) Evolution of the surface science of catalysis from single crystals to metal nanoparticles under pressure. *J. Chem. Phys.*, **128**, 182504.
12 Woodruff, D.P. and Delchar, T.A. (1994) *Modern Techniques of Surface Science*, 2nd edn, Cambridge University Press, Cambridge.
13 Niemantsverdriet, J. (2000) *Spectroscopy in Catalysis: An Introduction*, 2nd edn, John Wiley & Sons, Inc., New York.
14 Kolasinski, K.W. (2002) *Surface Science: Foundations of Catalysis and Nanoscience*, John Wiley & Sons, Inc., New York.
15 de Avila Arturo Soler-Illia, G.J., Sanchez, C., Lebeau, B., and Patarin, J. (2004) Chemical strategies to design textured materials: from microporous and mesoporous oxides to nanonetworks and hierarchical structures. *Chem. Rev.*, **102**, 4093–4138.
16 Thomas, J.M. (2008) Heterogeneous catalysis: enigmas, illusions, challenges, realities, and emergent strategies of design. *J. Chem. Phys.*, **128**, 182502.
17 Hellmann, A., Baerends, E.J., Biczysko, M., Bligaard, T., Christensen, C.H., Clary, D.G., Dahl, S., van Harrevelt, R., Honkala, K., Johnsson, H., Kroes, G.J., Luppi, M., Manthe, U., Norskov, J.K., Olsen, R.A., Rossmeisl, J., Skulason, E., Tautermann, C.S., Varandas, A.J.C., and Vincnet, J.K. (2006) Predicting catalysis: understanding ammonia synthesis from first-principles calculations. *J. Phys. Chem. B*, **110**, 17719–17735.
18 Norskov, J.K., Scheffler, M., and Toulhat, H. (2006) Density functional theory in surface science and heterogeneous catalysis. *MRS Bull.*, **31**, 669–674.
19 van Santen, R.A. and Neurock, M. (2006) *Molecular Heterogeneous Catalysis: A Conceptual and Computational Approach*, John Wiley & Sons, Inc., New York.
20 Heyden, A., Bell, A.T., and Keil, F.J. (2005) Efficient methods for finding transition states in chemical reactions: comparison of improved dimer method and partitioned rational function optimization method. *J. Chem. Phys.*, **123**, 224101.
21 Heyden, A., Bell, A.T., and Keil, F.J. (2005) Kinetic modeling of nitrous decomposition on Fe-ZSM-5 based on parameters obtained from first-principles calculations. *J. Catal.*, **233**, 26–35.
22 Brüggemann, T.C. and Keil, F.J. (2008) Theoretical investigation of the selective catalytic reduction of nitric oxide with ammonia of H-form zeolites. *J. Phys. Chem. C*, **112**, 17378–17387.
23 Brüggemann, T.C. and Keil, F.J. (2009) Theoretical investigation of the selective catalytic oxidation of ammonia of H-form zeolites. *J. Phys. Chem. C*, **113**, 13860–13876.
24 Brüggemann, T.C., Przybylski, M.-D., Balaji, S.P., and Keil, F.J. (2010) Theoretical investigation of the selective catalytic reduction of nitric dioxide with ammonia of H-form zeolites and the role of nitric and nitrous acids as intermediates. *J. Phys. Chem. C*, **114**, 6567–6587.
25 Masel, R.I. (1996) *Principles of Adsorption and Reaction on Solid Surfaces*, John Wiley & Sons, Inc., New York.
26 Masel, R.I. (2001) *Chemical Kinetics and Catalysis*, John Wiley & Sons, Inc., New York.

27 van Santen, R.A., van Leeuwen, P.W.N.M., Averill, B.A., and Moulijn, J.A. (1999) *Catalysis: An Integrated Approach*, 2nd edn, vol., 123, Studies in Surface Science and Catalysis, Elsevier, Amsterdam.
28 Groß, A. (2003) *Theoretical Surface Science: A Microscopic Perspective*, Springer, New York.
29 Rudzinski, W., Nieszporek, K., Moon, H., and Rhee, H.-K. (1994) Fundamentals of mixed-gas adsorption on heterogeneous solid surfaces. *Hetero. Chem. Rev.*, **1**, 275–308.
30 Mezaki, R. and Inoue, H. (1991) *Rate Equations of Solid-Catalyzed Reactions*, University of Tokyo Press, Tokyo.
31 Froment, G.F. and Bischoff, K.B. (1990) *Chemical Reactor Analysis and Design*, 2nd edn, John Wiley & Sons, Inc., New York.
32 Chorkendorff, I. and Niemantsverdriet, J.W. (2003) *Concepts of Modern Catalysis and Kinetics*, John Wiley & Sons, Inc., New York.
33 Nilsson, A., Petterson, L.G.M., and Norskov, J. (eds) (2007) *Chemical Bonding at Surfaces and Interfaces*, Elsevier, Amsterdam.
34 Lanzani, G., Martinazzo, R., Materzanini, G., Pini, J., and Tantardini, G.F. (2007) Chemistry at surfaces: from *ab initio* to quantum dynamics. *Theor. Chem. Acc.*, **117**, 805–825.
35 Ertl, G., Knoezinger, H., Schüth, F., and Weitkamp, J. (2008) *Handbook of Heterogeneous Catalysis*, 2nd edn, Wiley-VCH Verlag GmbH, Weinheim.
36 Somorjai, G.A. (2010) *Introduction to Surface Chemistry and Catalysis*, 2nd edn, John Wiley & Sons, Inc., New York.
37 Baker, J. (1986) An algorithm for the location of transition states. *J. Comput. Chem.*, **7**, 385–395.
38 Banerjee, A., Adams, N., Simons, J., and Shepard, R. (1985) Search for stationary points on surfaces. *J. Phys. Chem.*, **89**, 52–57.
39 Henkelman, G., Uberuaga, B.P., and Jonsson, H. (2000) A climbing image nudge elastic band method for finding saddle points and minimum energy paths. *J. Chem. Phys.*, **113**, 9901–9904.
40 Heyden, A., Peters, B., Bell, A.T., and Keil, F.J. (2005) A comprehensive DFT study on nitrous oxide decomposition over Fe-ZSM-5. *J. Phys. Chem. B*, **109**, 1857–1873.
41 Peters, B., Heyden, A., Bell, A.T., and Chakraborty, A. (2004) A growing string method for determining transition states: comparison to the nudged elastic band and string methods. *J. Chem. Phys.*, **120**, 7877–7886.
42 Sheppard, D., Terrell, R., and Henkelman, G. (2008) Optimization methods for finding minimum energy paths. *J. Chem. Phys.*, **128**, 134106.
43 Truhlar, D.G., Garrett, B.C., and Klippenstein, S.J. (1996) Current Status of Transition-State Theory. *J. Phys. Chem.*, **100**, 12771–12800
44 Broadbelt, L.J. and Snurr, R.Q. (2000) Applications of molecular modeling in heterogeneous catalysis research. *Appl. Catal. A*, **200**, 23–46.
45 Kjelstrup, S., Bedeaux, D., and Johannessen, E. (2006) *Elements of Irreversible Thermodynamics for Engineers*, 2nd edn, Tapir Academic Press, Trondheim.
46 Kohn, W. and Sham, L. (1965) Self-consistent equations including exchange and correlation effects. *Phys. Rev. A*, **14**, 1133–1138.
47 Parr, R.G. and Yang, W. (1989) *Density Functional Theory of Atoms and Molecules*, Oxford University Press, Oxford.
48 Kohanoff, J. (2006) *Electronic Structure Calculations for Solids and Molecules*, Cambridge University Press, Cambridge.
49 Martin, R. (2004) *Electronic Structure: Basic Theory and Practical Methods*, Cambridge University Press, Cambridge.
50 Kaxiras, E. (2003) *Atomic and Electronic Structure of Solids*, Cambridge University Press, Cambridge.
51 von Barth, U. (2004) Basic density functional theory: an overview. *Phys. Scr.*, **T109**, 9–39.
52 Langreth, D.C., Dion, M., Rydberg, H., Schröder, E., Hyldgaard, P., and Lundqvist, B.I. (2005) Van der Waals density functional theory with

applications. *Int. J. Quantum Chem.*, **101**, 599–610.

53 Thonhauser, T., Cooper, V.R., Li, S., Puzder, A., Hyldgaard, P., and Langreth, D.C. (2007) Van der Waals density functional: self-consistent potential and the nature of the van der Waals bond. *Phys. Rev. B*, **76**, 125112.

54 Helgaker, T., Jorgensen, P., and Olsen, J. (2002) *Molecular Electronic-Structure Theory*, John Wiley & Sons, Inc., New York.

55 Tuma, C. and Sauer, J. (2006) Treating dispersion effects in extended systems by hybrid MP2: DFT calculations – protonation of isobutene in zeolite ferrierite. *Phys. Chem. Chem. Phys.*, **8**, 3955–3965.

56 Hansen, N., Kerber, T., Sauer, J., Bell, A.T., and Keil, F.J. (2010) Quantum chemical modeling of benzene ethylation over H-ZSM-5 approaching chemical accuracy: a hybrid MP2:DFT study. *J. Am. Chem. Soc.*, **132**, 11525–11538.

57 Szabo, A. and Ostlund, N.S. (1996) *Modern Quantum Chemistry: An Introduction to Advanced Electronic Structure Theory*, Dover Publications, Mineola, NY.

58 Allen, M.P. and Tildesley, D.J. (1993) *Computer Simulation of Liquids*, Clarendon Press, Oxford.

59 Tuckerman, M.E. and Martyna, G.J. (2000) Understanding modern molecular dynamics: techniques and applications. *J. Phys. Chem. B*, **104**, 159–178.

60 Frenkel, D. and Smit, B. (2001) *Understanding Molecular Simulation*, 2nd edn, Academic Press, New York.

61 Leach, A. (2001) *Molecular Modelling*, Prentice hall, New York.

62 Rapaport, D.C. (2004) *The Art of Molecular Dynamics Simulations*, 2nd edn, Cambridge University Press, Cambridge.

63 Thijsen, J.M. (2007) *Computational Physics*, 2nd edn, Cambridge University Press, Cambridge.

64 Landau, D.P. and Binder, K. (2005) *A Guide to Monte Carlo Simulations in Statistical Physics*, 2nd edn, Cambridge University Press, Cambridge.

65 Raimondeau, S. and Vlachos, D.G. (2002) Recent developments on multiscale, hierarchical modeling of chemical reactors. *Chem. Eng. J.*, **90**, 3–23.

66 Vlachos, D.G. (2005) A review of multiscale analysis: examples from systems biology, materials engineering, and other fluid–surface interacting systems. *Adv. Chem. Eng.*, **30**, 1–61.

67 Berendsen, H.J.C. (2007) *Simulating the Physical World: Hierarchical Modeling from Quantum Mechanics to Fluid Dynamics*, Cambridge University Press, Cambridge.

68 Santiso, E. and Gubbins, K.E. (2004) Multi-scale molecular modeling of chemical reactivity. *Mol. Simul.*, **30**, 699–748.

69 Lerou, J.J. and Ng, K.M. (1996) Chemical reaction engineering: a multiscale approach to a multiobjective task. *Chem. Eng. Sci.*, **51**, 1595–1614.

70 Hansen, N., Krishna, R., van Baten, J.M., Bell, A.T., and Keil, F.J. (2010) Reactor simulation of benzene ethylation and ethane dehydrogenation catalyzed by ZSM-5: a multiscale approach. *Chem. Eng. Sci.*, **65**, 2472–2480.

71 Wheeler, A. (1951) Reaction rates and selectivity in catalyst pores. *Adv. Catal.*, **3**, 249–327.

72 Do, D.D. (1998) *Adsorption Analysis: Equilibria and Kinetics*, Imperial College Press, London.

73 Carman, P.C. (1937) Fluid flow through granular beds. *Trans. Inst. Chem. Eng.*, **17**, 150–166.

74 Johnson, M.F. and Stewart, W.E. (1965) Pore structure and gaseous diffusion in solid catalysts. *J. Catal.*, **4**, 248–252.

75 Wakao, N. and Smith, J.M. (1962) Diffusion in catalyst pellets. *Chem. Eng. Sci.*, **17**, 825–834.

76 Foster, R.N. and Butt, J.B. (1966) A computational model for the structure of porous materials employed in catalysis. *AIChE J.*, **12**, 180–185.

77 Mann, R. and Thomson, G. (1987) Deactivation of a supported zeolite catalyst: simulation of diffusion, reaction and coke deposition in a parallel bundle. *Chem. Eng. Sci.*, **42**, 555–563.

78 Epstein, N. (1989) On tortuosity and the tortuosity factor in flow and diffusion through porous media. *Chem. Eng. Sci.*, **44**, 777–779.

79 Sharratt, P.N. and Mann, R. (1987) Some observations on the variation of tortuosity with Thiele modulus and pore size distribution. *Chem. Eng. Sci.*, **42**, 1565–1576.

80 Wicke, E. and Kallenbach, R. (1941) Surface diffusion of carbon monoxide in active carbons. *Kolloidzeitschrift*, **97**, 135–151 (in German).

81 Feng, C. and Stewart, W.E. (1973) Practical models for isothermal diffusion and flow of gases in porous solids. *Ind. Eng. Chem. Fund.*, **12**, 143–147.

82 Wang, C.-T. and Smith, J.M. (1983) Tortuosity factors for diffusion in catalyst pellets. *AIChE J.*, **29**, 132–136.

83 Jackson, R. (1977) *Transport in Porous Catalysts*, Elsevier, Amsterdam.

84 Beeckmann, J.W. and Froment, G.F. (1980) Catalyst deactivation by site coverage and pore blockage. Finite rate of growth of the carbonaceous deposit. *Chem. Eng. Sci.*, **35**, 805–812.

85 Reyes, S. and Jensen, K.F. (1985) Estimation of effective transport coefficients in porous solids based on percolation concepts. *Chem. Eng. Sci.*, **40**, 1723–1734.

86 Stauffer, D. and Aharony, A. (1994) *Introduction to Percolation Theory*, 2nd edn (revised). Taylor & Francis, London.

87 Hoshen, J. and Kopelman, R. (1976) Percolation cluster distribution. I. Cluster multiple labeling technique and critical concentration algorithm. *Phys. Rev. B*, **24**, 3438–3445.

88 Sahimi, M., Gavalas, G.R., and Tsotsis, T.T. (1990) Statistical and continuum models of fluid solid reactions in porous media. *Chem. Eng. Sci.*, **45**, 1443–1502.

89 Sahimi, M. and Tsotsis, T.T. (2003) Molecular pore network models of nanoporous materials. *Physica B*, **338**, 291–297.

90 Sahimi, M. (1994) *Applications of Percolation Theory*, Taylor & Francis, London.

91 Fisher, M.E. (1971) The theory of critical point singularities, in *Critical Phenomena* (ed. M.S. Green) (Proceedings of the 1970 Enrico Fermi International School of Physics), Academic Press, New York, pp. 1–99.

92 Mohanty, K.K., Ottino, J.M., and Davis, H.T. (1982) Reaction and transport in disordered composite media: introduction of percolation concepts. *Chem. Eng. Sci.*, **37**, 905–924.

93 Voronoi, G. (1908) Novel applications of continuous parameters on the theory of quadratic forms. *J. Reine Angew Math.*, **134**, 198–287 (in French).

94 Winterfeld, P.H., Scriven, L.E., and Davis, H.T. (1981) Percolation and conductivity of random two-dimensional composites. *J. Phys. Chem.*, **14**, 2361–2376.

95 Okabe, A., Boots, B., and Sugihara, K. (1992) *Spatial Tessellations: Concepts and Applications of Voronoi Diagrams*, John Wiley & Sons, Inc., New York.

96 Reyes, S.C. and Iglesia, E. (1991) Effective diffusivities in catalyst pellets: new model porous structures and transport simulation techniques. *J. Catal.*, **129**, 457–472.

97 Drewry, H.P.G. and Seaton, N.A. (1995) Continuum random walk simulations of diffusion and reaction in catalyst particles. *AIChE J.*, **41**, 880–893.

98 Jerauld, G.R., Hatfield, J.C., Scriven, L.E., and Davis, H.T. (1984) Percolation and conduction on Voronoi and triangular networks: a case study in topological disorder. *J. Phys. C*, **17**, 1519–1529.

99 Hollewand, M.P. and Gladden, L.F. (1992) Modeling of diffusion and reaction in porous catalysts using a random three-dimensional network model. *Chem. Eng. Sci.*, **47**, 1761–1770.

100 Hollewand, M.P. and Gladden, L.F. (1992) Representation of porous catalysts using random pore networks. *Chem. Eng. Sci.*, **47**, 2757–2762.

101 Hollewand, M.P. and Gladden, L.F. (1995) Transport heterogeneity in porous pellets: I. PGSE NMR studies. *Chem. Eng. Sci.*, **50** (2), 309–326.

102 Hollewand, M.P. and Gladden, L.F. (1995) Transport heterogeneity in porous pellets: II. NMR imaging studies under transient and steady-state conditions. *Chem. Eng. Sci.*, **50** (2), 327–344.

103 Aris, R. (1975) *The Mathematical Theory of Diffusion and Reaction in Permeable*

Catalysts, vol. 2, Clarendon Press, Oxford.
104 Dullien, F.A.L. (1992) *Porous Media: Fluid Transport and Pore Structure*, 2nd edn, Academic Press, New York.
105 Torquato, S. (2002) *Random Heterogeneous Materials: Microstructure and Macroscopic Properties*, Springer, New York.
106 Mann, R. (1993) Development in chemical reaction engineering: issues relating to particle pore structures and porous materials. *Trans. IChem E (Part A)*, **71**, 551–562.
107 Sahimi, M. (2003) *Heterogeneous Materials. I. Linear Transport and Optical Properties*, Springer, New York.
108 Rolison, D.R. (2003) Catalytic nanoarchitectures: the importance of nothing and the unimportance of periodicity. *Science*, **299**, 1698–1701.
109 Keil, F.J. (1996) Modelling of phenomena within catalyst particles. *Chem. Eng. Sci.*, **51**, 1543–1567.
110 Keil, F.J. (1999) Diffusion and reaction in porous networks. *Catal. Today*, **53**, 245–258.
111 Keil, F. (1999) *Diffusion and Chemical Reactions in Gas/Solid Catalysis*, Springer, New York (in German).
112 Burganos, V.N. and Sotirchos, S.V. (1987) Diffusion in pore networks: effective medium theory and smooth field approximation. *AIChE J.*, **33**, 1678–1689.
113 Gregg, S.J. and Sing, K.S. (1982) *Adsorption, Surface Area and Porosity*, 2nd edn, Academic Press, London.
114 Zgrablich, G., Mendioroz, S., Daza, L., Pajares, J., Mayagoita, V., Rojas, F., and Conner, W.C. (1991) Effect of porous structure on the determination of pore size distribution by mercury porosimetry and nitrogen sorption. *Langmuir*, **7**, 779–785.
115 Conner, W.C., Christensen, S., Topsoe, H., Ferrero, M., and Pullen, A. (1994) The estimation of pore-network dimensions and structure: analysis of sorption and comparison with porosimetry. *Stud. Surf. Sci. Catal.*, **87**, 151–163.
116 Groen, J.C., Peffer, L.A.A., and Pérez-Ramirez, J. (2003) Pore size determination in modified micro- and macroporous materials. Pitfalls and limitations in gas adsorption data analysis. *Micropor. Mesopor. Mater.*, **60**, 1–17.
117 Seaton, N.A. (1991) Determination of the connectivity of porous solids from nitrogen sorption measurements. *Chem. Eng. Sci.*, **46**, 1895–1909.
118 Liu, H., Zhang, L., and Seaton, N.A. (1992) Determination of the connectivity of porous solids from nitrogen sorption measurements: II. Generalisation. *Chem. Eng. Sci.*, **47**, 4393–4404.
119 Naumov, S., Valiullin, R., Kärger, J., Pithcumani, R., and Coppens, M.-O. (2008) Tracing pore connectivity and architecture in nanostructured silica SBA-15. *Micropor. Mesopor. Mater.*, **110**, 37–40.
120 Calo, J.M. and Hall, P.J. (2004) The application of small angle scattering techniques to porosity characterization in carbons. *Carbon*, **42**, 1299–1304.
121 Adler, P.M. (1992) *Porous Media: Geometry and Transport*, Butterworth-Heinemann, Stoneham.
122 Yeong, C.L.Y. and Torquato, S. (1998) Reconstructing porous media. *Phys. Rev. E*, **57**, 495–506.
123 Yeong, C.L.Y. and Torquato, S. (1998) Reconstructing porous media. *Phys. Rev. E*, **58**, 224–233.
124 Meyerhoff, K. and Hesse, D. (1997) Determination of effective macropore diffusion coefficients by digital image processing. *Chem. Eng. Technol.*, **20**, 230–239.
125 Manwart, C. and Hilfer, R. (1999) Reconstruction of random media using Monte Carlo methods. *Phys. Rev. E*, **59**, 5596–5599.
126 Gelb, L.D. and Gubbins, K.E. (1998) Characterization of porous glasses: simulation models, adsorption isotherms, and the Brunauer–Emmettt–Teller analysis method. *Langmuir*, **14**, 2097–2111.
127 Gelb, L.D. and Gubbins, K.E. (1999) Characterization of porous glasses: simulation models, adsorption isotherms, and the

Brunauer–Emmettt–Teller analysis method. *Langmuir*, **15**, 305–308.
128 Mandelbrot, B.B. (1983) *The Fractal Geometry of Nature*, 2nd edn, Freeman, San Francisco.
129 Avnir, D. (ed.) (1989) *The Fractal Approach to Heterogeneous Chemistry*, John Wiley & Sons, Ltd, Chichester.
130 Coppens, M.-O. and Froment, G.F. (1994) Diffusion and reaction in a fractal catalyst pore: III. Application to the simulation of vinyl acetate production. *Chem. Eng. Sci.*, **49**, 4897–4907.
131 Coppens, M.-O. and Froment, G.F. (1995) Diffusion and reaction in a fractal catalyst pore: I. Geometrical aspects. *Chem. Eng. Sci.*, **50**, 1013–1026.
132 Coppens, M.-O. and Froment, G.F. (1997) The effectiveness of mass fractal catalysts. *Fractals*, **5**, 493–505.
133 Coppens, M.-O. (1999) The effect of fractal surface roughness on diffusion and reaction in porous catalysts: from fundamentals to practical applications. *Catal. Today*, **53**, 225–243.
134 Coppens, M.-O. (2005) Scaling up and down in a nature inspired way. *Ind. Eng. Chem.*, **44**, 5011–5019.
135 Coppens, M.-O. and Wang, G. (2009) Optimal design of hierarchically structured porous catalysts, in *Design of Heterogeneous Catalysts* (ed. U. Ozkan), John Wiley & Sons, Inc., New York.
136 Gutfraind, R., Sheintuch, M., and Avnir, D. (1991) Fractal and multifractal analysis of the sensitivity of catalytic reactions to catalyst structure. *J. Chem. Phys.*, **95**, 6100.
137 Havlin, S. and Ben-Avraham, D. (2002) Diffusion in disordered media. *Adv. Phys.*, **51** (1), 187–292.
138 Sheintuch, M. (2001) Reaction engineering principles of processes catalyzed by fractal solids. *Catal. Rev. Sci. Eng.*, **43**, 233–289.
139 Malek, K. and Coppens, M.-O. (2001) Effects of surface roughness on self- and transport diffusion in porous media in the Knudsen regime. *Phys. Rev. Lett.*, **87**, 125505.
140 Feng, C.F., Kostrov, V.V., and Stewart, W.E. (1974) Multi-component diffusion of gases in porous solids: models and experiments. *Ind. Eng. Chem. Fund.*, **13**, 5–9.
141 Rieckmann, C. and Keil, F.J. (1999) Simulation and experiment of multicomponent diffusion and reaction in three-dimensional networks. *Chem. Eng. Sci.*, **54**, 3485–3493.
142 Reyes, S.C., Sinfelt, J.H., and DeMartin, G.J. (2000) Diffusion in porous solids: the parallel contribution of gas and surface diffusion processes in pores extending from the mesoporous region into the microporous region. *J. Phys. Chem. B*, **104**, 5750–5761.
143 Skoulidas, A.I. and Sholl, D.S. (2001) Direct tests of the Darken approximation for molecular diffusion in zeolites using equilibrium molecular dynamics. *J. Phys. Chem. B*, **105**, 3151–3154.
144 Snyder, M.A. and Vlachos, D.G. (2005) The role of molecular interactions and interfaces in diffusion: transport diffusivity and evaluation of the Darken approximation. *J. Chem. Phys.*, **123**, 184707.
145 Skoulidas, A.I. and Sholl, D.S. (2003) Molecular dynamics simulations of self-diffusivities, corrected diffusivities, and transport diffusivities of light gases in four silica zeolites to assess influences of pore shape and connectivity. *J. Phys. Chem. A*, **107**, 10132–10141.
146 Kärger, J. and Ruthven, D.M. (1992) *Diffusion in Zeolites and Other Microporous Solids*, John Wiley & Sons, Inc., New York.
147 Cunningham, R.E. and Williams, R.J.J. (1980) *Diffusion in Gases and Porous Media*, Plenum Press, New York.
148 Evans, R.B., Watson, G.M., and Mason, E.A. (1961) Gaseous diffusion in porous media at uniform pressure. *J. Chem. Phys.*, **35**, 2076–2083.
149 Evans, R.B., Watson, G.M., and Mason, E.A. (1962) Gaseous diffusion in porous media. II: effects of pressure gradients. *J. Chem. Phys.*, **36**, 1894–1902.
150 Mason, E.A. and Malinauskas, A.P. (1983) *Gas Transport in Porous Media: The Dusty Gas Model*, Elsevier, Amsterdam.
151 Rothwell, L.B. (1963) Gaseous counterdiffusion in catalyst pellets. *AIChE J.*, **9**, 19–24.

152 Schneider, P. (1978) Multicomponent isothermal diffusion and forced flow of gases in capillaries. *Chem. Eng. Sci.*, **33**, 1311–1319.

153 Arnost, D. and Schneider, P. (1995) Dynamic transport of multicomponent mixtures of gases in porous solids. *Chem. Eng. J.*, **57**, 91–99.

154 Kerkhof, P.J.A.M. (1996) A modified Maxwell–Stefan model for transport through inert membranes: the binary friction model. *Chem. Eng. J.*, **64**, 319–343.

155 Kerkhof, P.J.A.M. and Geboers, M.A.M. (2005) Toward a unified theory of isotropic molecular transport phenomena. *AIChE J.*, **51**, 79–121.

156 Young, J.B. and Todd, B. (2005) Modelling of multi-component gas flows in capillaries and porous solids. *Int. J. Heat Mass Transf.*, **48**, 5338–5353.

157 Kjelstrup, S. and Bedeau, D. (2008) *Non-Equilibrium Thermodynamics of Heterogeneous Systems*, World Scientific, Singapore.

158 Düren, T., Keil, F.J., and Seaton, N.A. (2002) Molecular simulation of adsorption and transport diffusion of model fluids in carbon nanotubes. *Mol. Phys.*, **100**, 3741–3751.

159 Auerbach, S.M. (2000) Theory and simulation of jump dynamics, diffusion and phase equilibrium in nanopores. *Int. Rev. Phys. Chem.*, **19**, 155–198.

160 Keil, F.J., Krishna, R., and Coppens, M.-O. (2000) Modeling of diffusion in zeolites. *Rev. Chem. Eng.*, **16**, 71–197.

161 Sholl, D.S. (2006) Understanding macroscopic diffusion of adsorbed molecules in crystalline nanoporous materials via atomistic simulations. *Acc. Chem. Res.*, **39**, 403–411.

162 Dubbeldam, D. and Snurr, R. (2007) Recent developments in the molecular modeling of diffusion in nanoporous materials. *Mol. Simul.*, **33**, 305–324.

163 Smit, B. and Maesen, T. (2008) Molecular simulations of zeolites: adsorption, diffusion, and shape selectivity. *Chem. Rev.*, **108**, 4125–4184.

164 Dubbeldam, D., Beerdsen, E., Vlugt, T.J.H., and Smit, B. (2005) Molecular simulation of loading-dependent diffusion in nanoporous materials using extended dynamically corrected transition state theory. *J. Chem. Phys.*, **122**, 224712.

165 Beerdsen, E., Dubbeldam, D., and Smit, B. (2006) Understanding diffusion in nanoporous materials. *Phys. Rev. Lett.*, **96**, 044501.

166 Beerdsen, E., Dubbeldam, D., and Smit, B. (2006) Loading dependence of the diffusion coefficient of methane in nanoporous materials. *J. Phys. Chem. B*, **110**, 22754–22772.

167 Demontis, P. and Suffritti, G. (1997) Structure and dynamics of zeolites investigated by molecular dynamics. *Chem. Rev.*, **97**, 2845–2878.

168 Fritzsche, S., Wolfsberg, M., and Haberlandt, R. (2003) The importance of various degrees of freedom in the theoretical study of the diffusion of methane in silicalite-1. *Chem. Phys.*, **289**, 321–333.

169 Leroy, F., Rousseau, B., and Fuchs, A.H. (2004) Self-diffusion of *n*-alkanes in silicalite using molecular dynamics simulation: a comparison between rigid and flexible frameworks. *Phys. Chem. Chem. Phys.*, **6**, 775–783.

170 Zimmermann, N.E.R., Jakobtorweihen, S., Beerdsen, E., Smit, B., and Keil, F.J. (2007) In-depth study of the influence of host-framework flexibility on the diffusion of small gas molecules in one-dimensional zeolitic pore systems. *J. Phys. Chem. C*, **111**, 17370–17381.

171 Jakobtorweihen, S., Keil, F.J., and Smit, B. (2006) Temperature and size effects on diffusion in carbon nanotubes. *J. Phys. Chem. B*, **110**, 16332–16336.

172 Krishna, R. and van Baten, J.M. (2005) Diffusion of alkane mixtures in zeolites. Validating the Maxwell–Stefan formulation using MD simulations. *J. Phys. Chem. B*, **109**, 6386–6396.

173 Krishna, R. and van Baten, J.M. (2008) Onsager coefficients for binary mixture diffusion in nanopores. *Chem. Eng. Sci.*, **63**, 3120–3140.

174 Krishna, R. and van Baten, J.M. (2008) Insights into diffusion of gases in zeolites

gained from molecular dynamics simulation. *Micropor. Mesopor. Mater.*, **109**, 91–108.

175 Skoulidas, A.I., Sholl, D.S., and Krishna, R. (2003) Correlation effects in diffusion of CH_4/CF_4 mixtures in MFI zeolite. A study linking MD simulations with the Maxwell–Stefan formulation. *Langmuir*, **19**, 7977–7988.

176 Krishna, R., Paschek, D., and Baur, R. (2004) Modeling the occupancy dependence of diffusivities in zeolites. *Micropor. Mesopor. Mater.*, **76**, 233–246.

177 Rao, M.B. and Sircar, S. (1999) Thermodynamic consistency for binary gas adsorption equilibria. *Langmuir*, **15**, 7258–7267.

178 Krishna, R. and Baur, R. (2005) On the Langmuir–Hinshelwood formulation for zeolite catalysed reactions. *Chem. Eng. Sci.*, **60**, 1155–1166.

179 Ternel, B., Meskine, H., Reuter, K., Scheffler, M., and Metiu, H. (2007) Does phenomenological kinetics provide an adequate description of heterogeneous catalytic reactions? *J. Chem. Phys.*, **126**, 204711.

180 Corma, A., Llopis, F., Monton, J.B., and Weller, S. (1988) Comparison of models in heterogeneous catalysis for ideal and non-ideal surfaces. *Chem. Eng. Sci.*, **43**, 785–792.

181 Ostrovskii, V.E. (2004) Paradox of heterogeneous catalysis: paradox or regularity? *Ind. Eng. Chem. Res.*, **43**, 3113–3126.

182 Boudart, M. and Mariadassou, G.D. (1984) *Kinetics of Heterogeneous Catalytic Reactions*, Princeton University Press, Princeton, NJ.

183 Dumesic, J.A., Rudd, D.F., Aparacio, L.M., Rekoske, J.E., and Trevino, A.A. (1993) *The Microkinetics of Heterogeneous Catalysis*, American Chemical Society, Washington, DC.

184 Stoltze, P. (2000) Microkinetic simulation of catalytic reactions. *Prog. Surf. Sci.*, **65**, 65–150.

185 Garayhi, A. and Keil, F.J. (2001) Modeling of microkinetics in heterogeneous catalysis by means of frequency response techniques. *Chem. Eng. J.*, **82**, 329–346.

186 Keil, F.J. (2001) Development of microkinetic expressions by instationary methods. *Stud. Surf. Sci. Catal.*, **133**, 41–55.

187 Lynggaard, H., Andreasen, A., Stegelmann, C., and Stoltze, P. (2004) Analysis of simple kinetic models in heterogeneous catalysis. *Prog. Surf. Sci.*, **77**, 71–137.

188 Bligaard, T., Nørskov, J.K., Dahl, S., Matthiesen, J., Christensen, C.H., and Sehested, J. (2004) The Brønsted–Evans–Polanyi relation and the volcano curve in heterogeneous catalysis. *J. Catal.*, **224**, 206–217.

189 Lazo, C. and Keil, F.J. (2009) Thermodynamics from first principles: Phase diagram of oxygen adsorbed on Ni (111) and thermodynamic properties. *Phys. Rev. B*, **79**, paper 245418

190 Voter, A. (1997) A method for accelerating the molecular dynamics simulation of infrequent events. *J. Chem. Phys.*, **106**, 4665–4677.

191 Groot, R.D. and Warren, P.B. (1997) Dissipative particle dynamics: bridging the gap between atomistic and mesoscopic simulation. *J. Chem. Phys.*, **107**, 4423–4435.

192 Lowe, C.P. (1999) An alternative approach to dissipative particle dynamics. *Europhys. Lett.*, **47**, 145–151.

193 Damköhler, G. (1937) Influence of diffusion, fluid flow, and heat transport on the yield in chemical reactors, in *Der Chemieingenieur*, vol. 3 (Part 1) (eds A. Eucken and M. Jakob), Akademische Verlagsgesellschaft, Leipzig, pp. 359–485 (in German); English translation: *Int. Chem. Eng.* (1988), 28, 132–198.

194 Thiele, E.W. (1939) Relation between catalytic activity and size of particle. *Ind. Eng. Chem.*, **31**, 916–920.

195 Zeldovich, Ya. B. (1939) On the theory of reactions on powders and porous substances. *Acta Phys.*, **10**, 583–614.

196 Wicke, E. (1972) Physical phenomena in catalysis and in gas–solid surface reactions. *Adv. Chem. Ser.*, **109**, 183–208.

197 Schneider, P. (1975) Intraparticle diffusion in multicomponent catalytic

reactions. *Catal. Rev. Sci. Eng.*, **12**, 201–278.

198 Burghardt, A. (1986) Transport phenomena and chemical reactions in porous catalysts for multicomponent and multireaction systems. *Chem. Eng. Process.*, **21**, 229–244.

199 Luss, D. (1987) Diffusion–reaction interactions in catalyst pellets, in *Chemical Reaction and Reactor Engineering* (eds J.J. Carberry and A. Varma), Marcel Dekker, New York.

200 Rieckmann, C. and Keil, F.J. (1997) Multicomponent diffusion and reaction in three-dimensional networks: general kinetics. *Ind. Eng. Chem. Res.*, **36**, 3275–3281.

201 Wood, J., Gladden, L.F., and Keil, F.J. (2002) Modelling of diffusion and reaction accompanied by capillary condensation using three-dimensional pore networks. Part 2: dusty gas model and general reaction kinetics. *Chem. Eng. Sci.*, **57**, 3047–3059.

202 Deepak, P.D. and Bhatia, S.K. (1994) Transport in capillary network models of porous media: theory and simulation. *Chem. Eng. Sci.*, **49**, 245–257.

203 Sotirchos, S.V. and Burganos, V.N. (1988) Analysis of multicomponent diffusion in pore networks. *AIChE J.*, **34**, 1106–1118.

204 Bischoff, K.B. (1965) Effectiveness factors for general reaction forms. *AIChE J.*, **11**, 351–355.

205 Kulkarni, B.D., Jayaraman, V.K., and Doraiswamy, L.K. (1981) Effectiveness factors in bidispersed catalysts: the general nth order case. *Chem. Eng. Sci.*, **36**, 943–945.

206 Zhang, L. and Seaton, N.A. (1992) Prediction of the effective diffusivity in pore networks close to a percolation threshold. *AIChE J.*, **38**, 1816–1824.

207 Zhang, L. and Seaton, N.A. (1994) The application of continuum equations to diffusion and reaction in pore networks. *Chem. Eng. Sci.*, **49**, 41–50.

208 Weisz, P.B. and Hicks, J.S. (1962) The behaviour of porous catalyst particles in view of internal mass and heat diffusion effects. *Chem. Eng. Sci.*, **17**, 265–275.

209 Kaza, K.R., Villadsen, J., and Jackson, R. (1980) Intraparticle diffusion effects in the methanation reaction. *Chem. Eng. Sci.*, **35**, 17–24.

210 Keil, F.J., Hinderer, J., and Garayhi, A.R. (1999) Diffusion and reaction in ZSM-5 and composite catalysts for the methanol-to-olefines process. *Catal. Today*, **50**, 637–650.

211 Dogu, T. (1998) Diffusion and reaction in catalyst pellets with bidisperse pore size distribution. *Ind. Eng. Chem. Res.*, **37**, 2158–2171.

212 Goettmann, F. and Sanchez, C. (2007) How does confinement affect the catalytic activity of mesoporous materials. *J. Mater. Chem.*, **17**, 24–30.

213 Keil, F.J. and Coppens, M.-O. (2004) Dynamic Monte Carlo simulations of diffusion and reactions in zeolites, in *Computer Modeling of Microporous Materials* (eds C.R.A. Catlow, R.A. van Santen, and B. Smit), Elsevier, Amsterdam.

214 Turner, C.H., Brennan, J.K., Johnson, J.K., and Gubbins, K.E. (2002) Effect of confinement by porous materials on chemical reaction kinetics. *J. Chem. Phys.*, **116**, 2138–1248.

215 Turner, C.H., Brennan, J.K., Lisal, M., Smith, W.R., Johnson, J.K., and Gubbins, K.E. (2008) Simulation of chemical reaction equilibria by the reaction ensemble Monte Carlo method: a review. *Mol. Simul.*, **34**, 119–146.

216 Johnson, J.K., Panagiotopoulos, A.Z., and Gubbins, K.E. (1994) Reactive canonical Monte Carlo: a new simulation technique for reacting or associating fluids. *Mol. Phys.*, **81**, 717–733.

217 Smith, W.R. and Triska, B. (1994) The reaction ensemble method for the computer simulation of chemical and phase equilibria. *J. Chem. Phys.*, **100**, 3019–3027.

218 Hansen, N., Jakobtorweihen, S., and Keil, F.J. (2005) Reactive Monte Carlo and grand-canonical Monte Carlo simulations of the propene metathesis reaction system. *J. Chem. Phys.*, **122**, 164705.

219 Jakobtorweihen, S., Hansen, N., and Keil, F.J. (2006) Combining reactive and

configurational bias Monte Carlo: confinement influence on the propene metathesis reaction system in various zeolites. *J. Chem. Phys.*, **125**, 224709.
220 Smit, B. and Maesen, T.L.M. (2008) Towards a molecular understanding of shape selectivity. *Nature*, **451**, 671–678.
221 Gounaris, C.E., Wei, J., and Floudas, C.A. (2006) Rational design of shape selective separation and catalysis. *Chem. Eng. Sci.*, **61**, 7933–7962.
222 Hansen, N., Brüggemann, T., Bell, A.T., and Keil, F.J. (2008) Theoretical investigation of benzene alkylation with ethylene over H-ZSM-5. *J. Phys. Chem. C*, **112**, 15402–15411.
223 Hansen, N., Krishna, R., van Baten, J.M., Bell, A.T., and Keil, F.J. (2009) Analysis of diffusion limitation in the alkylation of benzene over H-ZSM-5 by combining quantum chemical calculations, molecular simulations, and continuum approach. *J. Phys. Chem. C*, **113**, 235–246.
224 Hegedus, L.L. and Pereira, C.J. (1990) Reaction engineering for catalyst design. *Chem. Eng. Sci.*, **45**, 2027–2044.
225 Hegedus, L.L. (1980) Catalyst pore structure by constrained nonlinear optimization. *Ind. Eng. Chem. Prod. Res. Dev.*, **19**, 533–537.
226 Keil, F.J. and Rieckmann, C. (1994) Optimization of three-dimensional catalyst pore structures. *Chem. Eng. Sci.*, **49**, 4811–4822.
227 El-Nafaty, U.A. and Mann, R. (1999) Support-pore architecture optimization in FCC catalyst particles using designed pore networks. *Chem. Eng. Sci.*, **54**, 3475–3484.
228 Prachayawarakorn, S. and Mann, R. (2007) Effects of pore assembly architecture on catalyst particle tortuosity and reaction effectiveness. *Catal. Today*, **128**, 88–99.
229 Wang, G., Johannessen, E., Kleijn, C.R., de Leeuw, S.W., and Coppens, M.-O. (2007) Optimizing transport in nanostructured catalysts: a computational study. *Chem. Eng. Sci.*, **62**, 5110–5116.
230 Wang, G. and Coppens, M.-O. (2008) Calculation of the optimal macropore size in nanoporous heterogeneous catalysts and its application to DeNO$_x$ catalysis. *Ind. Eng. Chem. Res.*, **37**, 3847–3855.
231 Wang, G., Groen, J.C., Yue, W., Zhou, W., and Coppens, M.-O. (2008b) Facile synthesis of ZSM-5 composites with hierarchical porosity. *J. Mater. Chem.*, **18**, 468–474.
232 Wang, G. and Coppens, M.-O. (2010) Rational design of hierarchically structured porous catalysts for autothermal reforming of methane. *Chem. Eng. Sci.*, **65**, 2344–2351
233 Gavriilidis, A., Varma, A., and Morbidelli, M. (1993) Optimal distribution of catalyst in pellets. *Catal. Rev. Sci. Eng.*, **35**, 399–456.
234 Davis, M.E. (2002) Ordered porous materials for emerging applications. *Nature*, **417**, 813–821.
235 Hartmann, S., Brandhuber, D., and Hüsing, N. (2007) Glycol-modified silanes: novel possibilities for the synthesis of hierarchically organized (hybrid) porous materials. *Acc. Chem. Res.*, **40**, 885–894.
236 Schüth, F. (2005) Engineered porous catalytic materials. *Annu. Rev. Mater. Res.*, **35**, 209–238.
237 Taguchi, A. and Schüth, F. (2005) Ordered mesoporous materials in catalysis. *Micropor. Mesopor. Mater.*, **77**, 1–45.
238 Yuan, Z.-Y. and Su, B.-L. (2006) Insights into hierarchically meso-macroporous structured materials. *J. Mater. Chem.*, **16**, 663–677.
239 Zeng, H.C. (2006) Synthetic architecture of interior space for inorganic nanostructures. *J. Mater. Chem.*, **16**, 649–662.
240 Zhao, X.S., Su, F., Yan, Q., Guo, W., Bao, X.Y., Lo, L., and Zhou, Z. (2006) Templating methods for preparation of porous structures. *J. Mater. Chem.*, **16**, 637–648.
241 Groen, J.C., Moulijn, J.A., and Perez-Ramirez, J. (2006) Desilication: on the controlled generation of mesoporosity in MFI zeolites. *J. Mater. Chem.*, **16**, 2121–2131.
242 Cejka, J. and Mintova, S. (2007) Perspectives of micro/mesoporous composites in catalysis. *Catal. Rev.*, **49**, 457–509.

243 Egeblad, K., Christensen, C.H., and Kustova, M. (2008) Templating mesoporous zeolites. *Chem. Mater.*, **20**, 946–960.

244 Tao, Y.S., Kanoh, H., Abrams, L., and Kaneko, K. (2006) Mesopore-modified zeolites: preparation, characterization, and applications. *Chem. Rev.*, **106**, 896–910.

245 Hoffmann, F., Cornelius, M., Morell, J., and Fröba, M. (2006) Silica-based mesoporous organic–inorganic hybrid materials. *Angew Chem., Int. Ed.*, **45**, 3216–3251.

246 Marquez-Alvarez, C., Zilkova, N., Perez-Pariente, J., and Cejka, J. (2008) Synthesis, characterization and catalytic applications of organized mesoporous aluminas. *Catal. Rev.*, **50**, 222–286.

247 Stair, P.C. (2008) Advanced synthesis for advancing heterogeneous catalysis. *J. Chem. Phys.*, **128**, 182507.

248 Centi, G. and Perathoner, S. (2008) Catalysis by layered materials: a review. *Micropor. Mesopor. Mater.*, **107**, 3–15.

249 Kitagawa, S., Kitaura, R., and Noro, S.-I. (2004) Functional porous coordination polymers. *Angew Chem., Int. Ed.*, **43**, 2334–2375.

250 van Donk, S., Janssen, A.H., Bitter, J.H., and de Jong, K.P. (2003) Generation, characterization, and impact of mesopores in zeolite catalysts. *Catal. Rev. Sci. Eng.*, **45**, 297–319.

251 Stein, A. (2003) Advances in microporous and mesoporous solids: highlights of recent progress. *Adv. Mater.*, **15**, 763–775.

252 Carberry, J.J. (1962) The micro–macro effectiveness factor for the reversible catalytic reaction. *AIChE J.*, **8**, 557–558.

253 Ertl, G. (2000) *Dynamics of Reactions at Surfaces* (eds B. Gates and H. Knoezinger), Academic Press, New York.

6
Modeling Porous Media Transport, Heterogeneous Thermal Chemistry, and Electrochemical Charge Transfer

Robert J. Kee and Huayang Zhu

6.1
Introduction

Electrochemical processes (e.g., fuel cells) depend upon the concerted interactions between fluid transport, ion and electron transport, thermal heterogeneous chemistry, and electrochemical charge transfer. Using solid oxide fuel cells (SOFC) as an example, this chapter presents and discusses the underpinning theory and computational solution.

Figure 6.1 illustrates an anode-supported tubular fuel cell having a gaseous fuel, such as synthesis gas (typically a mixture of H_2, CO, CH_4, H_2O, CO_2, and N_2), flowing in the tube and air flowing outside the tube. The heart of a fuel cell is the membrane electrode assembly, composed of an anode, a dense electrolyte, and a cathode. The most common SOFCs use a porous composite anode of nickel and yttria-stabilized zirconia (Ni–YSZ), a thin YSZ dense electrolyte, and a porous composite cathode of strontium-doped lanthanum manganite and YSZ (LSM–YSZ). The cells operate at elevated temperature around 800 °C.

During operation, gas-phase fuel species are transported through the anode pore spaces toward the dense electrolyte. The Ni serves three purposes. It acts as a heterogeneous catalyst to reform hydrocarbons (e.g., globally, $CH_4 + H_2O \rightleftharpoons CO + 3H_2$) and promotes water-gas shift chemistry (globally, $CO + H_2O \rightleftharpoons CO_2 + H_2$). It acts as an electrocatalyst that facilitates electrochemical charge transfer in regions near the dense electrolyte. The Ni is also the electronic conductor that carries electrons (products of the anodic charge transfer reactions) toward the anode current collection (shown in Figure 6.1 as an internal spiral wire). Charge transfer within the anode (globally, $H_2 + O^{2-} \rightleftharpoons H_2O + 2e^-$) takes place at the three-phase boundaries formed between electrode particles (Ni), electrolyte particles (YSZ), and the gas. The YSZ particles within the anode structure transport oxygen ions from the dense electrolyte and thus significantly expand the extent of the three-phase boundaries.

Modeling and Simulation of Heterogeneous Catalytic Reactions: From the Molecular Process to the Technical System,
First Edition. Edited by Olaf Deutschmann.
© 2012 Wiley-VCH Verlag GmbH & Co. KGaA. Published 2012 by Wiley-VCH Verlag GmbH & Co. KGaA.

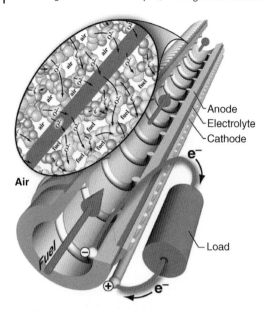

Figure 6.1 Illustration of a tubular anode-supported SOFC. The inset illustrates ion and electron transport at the particle (micron) scale.

The dense YSZ electrolyte membrane is a pure oxygen ion O^{2-} conductor, which has essentially no electronic conductivity. Oxygen ions are produced within the cathode by the electrochemical reduction of molecular oxygen (globally, $(1/2)O_2 + 2e^- \rightleftharpoons O^{2-}$). The electrons are supplied from the external load and the charge transfer occurs at the three-phase boundaries formed between LSM, YSZ, and air. The oxygen ions are conducted through the YSZ particles, across the dense YSZ electrolyte membrane, and into the composite anode structure, where they react electrochemically to oxidize the fuel. The electrons flow through the Ni from the anode to the external load.

Before proceeding, a few words are warranted to clarify the slightly alternative usage of the words anode, cathode, and electrolyte. The composite anode structure is a mixture of small particles of anode phase material (e.g., Ni) and electrolyte phase material (e.g., YSZ). In an "anode-supported" cell (e.g., Figure 6.1), the composite anode may be thick on the order of a millimeter. However, within this relatively thick porous structure, the anode and electrolyte particles are on the order of a micrometer in diameter. It is common to call the entire composite anode structure the anode. However, the fundamental charge transfer chemistry proceeds at the interfaces between individual particles of anode and electrolyte phases. Thus, depending upon the context, the word anode can have a different meaning. Analogous alternative nomenclature is relevant for the cathode and electrolyte particles that compose the composite cathode. Moreover, the current collection terminals are correctly called the anode and the cathode. The dense electrolyte that separates the composite anode and cathode structures is often simply called the electrolyte. However, as already noted, small electrolyte particles are essential in the composition of the composite electrodes.

Overall, a fuel cell uses the chemical potential that is available by oxidizing a fuel with air to produce electricity. The remainder of this chapter discusses the fundamental physical and chemical processes that are responsible for converting chemical potential energy to electricity.

6.2 Qualitative Illustration

Consider the highly idealized cell illustrated in Figure 6.2. The cathode is on the left and the anode is on the right. The cathode and the anode are both represented as two pure material, intersecting columns. One column is a pure electronic conductor (i.e., the electrode, such as a Ni anode or an LSM cathode) and the other column is a pure oxygen ion conductor (i.e., an electrolyte, such as YSZ). The columnar cathode and anode are separated by a pure oxygen ion-conducting electrolyte membrane. The cathode side is supplied with air and the anode side is supplied with a fuel such as hydrogen. Electrochemical charge transfer can proceed at the three-phase boundaries formed at the intersections between the ion-conducting and electron-conducting columns and the gas phase.

The SOFC illustrated in Figure 6.2 could operate, but very ineffectively. The reason is that the three-phase boundary (TPB) length is small compared to the characteristic dimensions of the electrode and electrolyte columns. Typical SOFC systems use intermixed submicron particles of electrode and electrolyte materials to dramatically increase TPB length (e.g., Figure 6.1). If the columns shown in Figure 6.2 were on the micron scale (or smaller) and millions of them arranged as a "forest" of intersecting columns, such a cell could potentially deliver very high performance. However, the

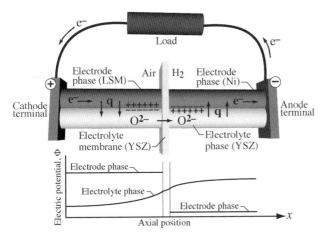

Figure 6.2 Idealized dual-column SOFC.

purpose here is to use the dual-column cell simply as the basis to illustrate aspects of SOFC electrochemistry and transport.

The cathode (positive terminal) is maintained at higher electric potential than the anode (the negative terminal). Because the electrode materials (e.g., Ni and LSM) are good electronic conductors, there is relatively little electric potential variation within the electrode columns. The electrolyte material (YSZ) is a pure oxygen ion conductor, but with relatively low ion conductivity. Therefore, as illustrated in the lower panel of Figure 6.2, relatively large electric potential gradients are needed to drive the ion flux. There is a very thin electrical double layer at the intersections between the electrode and electrolyte columns.

Along the axial length of the columns, electrochemical reactions transfer charge q between electrode and electrolyte. On the cathode side, charge is transferred from electrons in the electrode phase to oxygen ions in the electrolyte phase. The oxygen ions are transported through the dense electrolyte membrane to the anode side. The role of the anode is to transfer the charge associated with the oxygen ions to the electrons in the electrode. Note that in both cases, the negative charge is being transferred toward regions of more negative electric potential. Because the negative charge is naturally repelled by the negative electric potential, chemical energy is needed to overcome the barrier. Electrochemical charge transfer reactions accomplish this function. The chemical energy that is available from the oxidation of H_2 with O_2 to produce H_2O is used to produce electric current. As discussed in subsequent sections, the rates of charge transfer reactions depend upon the electric potential differences between phases.

The shapes of the electrolyte-phase electric potential profiles are the result of interactions between charge transport and charge transfer. Within the electrolyte membrane, the electric potential profile is linear, with the slope being directly proportional to the ion flux. Far from the electrolyte membrane, the charge transfer rates are low and the electrolyte electric potential profiles are nearly flat. Near the electrolyte membrane, the differences in electric potential between electrode and electrolyte are smaller than that when they are far from the electrolyte membrane. On the cathode side, the increasing electrolyte electric potential increases charge transfer rate. That is, transferring negatively charged electrons into the electrolyte, which is at a relatively more negative electric potential, is facilitated as the electrolyte electric potential becomes more positive. An analogous behavior takes place on the anode side, where the charge transfer rates are highest near the electrolyte membrane. A subsequent section provides a quantitative discussion of how the charge transfer rates depend on electric potential differences.

6.3
Gas-Phase Conservation Equations

Consider first the gas-phase transport and reaction processes within the porous electrodes. Assuming a continuum formulation, the gas-phase mass conservation equations within the porous electrodes can be written as

$$\frac{\partial\left(\phi_g \varrho Y_k\right)}{\partial t} + \nabla \cdot \mathbf{j}_k = \dot{s}_k W_k, \quad k = 1, \ldots, K_g, \tag{6.1}$$

where ϕ_g is the porosity, ϱ is the gas-phase mass density, Y_k and W_k are the gas-phase mass fractions and molecular weights for the K_g gas-phase species, respectively. Molar production rates of gas-phase species by heterogeneous chemistry (both thermal and electrochemical) are given as $\dot{s}_k$. The diffusive mass fluxes are represented as $\mathbf{j}_k$.

6.3.1
Gas-Phase Transport

The gas-phase species mass fluxes $\mathbf{j}_k$ through the porous structure due to the concentration and pressure gradients can be evaluated using the dusty gas model (DGM) [1, 2], which is expressed as an implicit relationship among the gas-phase species molar fluxes $\mathbf{J}_k$, molar concentrations $[X_k]$, concentration gradients, and the pressure p gradient as

$$\sum_{\ell \neq k} \frac{[X_\ell]\mathbf{J}_k - [X_k]\mathbf{J}_\ell}{[X_T] D_{k\ell}^e} + \frac{\mathbf{J}_k}{D_{k,\mathrm{Kn}}^e} = -\nabla[X_k] - \frac{[X_k]}{D_{k,\mathrm{Kn}}^e} \frac{B_g}{\mu} \nabla p, \tag{6.2}$$

where μ is the mixture viscosity and $[X_T] = p/RT$ is the total molar concentration. The mass fluxes $\mathbf{j}_k$ are related to the molar fluxes $\mathbf{J}_k$ as $\mathbf{j}_k = W_k \mathbf{J}_k$. $D_{k\ell}^e$ and $D_{k,\mathrm{Kn}}^e$ are the effective ordinary and Knudsen diffusion coefficients, respectively. Knudsen diffusion represents mass transport assisted by gas–wall collisions. The Knudsen diffusion coefficients depend upon the porous media microstructure, including porosity, average pore radius r_p, and tortuosity τ_g. The effective binary and Knudsen diffusion coefficients $D_{k\ell}^e$ and $D_{k,\mathrm{Kn}}^e$ can be evaluated as

$$D_{k\ell}^e = \frac{\phi_g}{\tau_g} D_{k\ell}, \quad D_{k,\mathrm{Kn}}^e = \frac{2}{3} \frac{r_p \phi_g}{\tau_g} \sqrt{\frac{8RT}{\pi W_k}}. \tag{6.3}$$

The ordinary multicomponent diffusion coefficients $D_{k\ell}$ and the mixture viscosities μ are determined from kinetic theory [3]. The permeability B_g can be evaluated from the Kozeny–Carman relationship as

$$B_g = \frac{\phi_g^3 d_p^2}{72 \tau_g (1-\phi_g)^2}, \tag{6.4}$$

where d_p is the particle diameter. Details about the computational implementation of the DGM are reported by Zhu et al. [2].

6.3.2
Chemical Reaction Rates

The molar production rates of gas-phase species due to heterogeneous thermal catalytic chemistry and electrochemical charge transfer reactions are represented by $\dot{s}_k$. Because the pore size within an electrode is usually comparable to the molecular mean free path length, there is relatively little probability for gas–gas collisions. Thus, the homogeneous gas-phase kinetics is usually negligible. However, if gas-phase chemistry is included, the right-hand side of Eq. (6.1) is extended as $(\dot{s}_k + \phi_g \dot{\omega}_k) W_k$, where $\dot{\omega}_k$ is the molar production rate of gas-phase species k via homogeneous chemistry. The production rates $\dot{s}_k$ are functions of temperature, gas composition, surface species coverages, and electric potential differences between the electrode and electrolyte phases. In addition to gas-phase species, the chemistry also depends on surface-adsorbed species. The temporal variations in site coverages θ_k of surface-adsorbed species are represented as

$$\frac{d\theta_k}{dt} = \frac{\dot{s}_k}{\Gamma}, \quad k = 1, \ldots, K_s, \tag{6.5}$$

where Γ is the available site density and K_s is the number of surface-adsorbed species. At a steady state, the net production rates of surface species vanish (i.e., $\dot{s}_k = 0$).

6.3.3
Boundary Conditions

Boundary conditions are needed to solve Eq. (6.1) within the porous electrodes. At the interfaces between the porous electrode and the external gas flow, the gas-phase composition is assumed to be that within the fuel or within the airflow. The gas-phase species fluxes at the interfaces with the dense electrolyte membrane vanish (i.e., $\mathbf{j}_k \cdot \mathbf{n} = 0$) because the dense electrolyte is impervious.

It is sometimes reasonable to assume that electrochemical charge transfer processes are limited to a vanishingly thin region at the electrode–electrolyte interface. If the charge transfer chemistry is assumed to be confined to the dense electrolyte interface, then the gas-phase species boundary condition can be written as a flux boundary condition that depends upon the charge transfer rate. That is, the species fluxes at the interfaces with the dense electrolyte membrane may be written as

$$\mathbf{j}_k \cdot \mathbf{n} = -\frac{W_k \nu_k}{n_e F} i_e, \tag{6.6}$$

where i_e is the current density, F is the Faraday's constant, ν_k are the stoichiometric coefficients in a global charge transfer reaction, and n_e is the number of electrons transferred. For hydrogen oxidation at the anode–electrolyte interface (i.e., $H_2 + O^{2-} \leftrightarrow H_2O + 2e^-$), $\nu_{H_2} = -1$, $\nu_{H_2O} = 1$, and $n_e = 2$. For oxygen reduction at the cathode–electrolyte interface (i.e., $O_2 + 4e^- \leftrightarrow 2O^{2-}$), $\nu_{O_2} = 1$ and $n_e = 4$. Although widely used in the literature, this approximation is not used for the results presented in this chapter.

6.4
Ion and Electron Transport

Charged species transport within electrode and electrolyte phases is an essential aspect of electrochemical processes. Generally speaking, molar flux of species $\mathbf{J}_k$

depends upon gradients in electrochemical potential $\tilde{\mu}_k$ as

$$\mathbf{J}_k = -\frac{\sigma_k}{(z_k F)^2}\nabla\tilde{\mu}_k = -\frac{\sigma_k}{(z_k F)^2}(\nabla\mu_k + z_k F\nabla\Phi). \tag{6.7}$$

In this equation, σ_k is a conductivity, z_k is the charge of species k, μ_k is the chemical potential, and Φ is the electric potential.

Consider the transport of oxygen ions O^{2-} within a YSZ electrolyte:

$$\mathbf{J}_{O^{2-}} = -\frac{\sigma_{O^{2-}}}{4F^2}\nabla\tilde{\mu}_{O^{2-}} = -\frac{\sigma_{O^{2-}}}{4F^2}(\nabla\mu_{O^{2-}} - 2F\nabla\Phi), \tag{6.8}$$

where $\sigma_{O^{2-}}$ is the oxygen ion conductivity in YSZ. Because of the high yttrium doping levels in a material such as YSZ, the chemical potential is essentially constant except within the *extremely thin* double layers at electrode–electrolyte interfaces. Thus, within the bulk of the electrolyte material (including porous structures), the ion flux depends primarily upon the electric potential gradient (i.e., the electric field within the electrolyte) as

$$\mathbf{J}_{O^{2-}} = \frac{\sigma_{O^{2-}}}{2F}\nabla\Phi. \tag{6.9}$$

Note that the negatively charged oxygen ion flux proceeds in the direction of positive electric field (i.e., from relatively low electric potential to relatively high electric potential).

Although representing ion flux as in Eq. (6.9) is appropriate for YSZ, flux transport in the form of Eq. (6.8) is more general. In other electrolyte materials (e.g., liquid organic electrolytes such as Li ion conductors in Li ion batteries), the ion conduction is driven primarily by chemical potential gradients. However, the subsequent discussion in this chapter is primarily concerned with the performance of composite electrodes such as Ni–YSZ.

A material such as YSZ has virtually no electronic conductivity. Nickel, of course, is an electron conductor, with virtually no oxygen ion conductivity. The chemical potential gradient of an electron in a metal conductor vanishes, so

$$\mathbf{J}_{e^-} = \frac{\sigma_{e^-}}{F}\nabla\Phi. \tag{6.10}$$

The electric *current* (A/cm^2), by usual definition, has the opposite sign, that is, current is proportional to the negative electric potential gradient. Thus,

$$\mathbf{i}_e = -\sigma_{e^-}\nabla\Phi. \tag{6.11}$$

A composite electrode such as Ni–YSZ conducts electrons in the Ni phase and oxygen ions in the YSZ phase. The electric potential fields within the two phases are usually different. At interfaces between the phases (i.e., between Ni and YSZ particles), charge transfer chemistry can proceed. That is, the charge associated with an O^{2-} in the YSZ can be transferred to produce an electron in the Ni. As discussed subsequently, the charge transfer process can be the result of multiple elementary reaction steps. The charge transfer rate depends upon the difference in electric potential between the phases.

In a composite cathode such as LSM–YSZ, the LSM acts primarily as an electronic conductor. Thus, the model is analogous to the Ni–YSZ. However, perovskite materials such as LSM typically also have some ion conductivity. Some doped perovskites, such as LSC, $(La_xSr_{1-x})CoO_{\delta-3}$, conduct both electrons and ions. Charge transport and charge transfer in such mixed ionic electronic conductors (MIECs) are the subject of much current research, but are not discussed further in this chapter.

6.5
Charge Conservation

Because cell performance depends strongly upon electric potential variations, the electrode and electrolyte electric potentials must be determined throughout the cell. Charge transport and charge transfer both depend upon spatial electric potential variations within the electron-conducting phase (e.g., the Ni phase in the anode and LSM in the cathode) and ion-conducting phase (e.g., the ceramic YSZ in the anode, cathode, and electrolyte). Quantitative models of these processes depend upon evaluating spatial variations in electric potentials. In a membrane–electrode assembly such as Ni–YSZ|YSZ|LSM–YSZ, three electric potential fields are relevant. They are the electric potential for the electron-conducting phase within the composite anode Φ_a, the electric potential for the electron-conducting phase within the cathode Φ_c, and the electric potential for the electrolyte phase Φ_e, which spans the anode, dense electrolyte, and cathode. The three electric potentials are governed by charge conservation equations. For steady-state operating conditions, the charge conservation equations may be stated as

$$\nabla \cdot (\sigma_e^e \nabla \Phi_e) = \begin{cases} \dot{s}_{a,e}, & \text{within anode,} \\ 0, & \text{within electrolyte,} \\ \dot{s}_{c,e}, & \text{within cathode,} \end{cases} \qquad (6.12)$$

$$\nabla \cdot (\sigma_a^e \nabla \Phi_a) = -\dot{s}_{a,e}, \quad \text{within anode,} \qquad (6.13)$$

$$\nabla \cdot (\sigma_c^e \nabla \Phi_c) = -\dot{s}_{c,e}, \quad \text{within cathode.} \qquad (6.14)$$

In these equations, σ_a^e and σ_c^e are the effective conductivities of the electron-conducting phases within the composite anode and cathode, respectively. The effective ion conductivity for the ion-conducting phase is σ_e^e. As discussed subsequently, the effective properties in a porous electrode structure (e.g., Figure 6.1) are quite different from the properties of the pure bulk materials. In Eq. (6.12), σ_e^e will have different values in the anode, cathode, and dense electrolyte. Within the dense electrolyte, the ion conductivity is that of the pure bulk material (e.g., YSZ).

Electrochemical transfer of charge between phases affects the electric potentials within the phases. In the continuum conservation equations, the net charge transfer rates are represented as source terms on the right-hand sides of Eqs (6.12)–(6.14). The nomenclature $\dot{s}_{m,e}$ is used where the phase m may be anode "a" or cathode "c." Charge transfer reactions usually involve both charged and uncharged species. The subscript "e" in $\dot{s}_{m,e}$ indicates the transfer of electrons, which are the only rates that are directly relevant to the charge conservation equations. A positive value of $\dot{s}_{m,e}$ indicates the transfer of negative charge into phase m, which tends to decrease the electric potential of the phase. Evaluation of $\dot{s}_{m,e}$ in terms of faradic electrochemical reactions is discussed subsequently. Note that the charge transfer source terms appear in the charge conservation equations for both participating phases, but with opposite signs. That is, all the charge that leaves one phase enters the partner phase.

6.5.1
Effective Properties

As can be anticipated from the inset in Figure 6.1, the effective conductivities of the porous composite electrodes can be significantly different from the pure material properties. Experimental observations with high-resolution microscopy provide clues to the details of particular composite structures [4]. Generally speaking, the ion conductor, the electron conductor, and the pores each occupy approximately one-third of the composite structure. Thus, on a simple geometric basis, one expects that the effective conductivity is no greater than one-third of the pure material conductivity. Because of necking between particles and possibly material inhomogeneities such as grain boundaries, the effective conductivities are usually even lower. Percolation theory provides an approach to estimate the effective properties of the composite electrodes [5].

In addition to effective conductivities, other effective properties of the composite electrodes are needed. For example, the effective area (per unit volume) of Ni is needed to evaluate catalytic chemistry. The three-phase boundary length (per unit volume) is needed to evaluate the electrochemical charge transfer rates.

6.5.2
Boundary Conditions

Boundary conditions are required to solve Eqs (6.12)–(6.14). At the interface between the cathode and the air compartment, the ionic fluxes vanish (i.e., $\mathbf{n} \cdot \sigma_e^e \nabla \Phi_e = 0$) and the cathode electric potential Φ_c can be set to be a reference electric potential as Φ_c^0. At the interface between the anode and the fuel compartment, the ionic fluxes vanish (i.e., $\mathbf{n} \cdot \sigma_e^e \nabla \Phi_e = 0$) and the anode electric potential Φ_a can be specified as Φ_a^0. The operating cell potential is $E_{\text{cell}} = \Phi_c^0 - \Phi_a^0$. The electron flux at the interface between the composite electrode structures and the dense electrolyte must vanish (i.e., $\mathbf{n} \cdot \sigma_a^e \nabla \Phi_a = 0$, and $\mathbf{n} \cdot \sigma_c^e \nabla \Phi_c = 0$) because the dense electrolyte membrane is assumed to be a purely ionic conductor.

6.5.3
Current Density and Cell Potential

The current density i_e (i.e., the charge flux) through the MEA structure can be evaluated as the ion flux through the dense electrolyte membrane. If the charge transfer processes are assumed to occur spatially throughout the composite electrodes, i_e may also be evaluated in terms of the charge transfer rates within the anode or the cathode structures as

$$i_e = \int_{x_{AF}}^{x_{EA}} \dot{s}_{a,e} dx = \int_{x_{EC}}^{x_{OC}} \dot{s}_{c,e} dx, \qquad (6.15)$$

where x_{AF} is the spatial location of the interface between the anode and the fuel compartment, x_{EA} is the interface between the electrolyte and the anode, x_{EC} is the interface between the electrolyte and the cathode, and x_{OC} is the interface between the cathode and the oxidizer (e.g., air) compartment.

6.6
Thermal Energy

There are important thermal consequences associated with charge transport and electrochemical charge transfer, with inefficiencies manifested as heat sources. For example, if a fuel cell has a conversion efficiency of 60%, then 40% of the chemical energy made available by oxidizing the fuel is realized as heat. In a practical SOFC operating at around 800 °C, it is the heat associated with these inefficiencies that maintains the operating temperature.

Ohmic heating is caused primarily by ion transport through the electrolyte materials. Generally speaking, chemical reactions can be endothermic (consuming heat) or exothermic (producing heat). Catalytic reforming chemistry, such as endothermic steam reforming of methane within composite anodes, contributes to cell cooling. Catalytic partial oxidation, as occurring when some oxygen is introduced into a fuel stream, contributes to cell heating. Charge transfer reactions are usually exothermic, which contributes to cell heating. Thus, all reactions, including the electrochemical charge transfer reactions, contribute to the energy balance. The effective resistance of charge transfer processes has been discussed by Zhu and Kee [6].

Despite the practical importance of thermal energy balances in practical fuel cell systems, the subject is not discussed further in this chapter. Instead, this chapter focuses upon chemical kinetics and transport. Interested readers may refer to literature that discusses thermal behaviors in more detail [7–10].

6.7
Chemical Kinetics

Electrochemical cell performance depends upon coupled interactions between thermal heterogeneous chemical kinetics and electrochemical charge transfer kinetics. The following sections discuss the formulation of chemical kinetics models and evaluation of reaction rates.

6.7.1
Thermal Heterogeneous Kinetics

Thermal chemistry plays at least two important roles. The first concerns the reforming or partial oxidation of hydrocarbons to produce H_2 and CO, which can then participate in electrochemical charge transfer reactions. This process usually proceeds heterogeneously on a catalyst surface such as Ni. The second concerns the direct role of heterogeneous chemistry in the charge transfer process. Surface adsorbates on both the electrode (e.g., Ni) and electrolyte (e.g., YSZ) surfaces can participate in thermal reactions and charge transfer reactions.

SOFC systems that use higher hydrocarbon fuels (e.g., diesel fuel) usually require an upstream steam reformer or a catalytic partial oxidation reactor to produce the syngas that is supplied to the fuel cell. However, it is often beneficial to retain some small hydrocarbons (particularly CH_4) in the reformate.

Within the porous composite anode structure, small hydrocarbons are further reformed to syngas, with the necessary steam being supplied as a product of the charge transfer chemistry. Internal reforming increases overall efficiency because the heat needed for reforming is available directly as a result of ohmic heating and other polarization losses. By retaining some CH_4 in a syngas fuel stream, endothermic steam reforming also assists the thermal management of the fuel cell stack.

Deutschmann and coworkers have developed an elementary heterogeneous reaction mechanism that describes Ni-based reforming kinetics within SOFC anodes. Although details are documented elsewhere [11, 12], Table 6.1 is included to emphasize the fundamental nature of the reaction chemistry. This mechanism, which involves 21 reversible elementary reactions among 6 gas-phase and 12 surface-adsorbed species, has been validated experimentally for both steam and dry (CO_2) reforming within Ni–YSZ anodes at 800 °C [11]. Because the elementary reaction mechanism describes all combinations of steam reforming, dry reforming, partial oxidation, and autothermal reforming, it is far more general than the global steam-reforming mechanisms that are often used in SOFC modeling. At least for methane at temperatures up to 800 °C, gas-phase chemistry within the channels and the anode pore space is usually assumed to be negligible [13].

Because fundamental formulations for heterogeneous thermal chemistry and approaches for evaluating reaction rates are well documented [3, 14], the underpinning theory is not discussed in this chapter. However, Section 6.7.2 discusses the formulation of electrochemical charge transfer chemistry using elementary reaction mechanisms.

The formation of carbon deposits is an important consideration in anode design. Deposits may be in the form of graphitic carbon (coke), which is known to be catalyzed by Ni. Deposits may also be polyaromatic hydrocarbons (similar to soot), which are likely formed via gas-phase chemistry [15]. Understanding the kinetics of deposit formation is difficult, which remains a largely unsolved problem. However,

Table 6.1 Heterogeneous reaction mechanism for CH_4 reforming on a Ni-based catalyst.

$H_2(g) + 2(Ni) \leftrightarrows H(Ni) + H(Ni)$
$O_2(g) + (Ni) + (Ni) \leftrightarrows O(Ni) + O(Ni)$
$CH_4(g) + (Ni) \leftrightarrows CH_4(Ni)$
$H_2O(g) + (Ni) \leftrightarrows H_2O(Ni)$
$CO_2(g) + (Ni) \leftrightarrows CO_2(Ni)$
$CO(g) + (Ni) \leftrightarrows CO(Ni)$
$O(Ni) + H(Ni) \leftrightarrows OH(Ni) + (Ni)$
$OH(Ni) + H(Ni) \leftrightarrows H_2O(Ni) + (Ni)$
$OH(Ni) + OH(Ni) \leftrightarrows O(Ni) + H_2O(Ni)$
$O(Ni) + C(Ni) \leftrightarrows CO(Ni) + (Ni)$
$O(Ni) + CO(Ni) \leftrightarrows CO_2(Ni) + (Ni)$
$HCO(Ni) + (Ni) \leftrightarrows CO(Ni) + H(Ni)$
$HCO(Ni) + (Ni) \leftrightarrows O(Ni) + CH(Ni)$
$CH_4(Ni) + (Ni) \leftrightarrows CH_3(Ni) + H(Ni)$
$CH_3(Ni) + (Ni) \leftrightarrows CH_2(Ni) + H(Ni)$
$CH_2(Ni) + (Ni) \leftrightarrows CH(Ni) + H(Ni)$
$CH(Ni) + (Ni) \leftrightarrows C(Ni) + H(Ni)$
$O(Ni) + CH_4(Ni) \leftrightarrows CH_3(Ni) + OH(Ni)$
$O(Ni) + CH_3(Ni) \leftrightarrows CH_2(Ni) + OH(Ni)$
$O(Ni) + CH_2(Ni) \leftrightarrows CH(Ni) + OH(Ni)$
$O(Ni) + CH(Ni) \leftrightarrows C(Ni) + OH(Ni)$

---

equilibrium behavior provides a means to estimate propensity to form carbon deposits [16].

Figure 6.3 is a ternary diagram showing the region where equilibrium solid carbon is stable. The deposit region depends upon composition and temperature. Hydrocarbons and most oxygenated fuels (e.g., alcohols) lie well into the deposit region. Syngas mixtures (i.e., mixtures of H_2 and CO) generally fall in the deposit-free region at typical SOFC operating temperatures. As steam and CO_2 are added to hydrocarbon fuels, the mixture tends toward the deposit-free region. The deposit-free region generally increases with increasing temperature.

6.7.2
Charge Transfer Kinetics

In a composite electrode, such as a Ni–YSZ anode or a LSM–YSZ cathode, the overall charge transfer process proceeds via a complex interaction of thermal chemistry and electrochemistry. Figure 6.4 and Table 6.2 represent a hydrogen oxidation mechanism developed by Goodwin *et al.* for a Ni–YSZ electrode [17]. Although the mechanism considers only hydrogen electro-oxidation, it involves a

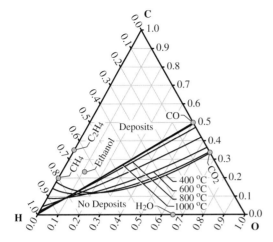

Figure 6.3 Ternary chart showing regions of equilibrium solid carbon formation as functions of temperature and composition.

relatively complex set of elementary reactions. This mechanism is used in this chapter as an example to facilitate the discussion of charge transfer chemistry. Elementary representations of charge transfer chemistry are the topic of significant current research. There are similar, but alternative, reaction mechanisms that are also under development [18–20]. However, the electrochemical formalism is essentially the same as is discussed below.

Full explanation of the mechanism in Table 6.2 is well beyond the scope of this chapter, but with details available in Ref. [17]. Nevertheless, a few comments on

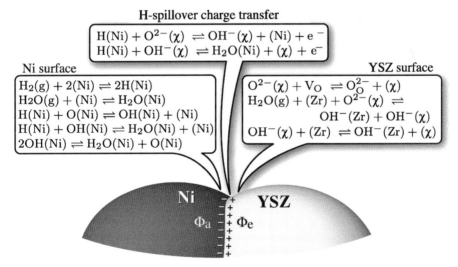

Figure 6.4 Chemical reactions on the electrode (Ni) and electrolyte (YSZ) surfaces and charge transfer reactions at the particle interface.

Table 6.2 Thermal and charge transfer chemistry in the vicinity of a Ni–YSZ three-phase boundary.

Thermal chemistry on Ni surface

$H_2(g) + 2(Ni) \leftrightarrows 2H(Ni)$
$H_2O(g) + (Ni) \leftrightarrows H_2O(Ni)$
$H(Ni) + O(Ni) \leftrightarrows OH(Ni) + (Ni)$
$H(Ni) + OH(Ni) \leftrightarrows H_2O(Ni) + (Ni)$
$2OH(Ni) \leftrightarrows H_2O(Ni) + O(Ni)$

Thermal chemistry on YSZ surface

$O_O^{2-} + (\chi) \leftrightarrows O^{2-}(\chi) + V_O$
$H_2O(g) + (Zr) + O^{2-}(\chi) \leftrightarrows OH^-(Zr) + OH^-(\chi)$
$OH^-(\chi) + (Zr) \leftrightarrows OH^-(Zr) + (\chi)$
$H_2(g) + (Zr) + O^{2-}(\chi) \leftrightarrows H^-(Zr) + OH^-(\chi)$

Charge transfer at the three-phase boundary

$H(Ni) + O^{2-}(\chi) \leftrightarrows OH^-(\chi) + (Ni) + e^-$
$H(Ni) + OH^-(\chi) \leftrightarrows H_2O(Ni) + (\chi) + e^-$

---

nomenclature are needed. The YSZ surface is composed of two site types, designated as Zr and χ. The site type is designated by the parentheses. For example, a hydroxyl ion on a χ site is written as $OH^-(\chi)$. An empty χ site is written as (χ). The gas-phase species are designated by (g). The site fractions of species (including the vacant site fraction) on each site type must sum to unity. The species V_O represents an oxygen vacancy in the bulk YSZ crystal lattice. The species O_O^{2-} represents a mobile oxygen ion within the bulk YSZ crystal lattice.

Electrochemical charge transfer reactions can generally be written as

$$\sum_{k=1}^{K} \nu'_{ki} \Lambda_k^{z_k} \rightleftharpoons \sum_{k=1}^{K} \nu''_{ki} \Lambda_k^{z_k}, \tag{6.16}$$

where $\Lambda_k^{z_k}$ is the chemical symbol of the kth species with charge z_k. The forward and backward stoichiometric coefficients for the kth species in the ith reaction are written as ν'_{ki} and ν''_{ki}, respectively.

Consider the following reversible reaction that is written in the anodic direction (i.e., producing electrons):

$$H(Ni) + OH^-(\chi) \rightleftharpoons H_2O(Ni) + (\chi) + e^-. \tag{6.17}$$

Two surface sites are involved: one is the Ni surface and the other is a χ site on the YSZ surface [17]. The anodic (forward) and cathodic (backward) stoichiometric coefficients are

$$\nu'_{H(Ni)} = 1, \quad \nu'_{OH^-(\chi)} = 1, \quad \nu''_{H_2O(Ni)} = 1, \quad \nu''_{e^-} = 1, \nu''_{(\chi)} = 1. \tag{6.18}$$

The H(Ni), H$_2$O(Ni), and (χ) are assumed not to carry a charge (i.e., $z = 0$) and the hydroxyl OH$^-$(χ) and the electron both carry a single negative charge (i.e., $z = -1$).

The net production rate for kth species resulting from the ith reaction $\dot{s}_{ki}$ can be represented in terms of the reaction rate of progress q_i as

$$\dot{s}_{ki} = \left(v''_{ki} - v'_{ki}\right) q_i = v_{ki} q_i, \tag{6.19}$$

where the net stoichiometric coefficients are defined as $v_{ki} \equiv v''_{ki} - v'_{ki}$. The rate of progress for each reaction can be evaluated as

$$q_i = k_{ai} \prod_{k=1}^{K} C_k^{v'_{ki}} - k_{ci} \prod_{k=1}^{K} C_k^{v''_{ki}}, \tag{6.20}$$

where C_k are the activities of the participating species. The activity of gas-phase species is its molar concentration $[X_k]$. The activity of surface-adsorbed species is the surface coverage $\Gamma_m \theta_{k,m}$, where $\theta_{k,m}$ is the site fraction for species k on the surface of site-type m and Γ_m is the total available surface site density for site-type m.

Charge transfer reactions are said to be anodic when producing electrons and cathodic when consuming electrons. By the convention adopted here, reactions are written such that the forward direction is anodic. The anodic (i.e., producing electrons) and cathodic (i.e., consuming electrons) rate expressions for ith reaction are written as

$$k_{ai} = k_{ai}^t \exp\left[-\beta_{ai} \frac{F}{RT} \sum_{k=1}^{K} v_{ki} z_k \Phi_k\right], \tag{6.21}$$

$$k_{ci} = k_{ci}^t \exp\left[+\beta_{ci} \frac{F}{RT} \sum_{k=1}^{K} v_{ki} z_k \Phi_k\right]. \tag{6.22}$$

The anodic and cathodic symmetry factors are β_{ai} and β_{ci}, respectively. For the elementary charge transfer reactions (i.e., those transferring a single electron), $\beta_{ai} + \beta_{ci} = 1$. The Faraday constant and the gas constant are represented as F and R, respectively, and T is the temperature.

Each species is assumed to have an associated charge z_k; charge-neutral species have $z_k = 0$. Each phase m (e.g., the Ni and YSZ in Eq. (6.17)) is assumed to be at an electric potential Φ_m. The rate expressions in Eqs (6.21) and (6.22) are written with the assumption that each charged species k is at the electric potential of the phase in which it exists. Thus, for convenience sake, in Eqs (6.21) and (6.22), the electric potential Φ_k is associated with the species and not directly with the phase. This also assists in the computational implementation.

The thermal component of the anodic (forward) rate expression is written in modified Arrhenius form as

$$k_{ai}^t = A_i T^{n_i} \exp\left(-\frac{E_i}{RT}\right), \tag{6.23}$$

where E_i is an activation energy, A_i is a pre-exponential factor, and n_i is a temperature exponent. To satisfy microscopic reversibility and maintain thermodynamic consistency, the thermal component of the cathodic (reverse) rate is related to the forward rate via the reaction's equilibrium constant K_i as

$$K_i = \frac{k_{ai}^t}{k_{ci}^t} = \exp\left(\frac{-\Delta G^0}{RT}\right), \tag{6.24}$$

where ΔG^0 is the change in standard-state free energy for the reaction. Evaluating ΔG^0, and hence the equilibrium constant, requires quantitative thermochemical properties for all species in the reaction.

When all species in a reaction are at the same electric potential, or when all participating species are charge neutral, the exponential factors in Eqs (6.21) and (6.22) become unity, and are thus irrelevant. In this case, the rate expressions revert to the purely thermal rate expressions.

Figure 6.5 represents potential energy surfaces that assist understanding the influence of electric potentials on charge transfer rates. The potential energy surface on the left represents the reactants and the one on the right represents the products. The electric potential difference between the electrode (here, the anode Ni) and the electrolyte (here, the YSZ) is written as $E_a = \Phi_a - \Phi_e$. The equilibrium electric potential difference E_a^{eq} is the electric potential difference at which the reaction proceeds at equal and opposite rates in the anodic and cathodic directions (illustrated as the dashed line). There is a potential energy barrier between the reactant and product states, which tends to be cusp-like for charge transfer reactions.

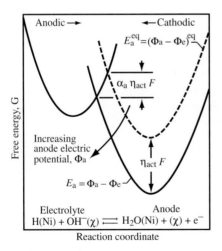

Figure 6.5 Potential energy surfaces to assist visualizing the effect of electric potential difference on charge transfer reaction rates.

When proceeding in the anodic (forward) direction, the charge transfer reaction illustrated in Figure 6.5 delivers electrons into the anode, which is at a lower electric potential than the electrolyte. As the electric potential of the anode is increased relative to the electrolyte (i.e., E_a increases), the barrier to the electron transfer is decreased. The negatively charged electron is naturally repelled from the negative electrode. Some of the chemical energy stored in the reactants is converted to electricity as electrons are delivered into the anode (i.e., the conduction band of the Ni). When the electric potential of the anode is increased (i.e., becomes less negative), the barrier to charge transfer decreases. The symmetry factors β are related to the magnitudes of the slopes of the potential energy surfaces at their crossing point. Because the slopes are typically similar, the symmetry factors for elementary reactions are usually near $\beta \approx 1/2$. More detailed discussion on formulating charge transfer rates can be found in Ref. [21] and in textbooks on electrochemistry [22, 23].

Consider the evaluation of E_a^{eq} for Eq. (6.17). For a reaction at equilibrium, $\sum v_k \tilde{\mu}_k = 0$. Thus,

$$\mu_{H(Ni)} + (\mu_{OH^-(\chi)} - F\Phi_\chi) - \mu_{H_2O(Ni)} - \mu_{(\chi)} + F\Phi_{Ni} = 0. \tag{6.25}$$

It follows directly that

$$E_a^{eq} = (\Phi_{Ni} - \Phi_\chi)^{eq} = \frac{1}{F}\left(\mu_{H_2O(Ni)} + \mu_{(\chi)} - \mu_{OH^-(\chi)} - \mu_{H(Ni)}\right). \tag{6.26}$$

Quantitatively evaluating E_a^{eq} requires evaluating the chemical potentials, which can be difficult for the surface adsorbates for which data are not readily available. Goodwin et al. [17] discuss the evaluation of thermodynamic properties for the reaction mechanism in Table 6.2.

The net species production rate $\dot{s}_k$ resulting from all the reactions I involving species k can be written as

$$\dot{s}_k = \sum_{i=1}^{I} \dot{s}_{ki} \Psi_{ik}, \tag{6.27}$$

where Ψ_{ik} are electrode microstructural factors that depend upon the reaction and the species involved [7]. Because the charge transfer reactions typically proceed at three-phase interfaces between electrode, electrolyte, and gas phases, the species production rates $\dot{s}_{ki}$ are usually written in terms of three-phase boundary length (mol/(cm s)). Such reactions usually involve species on different surface phases, electrons and ions within different phases, and possibly gas-phase species. For both gas-phase species and electrons, $\Psi_{ik} = \lambda_{TPB,i}^V$ is the three-phase boundary length per unit volume of porous electrode. For surface species, $\Psi_{ik} = \lambda_{TPB,i}^V/A_s$ is the ratio of the specific three-phase boundary length and the specific area of the surface on which the species exists. The total charge transfer rate per unit volume from all the electrochemical reactions can be represented as

$$\dot{s}_e = \sum_{i=1}^{I} \lambda_{\mathrm{TPB},i}^{\mathrm{V}} i_{e,i}, \tag{6.28}$$

where $i_{e,i} = n_{e,i} F q_i$ is the current density per unit length resulting from the ith electrochemical charge transfer reaction and $n_{e,i}$ is the number of electrons transferred.

6.7.3
Butler–Volmer Formulation

The foregoing discussion considers charge transfer chemistry in the context of elementary reactions. However, it is common to represent electrochemical charge transfer rates in the Butler–Volmer form. With some important assumptions concerning gas-phase equilibrium and a rate-limiting charge transfer step, the Butler–Volmer formulation can be derived from the elementary reactions.

The Butler–Volmer equation expresses the electrical current density i_e as

$$i_{\mathrm{BV},e} = i_0 \left[\exp\left(\frac{\alpha_a F \eta_{\mathrm{act}}}{RT} \right) - \exp\left(-\frac{\alpha_c F \eta_{\mathrm{act}}}{RT} \right) \right], \tag{6.29}$$

where i_0 is the exchange current density, α_a and α_c are anodic and cathodic symmetry factors, and η_{act} is the activation overpotential. The first term in the square brackets represents the anodic current density (producing electrons) and the second term represents the cathodic current density. At equilibrium electric potential, the chemistry proceeds at equal and opposite rates in the anodic and cathodic directions. The Butler–Volmer expression describes the net rate of a reversible reaction. The exchange current density represents the magnitude of the equal and opposite current densities at equilibrium electric potential. The value of i_0 depends upon the length of three-phase boundary, the effectiveness of electrocatalyst, the temperature, and the activities of the participating species. Because the Butler–Volmer formulation represents a global charge transfer process, the symmetry factors α are not the same as they would be for an elementary reaction β. Moreover, in general, $\alpha_a + \alpha_c \neq 1$ as would be the case for an elementary reaction.

The activation overpotential η_{act} is defined as

$$\eta_{\mathrm{act}} = E_a - E_a^{\mathrm{eq}} = (\Phi_a - \Phi_e) - (\Phi_a - \Phi_e)^{\mathrm{eq}}. \tag{6.30}$$

That is, $E_a = (\Phi_a - \Phi_e)$ is the electric potential difference between the electrode phase and the electrolyte phase, and $E_a^{\mathrm{eq}} = (\Phi_a - \Phi_e)^{\mathrm{eq}}$ is the electric potential difference that causes the charge transfer reaction to be equilibrated (i.e., proceeding in the anodic and cathodic directions at equal and opposite rates). When the anode electric potential Φ_a is increased relative to the adjoining electrolyte electric potential Φ_e, the activation overpotential η_{act} is increased by the same amount. As illustrated in Figure 6.5, the product-side potential energy surface is lowered by $\eta_{\mathrm{act}} F$ and the anodic energy barrier is lowered by $\alpha_a \eta_{\mathrm{act}} F$.

The preceding paragraph is written in the context of a reaction that is proceeding in the anodic direction (i.e., producing electrons). In this case, the activation overpotential is positive and the first term in Eq. (6.29) dominates. Considering the oxygen reduction at the cathode (globally, $(1/2)O_2 + 2e^- \rightleftharpoons O^{2-}$), the reaction proceeds in the cathodic direction (i.e., consuming electrons). In this case, η_{act} has a negative value and the second term in Eq. (6.29) dominates.

Assuming that H_2 is the fuel, Zhu et al. [2, 7] derived expressions for the net oxidation exchange current density as

$$i_{0,H_2} = i^*_{H_2} \frac{\left(p_{H_2}/p^*_{H_2}\right)^{(1-\alpha_a/2)} (p_{H_2O})^{\alpha_a/2}}{1 + \left(p_{H_2}/p^*_{H_2}\right)^{1/2}}, \qquad (6.31)$$

where p_k are the gas-phase partial pressures (measured in atmospheres). The parameter $p^*_{H_2}$ depends upon hydrogen adsorption/desorption rates and the parameter $i^*_{H_2}$ is assigned empirically to fit measured polarization data. Goodwin et al. [17] provide further validation that the modified Butler–Volmer model for the electrochemical oxidation of hydrogen provides a reasonable representation of charge transfer experiments. Zhu et al. [2, 7] also provide an expression for the cathode exchange current density as

$$i_{0,O_2} = i^*_{O_2} \frac{\left(p_{O_2}/p^*_{O_2}\right)^{\alpha_a/2}}{1 + \left(p_{O_2}/p^*_{O_2}\right)^{1/2}}. \qquad (6.32)$$

When casting a model in terms of the Butler–Volmer equation, the cell operating voltage E_{cell} is written in terms of the reversible (Nernst) potential E_{rev} and the summation of relevant overpotentials (all of which depend upon the current density i_e),

$$E_{cell} = E_{rev} - [\eta_{act,a}(i_e) + \eta_{ohm}(i_e) + |\eta_{act,c}(i_e)|]. \qquad (6.33)$$

The cell potential $E_{cell} = \Phi_c - \Phi_a$ is the electric potential difference between the cathode current collector and the anode current collector. The reversible potential is evaluated from the gas-phase compositions at the interface between the anode and the dense electrolyte and the interface between the cathode and the dense electrolyte as

$$E_{rev} = E^{eq}_c - E^{eq}_a = -\frac{\Delta G^0}{2F} + \frac{RT}{2F} \ln \frac{p_{H_2,a} p^{1/2}_{O_2,c}}{p_{H_2O,a}}, \qquad (6.34)$$

where the standard Gibbs free energy change for the overall global reaction (i.e., $H_2 + (1/2)O_2 \rightleftharpoons H_2O$) is evaluated as

$$\Delta G^0 = \mu^0_{H_2O} - \mu^0_{H_2} - \frac{1}{2}\mu^0_{O_2}, \qquad (6.35)$$

where μ^0_k are the species standard-state chemical potentials. The partial pressures of the H_2 and H_2O are evaluated within the porous anode (fuel electrode) at the interface with the dense electrolyte and the O_2 partial pressure is evaluated within the porous cathode (air electrode) at the interface with the dense electrolyte.

The ohmic overpotential is modeled as

$$\eta_{\text{ohm}} = i_e R_{\text{tot}}, \qquad (6.36)$$

where R_{tot} is usually dominated by the ion resistance through the dense electrolyte. However, other electrical resistances offered by materials interfaces and the interconnect materials also contribute. The so-called transport overpotentials can also be evaluated, which provide a measure of polarization associated with concentration variations through the porous electrode structures. However, for models that represent the species transport throughout the porous electrodes, there is no need to use transport overpotentials.

6.7.4
Elementary and Butler–Volmer Formulations

The incorporation of open-circuit voltage (OCV) is handled quite differently in an elementary reaction setting from that in a Butler–Volmer formulation. When using microscopically reversible elementary reactions, the OCV is an outcome of the model, not an input [19, 20]. However, a consistent set of species thermodynamic properties is required such that the OCV, which is easily measured experimentally, is predicted correctly. In the Butler–Volmer setting, the OCV (i.e., the Nernst potential) is established on the basis of gas-phase thermodynamic properties alone (e.g., Eq. (6.34)). With certain simplifying assumptions concerning equilibrium, the thermodynamic properties of all surface and bulk species cancel out of the OCV calculation. When the needed assumptions are valid, calculating the reversible potential from the gas-phase species has advantages because the gas-phase thermodynamic properties are usually well known. However, there are limitations to predicting OCV from stable gas-phase species alone.

If the gas-phase is not in equilibrium (e.g., in the case of rate-limiting gas-phase chemical kinetics), then equations such as Eq. (6.34) cannot predict the reversible potential accurately. If there are competing charge transfer reactions, then it is difficult to establish an unambiguous reversible potential based upon gas-phase thermodynamics alone. For example, in a global sense, assume that anodic charge transfer can proceed as $CO + O^{2-} \rightleftharpoons CO_2 + 2e^-$, operating in parallel with hydrogen-based charge transfer. In this case, using the equivalent of Eq. (6.34) would predict a different OCV for the H_2 and CO charge transfer processes. However, in practice, there is no ambiguity about OCV: it is the measured voltage at open circuit. The elementary reaction approach suffers no inherent limitations for competing charge transfer pathways or nonequilibrated processes. However, it does require consistent thermodynamic properties for all species, including surface adsorbates and charged species, which are difficult to measure or otherwise establish.

6.8 Computational Algorithm

Because computational solution methods are not central to this chapter, only a brief summary is included here. However, more details about solution algorithms and software implementation are available [2, 6, 7].

Equations (6.1), (6.5) and (6.12)–(6.14) form a coupled nonlinear system of equations to compute profiles of gas-phase mole fractions X_k, surface coverages θ_k, and electric potentials Φ_m throughout the cell, which all depend upon transport and chemistry processes. At a steady state, the system of equations is an elliptic boundary value problem. After discretizing the spatial operators using a finite volume approach, the computational problem becomes one of solving a large system of nonlinear algebraic equations. Because of nonlinearity and stiffness, solution via a modified Newton iteration can be problematic [3]. Therefore, a hybrid time marching approach is used. Solving the physically transient problem is more stable and reliable, but usually more computationally intensive. After discretization of the spatial operators, the transient problem forms a system of differential algebraic equations [24]. The results shown in this chapter are the result of solving the transient model to a steady state using the Limex software [25].

6.9 Button Cell Example

Button cells, as illustrated in Figure 6.6, are used widely in the laboratory environment for developing and characterizing fuel cell membrane electrode assemblies [10, 26]. These small cells are usually operated within a furnace that maintains nearly uniform temperature.

The illustrative results here consider a high-performance LSM–YSZ|YSZ|Ni–YSZ button cell that is intended to represent one reported by Jiang and Virkar [26]. Table 6.3 shows physical dimensions and model parameters. The LSM–YSZ cathode is 50 μm thick, the YSZ electrolyte is 20 μm thick, and the Ni–YSZ anode is 550 μm thick. The nominal particle diameters for both the porous anode and cathode are 1 μm. The anode-specific three-phase boundary length is set to be 2.4×10^9 cm^{-2}, and the specific effective Ni catalyst area is set to be 1080 cm^{-1}. Evaluation of the effective conductivities within the composite electrodes is discussed in Ref. [7].

The anode thermal chemistry is represented using the reaction mechanism shown in Table 6.1. The reaction rate expressions and species thermodynamic properties, which are reported in earlier literature [11, 12], are not repeated here. The anode electrochemistry is represented by the elementary reaction mechanism shown in Table 6.2. The cathode electrochemistry is based upon elementary reaction steps, but is implemented here using a Butler–Volmer formulation [2, 7].

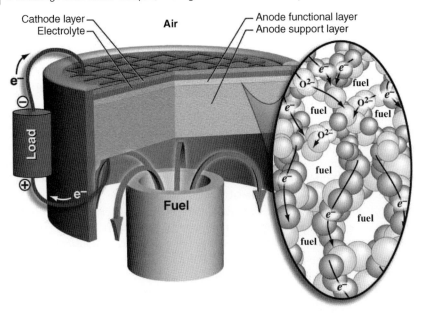

Figure 6.6 An illustration of the button cell.

6.9.1
Polarization Characteristics

Overall polarization behavior of a fuel cell is usually represented in terms of cell voltage and power density as functions of current density. Figure 6.7 shows polarization characteristics for the button cell as predicted by the model. The figure shows results for two fuel streams, with air being the oxidizer in both cases. One fuel stream is humidified H_2 (97% H_2 and 3% H_2O) and the other is a particular syngas mixture (35.3% H_2, 1.3% CO, 41.3% H_2O, 7.9% CO_2, and 14.2% CH_4). This particular syngas mixture is the equilibrium result of methane steam reforming with a steam–carbon ratio of 2.5 at 500 °C and atmospheric pressure. The results assume isothermal operation at 800 °C and atmospheric pressure. As should be anticipated, the cell delivers higher power with the humidified hydrogen than it does with the syngas mixture.

6.9.2
Electric Potentials and Charged Species Fluxes

Figure 6.8 shows profiles of electric potentials and current densities for the cell operating with the syngas mixture and at two different cell potentials. The abscissa is the position within the cell as measured from the interface between the fuel compartment and the composite anode. In both cases, the electric potential at the cathode terminal is set to be 0 V. Because of high electrical conductivity of the

Table 6.3 Parameters for modeling the MEA structure.

Parameters	Value	Units
Anode		
Thickness (L_a)	550	μm
Porosity (ϕ_g)	0.35	
Ni volume fraction (ϕ_{Ni})	0.23	
YSZ volume fraction (ϕ_{YSZ})	0.42	
Tortuosity (τ_g)	4.50	
Ni particle radius (r_{Ni})	0.50	μm
YSZ particle radius (r_{YSZ})	0.50	μm
Specific catalyst area (A_s)	1080.0	cm^{-2}
Specific TPB length (λ_{TPB}^V)	2.40E9	cm^{-2}
Cathode		
Thickness (L_c)	50	μm
Porosity (ϕ_g)	0.35	
LSM volume fraction (ϕ_{LSM})	0.31	
YSZ volume fraction (ϕ_{YSZ})	0.34	
Tortuosity (τ_g)	4.00	
LSM particle radius (r_{LSM})	0.625	μm
YSZ particle radius (r_{YSZ})	0.625	μm
Exchange current factor (i_{ref,O_2}^*)	5.60E4	A/cm^3
Activation energy (E_{a,O_2})	130.0	kJ/mol
Reference temperature (T_{ref})	800.0	°C
Anodic symmetry factor (α_a)	0.75	
Cathodic symmetry factor (α_c)	0.25	
Electrolyte		
Thickness (L_{el})	20	μm

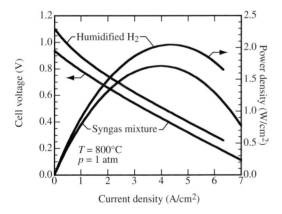

Figure 6.7 Predicted polarization characteristics for the button cell operating on a humidified H_2 and on a syngas fuel mixture at 800 °C and atmospheric pressure.

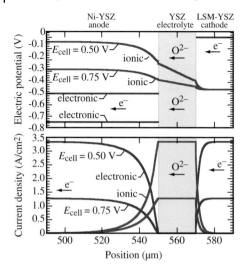

Figure 6.8 Comparison of the electric potential and current density profiles near the dense electrolyte for the button cell operating at the cell voltage of 0.5 and 0.75 V.

electrode materials in the composite electrodes, the small electric potential variations within the electrode phases are not apparent in the plots. Because the YSZ is a pure ion conductor, the electrode electric potential is not relevant within the dense electrolyte. The arrows indicate the direction of electron and ion flux, from cathode toward the anode.

Because of relatively low oxygen ion conductivity in YSZ, the electric potential variations in the electrolyte phase are significant. Within the 20 μm thick dense electrolyte, the electric potential variation is linear (i.e., uniform gradient). The signs of the gradients within the YSZ are such that the negatively charged oxygen ions are transported from regions of relatively low electric potential toward regions of relatively high electric potential, which is consistent with Eq. (6.9). The linear electric potential within the dense electrolyte is a result of the YSZ being a pure ion conductor and a uniform ion current through dense electrolyte. Electrons enter from the external load into the electron-conducting phase (LSM) of the composite cathode (Figure 6.6). As electrons are transported through the electrode phase toward the dense electrolyte, they participate in charge transfer reactions. These reactions transfer charge from electrons in the electrode phase (LSM particles) to oxygen ions in the electrolyte phase (YSZ particles). The local charge transfer rate depends upon both the local electric potential difference between the phases and the activities of participating species, temperature, TPB length, and so on.

The lower panel of Figure 6.8 shows that within the composite cathode, electron current density in the electrode phase is gradually converted to ion current density in the electrolyte phase. In this example, the cathodic charge transfer occurs over a distance of about 10 μm. Because the dense YSZ electrolyte is a pure ion conductor, no electrons emerge from the composite cathode and enter the dense electrolyte. Thus, the electron current density must vanish at the intersection between the composite cathode and the dense electrolyte. For the same reason, only ion current density can proceed through the dense electrolyte. Because the charged species transport through the membrane electrode assembly must be everywhere the same, the sum of the ion and electron current densities must be everywhere the same.

On the anode side, ion current within the electrolyte phase (YSZ) is gradually transferred to electron current in the electrode phase (Ni). In this example, the anodic charge transfer process occurs over a distance of about 30 μm. The electrons leave the anode structure and are delivered to the external load (Figure 6.6). Because the electron current collectors and the wiring to the external load are pure electron conductors, the ion current must vanish at the interface between the composite anode and the fuel compartment.

Charge transfer rates are highest near the dense electrolyte interfaces. This behavior is closely coupled with the shapes of electric potential profiles. On the cathode side, negative charge is being transferred from electrons in the electrode phase to oxygen ions in the electrolyte phase, which is at a relatively more negative electric potential. Near the dense electrolyte interface, the electric potential of the electrolyte phase increases (i.e., becomes less negative), which facilitates the cathodic charge transfer rate. Such behavior is discussed qualitatively in the context of Figure 6.2.

On the anode side, negative charge is being transferred from oxygen ions in the electrolyte phase to electrons in the electrode phase, which is at a relatively more negative electric potential. Near the dense electrolyte interface, the electric potential of the electrolyte phase decreases. As illustrated with the potential energy surfaces in Figure 6.5, the charge transfer rate increases as the electrode electric potential increases relative to the electrolyte electric potential.

Although the charge transfer rates are not shown in Figure 6.8, it is evident that far from the interfaces between the dense electrolyte and the composite electrodes, the charge transfer rate vanishes. Far from the dense electrolyte, the current density within the composite electrodes is entirely electronic. The electric potential gradients in the electrode phases are very small because the electronic conductivities are high. The fact that the electrolyte electric potential profiles are flat (i.e., zero gradient) indicates that there is no charge transfer. This behavior is easily understood in the context of Eqs (6.13) and (6.14).

As a practical matter, the electrode electric potentials can be controlled independently during the fuel cell operation by adjusting the external load resistance. However, the electrolyte electric potential profiles cannot be independently controlled. The electrolyte electric potential profiles depend upon cell geometry, particle sizes, electrode and electrolyte materials, three-phase boundary lengths, thermal and electrochemical reaction rates, fuel and oxidizer composition, and so on. The electrolyte electric potential profiles must be determined such that the coupled set of conservation equations are satisfied, including the contributing reaction rates and transport processes. For a given operating cell voltage, the current density depends upon all the physical and chemical processes that contribute to polarization within the cell.

As is evident from the polarization curves (Figure 6.7), reducing cell potential results in increasing current density. This behavior is also evident in Figure 6.8, with significantly higher current densities when $E_{cell} = 0.5$ V than when $E_{cell} = 0.75$ V. The polarization curves in Figure 6.7 are produced by solving the detailed chemistry transport model over a range of operating voltages. Thus, the magnitudes of the current densities shown in the lower panel of Figure 6.8 are seen to be consistent with those in Figure 6.7 at cell potentials of 0.5 and 0.75 V. Because of the higher current densities, the electric potential gradients (Figure 6.8, upper panel) are greater at 0.5 V than they are at 0.75 V.

6.9.3
Anode Gas-Phase Profiles

Figure 6.9 illustrates gas-phase mole fraction profiles within the button cell composite anode. The cell is operating with the syngas fuel mixture and is isothermal at 800 °C. Profiles are shown for two operating voltages. Because H_2 is consumed by charge transfer chemistry near the dense electrolyte interface, its profile has a negative slope. Because H_2O is a product of the charge transfer chemistry, its profile has a positive slope. The magnitudes of the gradients are higher at $E_{cell} = 0.5$ V than they are at $E_{cell} = 0.75$ V because the lower cell voltage causes higher charge transfer rates. The gas-phase diffusive processes within the pore spaces cause H_2 to be transported from the fuel compartment toward the dense electrolyte and H_2O to be transported out to the fuel compartment.

As methane is transported from the fuel compartment into the anode structure, it is catalytically reformed by available steam and CO_2. Thus, the CH_4 mole fraction

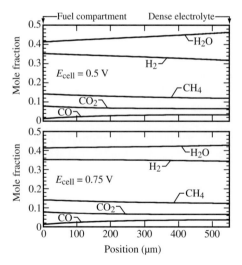

Figure 6.9 The gas-phase mole fraction profiles within the pore spaces of a Ni–YSZ button cell anode structure. The cell is operating with a syngas fuel mixture at operating voltages of $E_{cell} = 0.5$ V and $E_{cell} = 0.75$ V. The cell is operating isothermally at 800 °C and atmospheric pressure.

decreases toward the dense electrolyte, with CO and H_2 being formed. As evidenced by the relatively shallow slopes of the CH_4 profiles, the reforming rates for the circumstances shown in Figure 6.9 are relatively slow. Nevertheless, the effects of reforming are shown by the increases in CO. Some of the CO that is produced is shifted to H_2 and CO_2 by water-gas shift processes. The production of H_2 by the reforming chemistry partially offsets the H_2 consumption by charge transfer chemistry. If reforming were not active, the H_2 and H_2O gradients would be larger than those shown in Figure 6.9.

6.9.4
Anode Surface Species Profiles

To emphasize the elementary nature of the chemical kinetics (Tables 6.1 and 6.2), Figure 6.10 shows the profiles of surface site fractions within the Ni–YSZ anode. Models that use global reforming and charge transfer chemistry do not consider any surface-adsorbed species. The simulations shown here do not attempt to resolve site coverages on the nanoscale around individual electrode–electrolyte particle interfaces. The site coverages are averages that are resolved at the continuum level through the electrode thickness. However, models that were developed to analyze pattern anode behavior suggest that this is a good approximation [17].

Figure 6.10 shows relatively flat site coverage profiles for most adsorbates. The empty Ni site fraction decreases from the anode fuel compartment toward the dense electrolyte, with the sites being filled primarily as CO(Ni). According to Goodwin et al. [17], the χ site plays a major role in the elementary charge transfer chemistry. In region near dense electrolyte interface, the relatively large variation in hydroxyl $OH^-(\chi)$ is central to the charge transfer process.

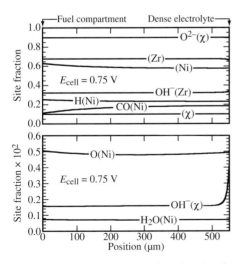

Figure 6.10 Site fraction profiles for selected surface species within the Ni–YSZ anode structure when operating with the syngas fuel mixture at 800 °C and $E_{cell} = 0.75$ V. The upper panel shows species with relatively high coverages and the lower panel shows species that have low coverage.

6.9.5
Applicability and Extensibility

A particular button cell example is used here to illustrate aspects of cell behavior. Different cell architectures, materials, and operating conditions will produce different results. For example, a recent study by Pillai et al. used alternative button cells architectures, including barrier layers, to study the influence of internal reforming on cell stability [10]. This study was also concerned with temperature variations within the cell. Models similar to the one discussed in this chapter were used to assist the interpretation of experimental observations.

The button cell example in this chapter, which concentrates on chemical and transport processes, is formulated as an isothermal one-dimensional problem. A study by Goldin et al. [27] develops a three-dimensional computational fluid dynamics to explore geometric effects and thermal effects in a button cell. However, the underpinning chemistry and transport models are similar to those in this chapter.

Simulating practical cells, such as the tubular cell shown in Figure 6.1, requires models that can represent at least two-dimensional behavior [2, 8, 9, 28]. As fuel flows through the tube, it is consumed by reforming and electrochemistry and diluted by reaction products. Thus, the chemical composition of the gas varies greatly along the tube length. Depending upon the operating conditions, the tube wall and gas temperatures may also vary considerably. The underpinning chemistry and transport approaches discussed in this chapter remain entirely applicable, but must be incorporated in conservation equations with two or three spatial dimensions.

6.10
Summary and Conclusions

Most SOFC models reported in the literature use global reactions to represent the methane steam reforming within composite anodes and the global Butler–Volmer representations of the charge transfer kinetics. Such models are widely used, providing practical benefits in understanding and interpreting performance and assisting design optimization. However, this chapter develops a different approach, which is more firmly grounded in fundamental chemistry. For example, because the elementary catalytic reaction mechanism (Table 6.1) describes all combinations of steam reforming, dry reforming, partial oxidation, and autothermal reforming, it is far more powerful than the global steam reforming mechanisms that are often used in SOFC modeling. The elementary charge transfer chemistry (Table 6.2) leads to overall reaction orders and rates that are more general and extensible than Butler–Volmer formulation with empirical reaction orders in the exchange current densities. This chapter discusses the formalism needed to develop models based upon elementary chemistry and transport. However, even with the formalism in hand, there remain great challenges in developing and validating the needed reaction mechanisms. These challenges are made even greater as practical systems seek to use higher hydrocarbon fuels (including impurities such as sulfur) and new electrode and electrolyte materials and cell architectures are developed.

Acknowledgment

The effort was supported by a DoD Research Tools Consortium (RTC) program administered by the Office of Naval Research under Grant N00014-05-1-0339.

List of Symbols

A_i	prefactor of the Arrhenius form of the i reaction
A_s	specific surface area of the active catalysts, cm^{-1}
B_g	permeability, cm^2
C_{dl}	double-layer capacitance, F
$\tilde{C}_k$	species activities, mol/cm^3
$D^e_{k,Kn}$	effective Knudsen diffusion coefficients, cm^2/s
$D_{k\ell}$	binary diffusion coefficients, cm^2/s
$D^e_{k\ell}$	effective binary diffusion coefficients, cm^2/s
d_p	particle diameter, cm
E_a	electric potential difference in the anode, V
E_c	electric potential difference in the cathode, V
E_{cell}	cell potential, V
E_{c,O_2}	activation energy for O$_2$ reduction, J/mol
E_a^{eq}	equilibrium electric potential difference in the anode, V
E_c^{eq}	equilibrium electric potential difference in the cathode, V
E_i	activation energy of the ith reaction, J/mol
E_{rev}	reversible cell potential, V
F	Faraday constant, C/mol
G^0	Gibbs free energy, J/mol
I	number of reaction steps
i_e	net current density, A/cm^2
$i_{e,BV}$	charge transfer rate in Butler–Volmer equation, A/cm^2
$i_{e,i}$	current density per unit length from the ith reaction, A/cm
i_0	exchange current density, A/cm^2
i_{0,H_2}	exchange current density for H$_2$ oxidation, A/cm^2
i_{0,O_2}	exchange current density for O$_2$ reduction, A/cm^2
$i^*_{H_2}$	parameter in the expression of i_{0,H_2}, A/cm^2
$i^*_{O_2}$	parameter in the expression of i_{0,O_2}, A/cm^2
J_e	molar flux of electrons, mol/(cm^2 s)
J_k	species mole flux, mol/(cm^2 s)
j_k	gas-phase species mass flux, g/(cm^2 s)

$J_{O^{2-}}$	molar flux of oxygen ions, mol/(cm² s)
K_g	number of gas-phase species
K_i	equilibrium constant of the ith reaction
K_s	number of surface-adsorbed species
k_{ai}	anodic rate constant of the ith reaction
k_{ai}^t	anodic thermal rate constant of the ith reaction
k_{ci}	cathodic rate constant of the ith reaction
k_{ci}^t	cathodic thermal rate constant of the ith reaction
L_a	anode thickness, cm
L_c	cathode thickness, cm
L_{el}	electrolyte thickness, cm
$\mathbf{n}$	surface-normal unit vector
n_e	number of charge transferred in the reactions
$n_{e,i}$	number of charge transferred in the ith reaction
n_i	exponent of the Arrhenius form of the i reaction
p	pressure, dyn/cm²
p_{H_2}	partial pressure of H_2, atm
p_{H_2O}	partial pressure of H_2O, atm
p_{O_2}	partial pressure of O_2, atm
$p_{H_2}^*$	parameter in the expression of i_{0,H_2}, atm
$p_{O_2}^*$	parameter in the expression of i_{0,O_2}, atm
q	charge, C
q_i	rate of progress for the ith reaction, mol/(cm³ s)
R	universal gas constant, J/(mol K)
R_{tot}	total ohmic resistance, Ω cm²
r_{Ni}	radius of the Ni particles, cm
r_{LSM}	radius of the LSM particles, cm
r_p	pore radius, cm
r_{YSZ}	radius of the YSZ particles, cm
$\dot{s}_{a,e}$	Faradic charge transfer rate within the anode, A/cm³
$\dot{s}_{c,e}$	Faradic charge transfer rate within the cathode, A/cm³
$\dot{s}_e$	charge transfer rate per unit volume, A/cm³
$\dot{s}_k$	molar production rate by heterogeneous reactions for gas-phase species, mol/(cm³ s), or for surface species, mol/(cm² s)
$\dot{s}_{ki}$	net production rate of the ith reactions, mol/(cm² s)
$\dot{s}_{m,e}$	Faradic charge transfer rate for phase m, A/cm³
T	temperature, K
t	time, s
W_k	species molecular weight, g/mol
$\bar{W}$	mean molecular weight, g/mol
X_k	gas-phase species mole fractions
$[X_k]$	gas-phase species mole concentrations, mol/cm³
$[X_T]$	total gas-phase mole concentrations, mol/cm³
x	spatial coordinate, cm
x_{AF}	position of the anode–fuel compartment interface, cm

x_{EA}	position of the anode–electrolyte interface, cm
x_{EC}	position of the cathode–electrolyte interface, cm
x_{OC}	position of the cathode–air compartment interface, cm
Y_k	gas-phase species mass fractions
z_k	number of charges of the kth species

6.10.1
Greek Letters

α_a	anodic symmetric factor in the BV equation
α_c	cathodic symmetric factor in the BV equation
β_{ai}	anodic symmetric factor of the ith reaction
β_{ci}	cathodic symmetric factor of the ith reaction
Γ	surface site density, mol/cm^2
γ_i	sticking coefficients
η_{act}	local activation overpotential, V
$\eta_{act,a}$	local activation overpotential within the anode, V
$\eta_{act,c}$	local activation overpotential within the cathode, V
η_{ohm}	ohmic overpotential, V
θ_k	site fractions of surface-adsorbed species
λ_{TPB}^V	TPB length per unit volume, cm^{-2}
σ_a^e	effective electric conductivity of anode phase, S/cm
σ_c^e	effective electric conductivity of cathode phase, S/cm
σ_e^e	effective electric conductivity of electrolyte phase, S/cm
$\sigma_{O^{2-}}$	electric conductivity of oxygen ions, S/cm
σ_{e^-}	electric conductivity of electrons, S/cm
μ	gas-phase viscosity, g/(cm s)
μ_{H_2}	chemical potentials of H_2, J/mol
μ_{H_2O}	chemical potentials of H_2O, J/mol
μ_k	species chemical potentials, J/mol
$\tilde{\mu}_k$	species electrochemical potentials, J/mol
μ_{O_2}	chemical potentials of O_2, J/mol
$\mu_{O_a^{2-}}$	chemical potentials of O^{2-} within the anode, J/mol
$\mu_{O_c^{2-}}$	chemical potentials of O^{2-} within the cathode, J/mol
$\mu_{H_2}^0$	standard-state chemical potential of H_2, J/mol
$\mu_{H_2O}^0$	standard-state chemical potential of H_2O, J/mol
μ_k^0	species standard-state chemical potentials, J/mol
$\mu_{O_2}^0$	standard-state chemical potential of O_2, J/mol
ν_k	reaction stoichiometry
ν'_{ki}	forward stoichiometric coefficients
ν''_{ki}	backward stoichiometric coefficients
ϱ	gas-phase mass density, g/cm^3
τ_g	tortuosity of the gas phase
Φ	electric potential, V

Φ_a anode electric potential, V
Φ_c cathode electric potential, V
Φ_a^0 electric potential at the anode–fuel interface, V
Φ_c^0 electric potential at the cathode–air interface, V
Φ_{ed} electric potential in the electron-conducting phase, V
Φ_{el} electric potential in the ion-conducting phase, V
$\Phi_{e,a}$ electrolyte electric potential within the anode, V
$\Phi_{e,c}$ electrolyte electric potential within the cathode, V
Φ_a^{eq} equilibrium anode electric potential, V
Φ_c^{eq} equilibrium cathode electric potential, V
$\Phi_{e,a}^{eq}$ equilibrium electrolyte electric potential in the anode, V
$\Phi_{e,c}^{eq}$ equilibrium electrolyte electric potential in the cathode, V
Φ_m electric potential of phase m, V
Φ_k phase electric potential where the kth species belongs, V
ϕ_g porosity
ϕ_{LSM} volume fraction of LSM particles
ϕ_{Ni} volume fraction of Ni particles
ϕ_{YSZ} volume fraction of YSZ particles
$\Lambda_k^{z_k}$ chemical symbol for the kth species with charge z_k
Ψ_{ik} geometric factor

References

1 Mason, E.A. and Malinauskas, A.P. (1983) *Gas Transport in Porous Media: The Dusty-Gas Model*, Elsevier, New York.

2 Zhu, H., Kee, R.J., Janardhanan, V.M., Deutschmann, O., and Goodwin, D.G. (2005) Modeling elementary heterogeneous chemistry and electrochemistry in solid-oxide fuel cells. *J. Electrochem. Soc.*, **152**, A2427–A2440.

3 Kee, R.J., Coltrin, M.E., and Glarborg, P. (2003) *Chemically Reacting Flow: Theory and Practice*, John Wiley & Sons, Inc., Hoboken, NJ.

4 Wilson, J.R., Kobsiriphat, W., Mendoza, R., Chen, H.Y., Hiller, J.M., Miller, D.J., Thornton, K., Voorhees, P.W., Adler, S.B., and Barnett, S.A. (2006) Three-dimensional reconstruction of a solid-oxide fuel-cell anode. *Nat. Mater.*, **5**, 541–544.

5 Chen, D., Lin, Z., Zhu, H., and Kee, R.J. (2009) Percolation theory to predict effective properties of solid oxide fuel-cell composite electrodes. *J. Power Sources*, **191**, 240–252.

6 Zhu, H. and Kee, R.J. (2006) Modeling electrochemical impedance spectra in SOFC button cells with internal methane reforming. *J. Electrochem. Soc.*, **153**, A1765–A1772.

7 Zhu, H. and Kee, R.J. (2008) Modeling distributed charge-transfer processes in SOFC membrane electrode assemblies. *J. Electrochem. Soc.*, **155**, B175–B729.

8 Zhu, H. and Kee, R.J. (2007) The influence of current collection on the performance of tubular anode-supported SOFC cells. *J. Power Sources*, **169**, 315–326.

9 Kee, R.J., Zhu, H., Sukeshini, A.M., and Jackson, G.S. (2008) Solid oxide fuel cells: operating principles, current challenges, and the role of syngas. *Combust. Sci. Technol.*, **180**, 1207–1244.

10 Pillai, M., Lin, Y., Zhu, H., Kee, R.J., and Barnett, S.A. (2009) Stability and coking of direct-methane solid oxide fuel cells: effect of CO_2 and air additions. *J. Power Sources*, **195** (1), 271–279.

11 Hecht, E.S., Gupta, G.K., Zhu, H., Dean, A.M., Kee, R.J., Maier, L., and Deutschmann, O. (2005) Methane reforming kinetics within a Ni–YSZ SOFC anode. *Appl. Catal. A*, **295**, 40–51.

12 Janardhanan, V.M. and Deutschmann, O. (2006) CFD analysis of a solid oxide fuel cell with internal reforming: coupled interactions of transport, heterogeneous catalysis and electrochemical processes. *J. Power Sources*, **162**, 1192–1202.

13 Gupta, G.K., Hecht, E.S., Zhu, H., Dean, A.M., and Kee, R.J. (2006) Gas-phase reactions of methane and natural-gas with air and steam in non-catalytic regions of a solid-oxide fuel cell. *J. Power Sources*, **156**, 434–447.

14 Coltrin, M.E., Kee, R.J., and Rupley, F.M. (1991) Surface Chemkin: a generalized formalism and interface for analyzing heterogeneous chemical kinetics at a gas–surface interface. *Int. J. Chem. Kinet.*, **23**, 1111–1128.

15 Sheng, C.Y. and Dean, A.M. (2004) The importance of gas-phase kinetics within the anode channel of a solid-oxide fuel cell. *J. Phys. Chem.*, **108**, 3772–3783.

16 Sasaki, K. and Teraoka, Y. (2003) Equilibria in fuel cell gases. *J. Electrochem. Soc.*, **150**, A878–A888.

17 Goodwin, D.G., Zhu, H., Colclasure, A.M., and Kee, R.J. (2009) Electrochemical charge-transfer processes for hydrogen–steam mixtures on Ni–YSZ pattern anodes. *J. Electrochem. Soc.*, **156**, B1004–B1021.

18 Vogler, M., Bieberle-Hütter, A., Gauckler, L., Warnatz, J., and Bessler, W.G. (2009) Modelling study of surface reactions, diffusion, and spillover at a Ni/YSZ patterned anode. *J. Electrochem. Soc.*, **156**, B663–B672.

19 Bessler, W.G., Gewies, S., and Vogler, M. (2007) A new framework for physically based modeling of solid oxide fuel cells. *Electrochim. Acta*, **53**, 1782–1800.

20 Bessler, W.G., Warnatz, J., and Goodwin, D.G. (2007) The influence of equilibrium potential on the hydrogen oxidation kinetics of SOFC anodes. *Solid State Ionics*, **177**, 3371–3383.

21 Kee, R.J., Zhu, H., and Goodwin, D.G. (2004) Solid-oxide fuel cells with hydrocarbon fuels. *Proc. Combust. Inst.*, **30**, 2379–2404.

22 Bockris, J.O., Reddy, A.K.N., and Gamboa-Aldeco, M. (2000) *Modern Electrochemistry: Fundamentals of Electrodics*, 2nd edn, Kluwer Academic/Plenum Publishers, New York.

23 Bard, A.J. and Faulkner, L.R. (2000) *Electrochemical Methods: Fundamentals and Applications*, 2nd edn, John Wiley & Sons, Inc., New York.

24 Asher, U.M. and Petzold, L.R. (1998) *Computer Methods for Ordinary Differential Equations and Differential-Algebraic Equations*, SIAM, Philadelphia, PA.

25 Deuflhard, P., Hairer, E., and Zugck, J. (1987) One-step and extrapolation methods for differential-algebraic systems. *Numer. Math.*, **51**, 501–516.

26 Jiang, Y. and Virkar, A.V. (2003) Fuel composition and diluent effect on gas transport and performance of anode-supported SOFCs. *J. Electrochem. Soc.*, **150**, A942–A951.

27 Goldin, G., Zhu, H., Kee, R.J., Bierschenk, D., and Barnett, S.A. (2009) Multidimensional flow, thermal, and chemical behavior in solid-oxide fuel cell button cells. *J. Power Sources*, **187**, 123–135.

28 Zhu, H., Colclasure, A.M., Kee, R.J., Lin, Y., and Barnett, S.A. (2006) Anode barrier layers for tubular solid-oxide fuel cells with methane fuel streams. *J. Power Sources*, **161**, 413–419.

7
Evaluation of Models for Heterogeneous Catalysis
John Mantzaras

7.1
Introduction

The progress in many industrial devices employing heterogeneous catalysis, such as chemical synthesis, exhaust gas cleanup, and heat/power generation plants, crucially depends on key advances in multidimensional modeling necessary for reactor design. The numerical models should entail detailed description of the heterogeneous (catalytic) and low-temperature homogeneous (gas-phase) kinetics, as well as of the underlying interphase and intraphase fluid transport. There is currently a pressing need for hetero-/homogeneous chemical reaction schemes valid over broad ranges of operating conditions, which for the heat/power generation systems alone encompass pressures from 1 bar in industrial boilers and household burners [1] to 5 bar in microreactors [2] and to 20 bar in large stationary gas turbines [3]. Moreover, the temperatures in industrial catalytic applications range from a few hundred degrees up to 1400 K.

Recent developments in surface science measuring techniques, as briefly discussed in Section 7.2, have assisted the construction of detailed catalytic reaction mechanisms by probing key elementary reaction pathways and identifying pertinent surface species. As most of the surface science data refer to ultrahigh vacuum (UHV) and to single-crystal surfaces, the extension of the resulting reaction mechanisms to realistic pressures and technical catalysts (polycrystalline surfaces) is not warranted. This necessitates additional validation tools, which are mainly based on measurements of gas-phase thermoscalars, either average or spatially resolved, in suitable laboratory-scale reactors.

In an effort to bridge the pressure and materials gap between surface science and technical heterogeneous catalysis, *in situ* spatially resolved measurements of major and minor gas-phase species concentrations over the catalyst boundary layer using spontaneous Raman and laser-induced fluorescence (LIF), respectively, have fostered fundamental hetero/homogeneous kinetic studies at elevated pressures and temperatures, realistic reactant compositions, and technical catalyst surfaces [4–7]. It is demonstrated in the forthcoming sections that the combination of *in situ* measurements with multidimensional modeling can lead to the development and refinement of catalytic and gas-phase chemical reaction schemes, which are valid at industrially relevant operating conditions. It is emphasized that even for the large surface-to-volume ratios typical to catalytic reactors, gas-phase chemistry cannot always be ignored, particularly at elevated pressures [7, 8]. The methodology for the validation of hetero/homogeneous chemical reaction schemes is presented in Section 7.3, first for combustion of fuel-lean methane/air, propane/air, and hydrogen/air mixtures over platinum at pressures of up to 16 bar [5–14] and then for fuel-rich catalytic combustion of methane over rhodium [15, 16] at pressures of up to 10 bar.

Fluid transport, either from the bulk of the gas to the outer catalyst surface (interphase transport) or within the catalyst porous structure (intraphase transport) is dealt in Section 7.4. Transport is coupled to the hetero/homogeneous chemistry, necessitating its accurate description in multidimensional codes, so as to avoid potential falsification of chemical kinetics. Turbulent transport, in particular, is an issue of interest in gas turbines using the catalytically stabilized thermal combustion (CST) methodology [17]. Finally, new modeling directions that include discrete lattice Boltzmann (LB) methods for intraphase transport are presented. Measurements at the microscales relevant to intraphase transport are challenging, such that the appropriate validation tools include not only experiments but also direct simulation Monte Carlo (DSMC) methods.

7.2
Surface and Gas-Phase Diagnostic Methods

7.2.1
Surface Science Diagnostics

Surface science measuring techniques have considerably advanced in the past years [18]. Briefly, optical sum-frequency generation (SFG) vibrational spectroscopy, high-pressure scanning tunnel microscopy (STM), and photoelectron emission microscopy (PEEM) have been applied to investigate elementary surface processes in heterogeneous catalysis [19–21]. Nonetheless, the bulk of these techniques have been applied at UHV, or at best for millibar pressures, and on well-defined single-crystal surfaces.

A notable example is the SFG vibrational spectroscopy study in Kissel-Osterrieder *et al.* [22], which has shown that carbon monoxide can dissociate on polycrystalline platinum at a pressure of 20 mbar, a reaction pathway not observed in earlier UHV experiments. *In situ* surface diagnostics techniques allowing investigations of catalytic reactions under technically relevant pressures and temperatures, on realistic catalysts and mixture compositions, need to be further advanced to become a main research tool for practical systems. As a result, the evaluation of kinetic models discussed in the following sections is largely based on gas-phase measurements over the catalyst surface.

7.2.2
In Situ Gas-Phase Diagnostics

Given the limitations of surface science measurements, the development of catalytic reaction mechanisms relies heavily on additional experiments carried out in a variety of laboratory-scale reactors. Such measurements aim at introducing appropriate kinetic rate modifications so as to bridge the pressure and materials gap of the largely UHV surface science experiments.

Various reactor designs have been established in recent years for assessing surface kinetics. An extensively used configuration is the nearly isothermal, low-temperature (typically less than 600 °C), gradientless (in the cross-flow direction) microflow catalytic reactor, which operates with highly diluted fuels in order to minimize the combustion temperature rise. Although the modeling of such reactors is straightforward, the isothermal operation may hinder the description of certain reaction pathways that are thermally controlled. More important, high temperature studies are largely precluded since they require very large gas hourly spatial velocities (GHSV) to remove the resulting transport limitations. Recent improvements in microreactor technology are the annular [23, 24] and the linear designs [25], where the reactants flow through submillimeter narrow channels leading to very high velocities, thus alleviating transport limitations. A general disadvantage of microflow reactors is that they cannot account for gas-phase chemistry contributions, thus hindering hetero/homogeneous combustion studies that are important for high-pressure applications.

Another popular device is the stagnation flow reactor, which has provided numerous data on catalytic ignition and extinction, steady fuel conversion, and product selectivity under realistic temperatures and mixture compositions [26–30]. Contrary to microflow reactors, stagnation flow reactors can tolerate large gas-phase gradients. This is because modeling with well-established 1D stagnation flow reactive codes [31] is nowadays straightforward, even when complex chemical reaction schemes and detailed transport are included. Gas-phase thermoscalar measurements in both microflow and stagnation flow reactors are typically global and not spatially resolved. The size limitations of microflow reactors allow the measurement of only the inlet and outlet species composition and the overall reactor temperature. Stagnation-flow experiments have provided global data (total fuel conversion, product selectivity, autothermal behavior, ignition/extinction temperatures), even though the geometry itself is very amenable to spatially resolved measurements. Spatially resolved experiments in stagnation flow reactors have advanced recently, providing species boundary layer profiles with intrusive [30] and nonintrusive [32] measuring techniques. The specific advantage of spatially resolved measurements is discussed next within the context of optically accessible reactors.

A new configuration suitable for kinetic studies is the high-pressure, high-temperature, optically accessible channel-flow reactor shown in Figure 7.1 [5, 7–14, 33]. The assembly consists of a rectangular reactor, which forms a liner in a cylindrical tank that provides pressurization up to 20 bar. The reactor comprises two horizontal Si[SiC] ceramic plates (300 mm long and 104 mm wide, placed 7 mm

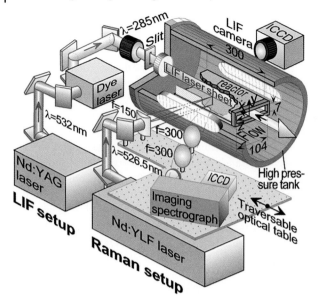

Figure 7.1 Catalytic channel-flow reactor and optical arrangement of the Raman/LIF setup [11]. All distances are in millimeters.

apart) and two 3 mm thick vertical quartz windows. The inner Si[SiC] surfaces are coated with the catalyst of interest, while the surface temperatures of both plates are monitored with thermocouples positioned along the x–y streamwise symmetry plane and embedded 0.9 mm beneath the catalyst. Optical accessibility is maintained from both reactor sides via two high-pressure quartz windows on the tank, while a counterflow streamwise optical access is also available through a rear tank flange. *In situ*, nonintrusive, laser-based spectroscopic techniques are applied, which are shown in their most recent configuration [11, 13, 14] in Figure 7.1. The measurements include 1D spontaneous Raman measurements of major gaseous species concentrations and temperature across the entire 7 mm channel height using a dedicated Nd:YLF laser, and planar laser-induced fluorescence of trace species along the x–y symmetry plane using a tunable dye laser pumped by an Nd:YAG laser. Probed trace species with LIF are typically OH and CH_2O in fuel-lean and fuel-rich hydrocarbon combustion, respectively. The aforementioned kinetic studies are performed at steady and laminar flow conditions. Fluid mechanical transport and, in particular, turbulent transport, where the modeling uncertainty becomes an issue, can also be assessed with particle image velocimetry (PIV) [34]. As a result, both chemistry and interphase transport can be investigated in this configuration.

The reactor of Figure 7.1 has a number of advantages, stemming either from the geometry itself or from the *in situ* nature of the measurements. Since large cross-flow gradients can be tolerated, catalytic reactivity studies can be performed at high surface temperatures, for example, up to 1400 K, without the need of excessively large GHSV. In addition, the reactor is suitable for investigating the hetero/homogeneous chemistry coupling because gaseous combustion is contained in the catalytic channel rather than in a follow-up burnout zone. On the other hand, the added benefits come at an increasing experimental and numerical cost. Multidimensional reactive computational fluid dynamics (CFD) models [35, 36] are required to simulate the *in situ* experiments.

7.3
Evaluation of Hetero/Homogeneous Chemical Reaction Schemes

7.3.1
Fuel-Lean Combustion of Methane/Air on Platinum

The methodology for the evaluation of catalytic and gaseous kinetics is demonstrated using as benchmark the combustion of fuel-lean methane/air mixtures (equivalence ratio $\varphi < 0.4$) over polycrystalline platinum, at pressures of up to 16 bar. The catalytic reactivity is firstly assessed by comparing Raman measurements and predicted species boundary layer profiles. Homogeneous combustion is subsequently evaluated using both Raman and OH-LIF measurements.

7.3.1.1 Heterogeneous Kinetics
The assessment of the catalytic reactivity from the Raman data is illustrated in Figure 7.2. The first step involves delineation of the reactor extent over which the gaseous pathway has a negligible contribution to the total conversion of methane. Figure 7.2a provides the catalytic (C) and gas-phase (G) methane axial conversion rates at 14 bar; the conversions were computed using the elementary heterogeneous and homogeneous reaction schemes of Deutschmann *et al.* [37] and Warnatz *et al.* [38], respectively. The aptness of the latter mechanism will be addressed in Section 7.3.1.2. The volumetric G conversion rate in Figure 7.2a has been integrated across the 7 mm channel height in order to facilitate comparisons with the surface C conversion rate. The location of appreciable gaseous methane conversion, marked as x_{ag} in Figure 7.2a and defined as the far upstream position where G amounts to 5% of C, is not related to the onset of homogeneous ignition. It will be shown next that the appreciable gaseous methane conversion is not always associated with a thermal runaway and hence with a significant exothermicity.

The measured transverse species profiles are compared against the simulations (Figure 7.2b and c) only at axial locations $x < x_{ag}$ so as to avoid potential falsification of the catalytic kinetics by the gaseous chemistry. Cardinal in such comparisons is the realization of kinetically controlled catalytic conversion, which is manifested in Figure 7.2b and c by the nonzero concentrations of methane at both channel walls ($y = 0$ and 7 mm). The good agreement between measurements and predictions in Figure 7.2b and c, especially in the profiles of the limiting reactant (methane), reflects the aptness of the employed catalytic scheme. Detailed studies [7] have further shown that the employed catalytic scheme from Ref. [37] is appropriate over the pressure and wall temperature ranges of 1 bar $\leq p \leq$ 16 bar and 780 K $\leq T_w \leq$ 1250 K, respectively.

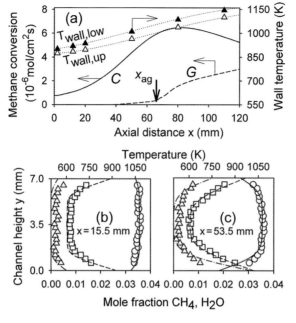

Figure 7.2 Methane/air combustion over platinum in the reactor of Figure 7.1. (a) Predicted axial profiles of the catalytic (C) and gaseous (G) methane conversions for $p = 14$ bar, $\varphi = 0.35$, $T_{IN} = 606$ K, and $U_{IN} = 0.55$ m/s. Adapted from Ref. [7]. The arrow marked x_{ag} defines the onset of appreciable gaseous contribution. Measured upper wall temperatures (open triangles) and lower wall temperatures (filled triangles). (b and c) Predicted (lines) and Raman-measured (symbols) transverse profiles: methane (circles), water (triangles), and temperature (squares).

Additional comparison of the measurements with other catalytic reaction schemes has revealed [7] that the proper performance of any surface mechanism reflects its capacity to capture the decrease in surface free site coverage with rising pressure that, in turn, restrains the rate of increase of the catalytic reactivity with increasing pressure. The catalytic reaction rate of methane has an overall pressure dependence p^{1-n}, where $0 < 1 - n < 1$ and the exponent n is, in reality, a function of pressure. The high-pressure validity of the employed catalytic mechanism [37] is due to the inclusion of a methane adsorption rate that has an order of 2.3 with respect to the platinum free site coverage. Attempts to fit the results of the detailed heterogeneous scheme to a global step have shown that there exists no constant pressure exponent and effective activation energy over the temperature and pressure ranges of interest [7]. Even so, the following global step has been proposed [7], yielding the best fit to computations with the validated detailed scheme:

$$\dot{s}_{CH_4} = A(p/p_0)^{-n} \exp(-E_a/RT_w)[CH_4]_w, \tag{7.1}$$

where $[CH_4]_w$ is the concentration of methane at the gas–wall interface, $A = 1.27 \times 10^5$ cm/s, $n = 0.53$, $E_a = 84$ kJ/mol and $p_0 = 1$ bar. Given the linear dependence of

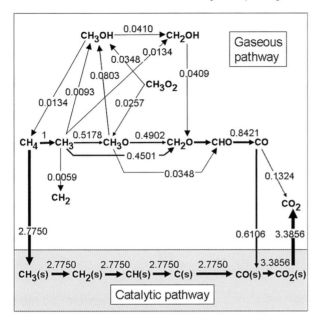

Figure 7.3 Heterogeneous and homogeneous carbon fluxes (only fluxes greater than 0.004 are shown) for the case in Figure 7.2 at $x = 95$ mm. Adapted from Ref. [7].

$[CH_4]_w$ on pressure, the overall reaction pressure order in Eq. (7.1) is 0.47. The global step of Eq. (7.1) reproduces in ±15% results of the validated detailed mechanism [37] over the pressure range of 1 bar $\leq p \leq$ 16 bar. For more accurate predictions, a reduced skeletal catalytic reaction scheme has also been constructed in Ref. [7]. The negative pressure exponent $(-n)$ in Eq. (7.1) is crucial in numerical models used for the design of natural gas-fueled turbine catalytic reactors, as shown in Ref. [39].

The hetero/homogeneous carbon flux diagram in Figure 7.3, referring to the conditions in Figure 7.2, indicates that methane is converted in parallel by both reaction pathways. Methane is oxidized homogeneously to CO, which is adsorbed very efficiently on platinum due to its high sticking coefficient. The adsorbed surface CO(s) is then oxidized catalytically to CO_2(s) that further desorbs to the gas phase. The gaseous hydrocarbon combustion can generally be described by two sequential steps: first, the incomplete oxidation of fuel to CO and, second, the main exothermic reaction of CO to CO_2 [40]. The catalytic pathway clearly inhibits the onset of homogeneous ignition by depriving CO from the gas phase. Nonetheless, even in the absence of homogeneous ignition, the gaseous pathway can amount to appreciable methane conversion, especially at higher pressures, through the incomplete oxidation of CH_4 to CO. The regimes of importance for the gaseous pathway are delineated in the parametric plot of Figure 7.4, which was constructed using a surface perfectly stirred reactor (SPSR) model [41] with the hetero/homogeneous chemical reaction schemes of Deutschmann/Warnatz [37, 38]. To the left of each line of fixed pressure and temperature, gaseous chemistry cannot be neglected as it accounts for

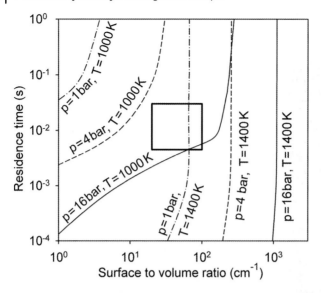

Figure 7.4 Delineation of regimes of significant gaseous chemistry contribution in fuel-lean methane/air combustion over Pt. Adapted from Ref. [7]. To the left of each line of constant pressure and temperature, gas-phase chemistry cannot be neglected. The rectangle in the center of the graph denotes operating conditions for CST-based turbine reactors.

more than 5% of the total methane conversion. The rectangle at the center denotes the operating regimes for turbines using the CST combustion methodology [17]. It is evident that gaseous chemistry cannot be neglected in numerical modeling at turbine-relevant conditions, meaning $p \geq 16$ bar and $T \geq 1000$ K. The implication for models used for high-pressure catalytic reactor design is that heterogeneous chemistry alone may not suffice, thus requiring a more complex hetero/homogeneous chemistry description.

7.3.1.2 Gas-Phase Kinetics

The combination of LIF and Raman data is well suited for the validation of gas-phase reaction schemes since it permits simultaneous investigation of both reaction pathways [5, 7, 8, 11–15]. An assessment of the catalytic processes preceding the onset of homogeneous ignition is essential, despite the validation of the catalytic scheme [37] in the foregoing section. The higher surface temperatures required to achieve homogeneous ignition may cause partial or total catalyst deactivation, resulting in a near-wall fuel excess that would in turn accelerate the onset of gas-phase ignition and thus falsify the gaseous kinetics. This problem is further compounded by the existence of multiple combinations of catalytic and gaseous reactivities yielding exactly the same homogeneous ignition distance [42]. In this respect, the Raman measurements remove any uncertainties associated with the catalytic pathway that may affect the evaluation of gaseous chemistry.

Homogeneous kinetics have been studied over the ranges of 1 bar $\leq p \leq 16$ bar and 1050 K $\leq T_w \leq$ 1430 K [8–10, 43]. Measured and predicted 2D distributions of the OH

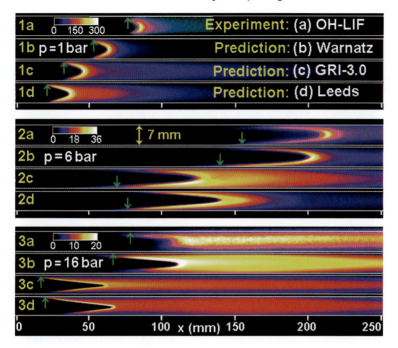

Figure 7.5 LIF-measured (a) and numerically predicted (b–d) OH maps in methane/air combustion over Pt: (1) $p = 1$ bar, $\varphi = 0.31$, $T_{IN} = 754$ K, (2) $p = 6$ bar, $\varphi = 0.36$, $T_{IN} = 569$ K, and (3) $p = 16$ bar, $\varphi = 0.40$, $T_{IN} = 643$ K. Adapted from Refs [8, 9]. Predictions with the catalytic scheme of Deutschmann et al. [37] and the gas-phase schemes of (b) Warnatz et al. [38], (c) GRI-3.0 [44], and (d) Leeds [45]. The arrows denote the onset of homogeneous ignition. The color bars provide the predicted OH in ppmv using the Warnatz mechanism [38].

radical are compared in Figure 7.5. The position of homogeneous ignition (x_{ig}), shown with the arrows in Figure 7.5, has been defined in both predictions and experiments as the far upstream location where OH reaches 5% of its maximum value in the reactor. Raman-measured and predicted transverse species profiles for case 3 (16 bar) in Figure 7.5 are shown in Figure 7.6, indicating a correct prediction of the catalytic processes in the gaseous induction zone (Figure 7.6a and b); it is here evident that the catalytic methane conversion is close to the mass transport limit. The gaseous scheme of Warnatz et al. [38] reproduces the measured homogeneous ignition position and the subsequent flame sweep angles for $p \geq 6$ bar (Figure 7.5). When used in conjunction with the catalytic scheme of Deutschmann et al. [37], it also reproduces the gas-phase contribution in the combined hetero/homogeneous conversion zone $x_{ag} \leq x \leq x_{ig}$ (see, for example, Figure 7.6b). In contrast, the mechanisms from GRI-3.0 [44] and Leeds [45] underpredict considerably the measured x_{ig} at all pressures. Ignition delays have been computed as $\tau_d = \tau(x_{ig})$ ($\tau(x) = \int_0^x dx/\bar{U}(x)$ where $\bar{U}(x)$ is the local mean axial velocity) using the three gaseous schemes and their resulting differences were up to 100 ms. Since turbine catalytic reactors have

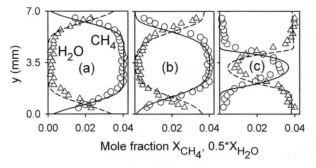

Figure 7.6 Measured (symbols) and predicted (lines) transverse profiles of methane and water mole fractions for case 3 in Figure 7.5: (a) $x=15.5\,mm$, (b) $x=53.5\,mm$, and (c) $x=93.5\,mm$. Predictions using the Deutschmann/Warnatz reaction schemes [37, 38].

typical residence times shorter than 30 ms, it is apparent that the burner design would be greatly impacted by the proper choice of gas-phase reaction mechanism.

Comprehensive analysis of the three employed gaseous mechanisms has revealed [8] that the differences in Figure 7.5 are a result of the low equivalence ratios and temperatures. The first factor bears the direct impact of the catalytic pathway, which consumes methane such that the effective equivalence ratio at the x_{ig} location is considerably lower than the already low inlet value. The second factor reflects the particular operating conditions of catalytic systems and it impacts the relative contribution of the low- and high-temperature oxidation routes of methane [8]. The capacity of the various schemes to reproduce x_{ig} is linked to their aptness in correctly predicting ignition delay times. Conversely, the postignition behavior of the gaseous schemes (e.g., the flame sweep angles in Figure 7.5) reflects their ability to reproduce propagation characteristics. The former quantity is certainly more important than the latter since the onset of gaseous combustion is deemed detrimental to the catalyst and reactor integrity and numerical models should be able to predict the likelihood of such an event.

Since the scheme of Warnatz et al. [38] is suitable over 6 bar $\leq p \leq$ 16 bar (Figure 7.5), it provides a suitable platform for the construction of a modified mechanism valid over the extended range of 1 bar $\leq p \leq$ 16 bar. Sensitivity analysis reveals that at $p=1$ bar, the homogeneous ignition is particularly sensitive to the chain branching step $CHO + M = CO + H + M$; while at $p=16$ bar, the corresponding sensitivity is weak. Friedrichs et al. [46] have recently suggested slower rates for this reaction. To improve the predictions of Figure 7.5 at $p<6$ bar, the preexponential of this reaction was reduced by a factor of 1.8 in the scheme of Warnatz et al. [38]. The altered scheme provided very good agreement with the LIF measurements over 1 bar $\leq p \leq$ 16 bar [8], as seen in Figure 7.7.

The hetero/homogeneous chemistry coupling in terms of radicals and major species is now briefly addressed using the schemes of Deutschmann and modified Warnatz. It has been shown that the catalyst itself is a poor source of OH radicals so as to meaningfully affect the gaseous pathway at conditions away from homogeneous ignition [7]. However, at the late stages of the induction zone or at postignition

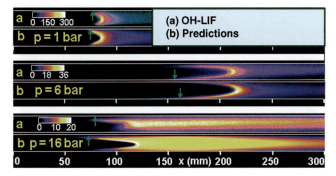

Figure 7.7 LIF-measured (a) and numerically predicted (b) OH maps in methane/air combustion over Pt using the catalytic scheme of Deutschmann *et al.* [37] and the amended gas scheme of Warnatz *et al.* [38]. Adapted from Ref. [8].

positions, the catalyst turns to a sink of OH radicals. This behavior determines an overall mild inhibition of homogeneous ignition on radical (OH as well as O and H) reactions: for example, x_{ig} decreases by 10% when removing the radical adsorption/desorption reactions. In addition, it has been shown [6, 43] that the catalytically formed major product H_2O promotes chemically the onset of homogeneous ignition in methane combustion.

7.3.2
Fuel-Lean Combustion of Propane/Air on Platinum

The Raman/LIF methodology has been further applied to the hetero/homogeneous combustion of fuel-lean ($\varphi \leq 0.43$) propane/air mixtures over polycrystalline platinum at pressures of up to 7 bar and temperatures $720\,\mathrm{K} \leq T_w \leq 1280\,\mathrm{K}$ [11]. Such moderate pressures are of particular interest in microreactors used for portable power generation [2]. Key issue is the determination of the pressure dependence of the catalytic reactivity. Computations were performed using the atmospheric pressure global catalytic mechanism of Garetto *et al.* [47], which has been augmented with a pressure correction factor $(p/p_0)^{-n}$ in a fashion similar to the methane expression in Eq. (7.1):

$$\dot{s}_{C_3H_8} = A(p/p_0)^{-n} T_w^{1.15} \exp(-E_a/RT_w) [C_3H_8]_w^a, \tag{7.2}$$

where $A = 93.2\,\mathrm{K}^{-1.15}\,\mathrm{cm}^{1.45}/(\mathrm{mol}^{0.15}\,\mathrm{s})$, $E_a = 71.128\,\mathrm{kJ/mol}$, and $a = 1.15$. Raman-measured and predicted transverse profiles of propane and water mole fractions are illustrated in Figure 7.8. The comparisons pertain to axial positions upstream of the onset of appreciable gas-phase propane conversion, in a manner analogous to the methane studies of Section 7.3.1.1. Given that detailed heterogeneous mechanisms for fuel-lean combustion of higher hydrocarbons are still under construction, global steps such as Eq. (7.2) offer at this stage the only possibility for much needed engineering calculations. A value of $n = 0.4 \pm 0.03$ in Eq. (7.2) yielded the best fit to the experimentally acquired boundary layer propane profiles over the entire pressure

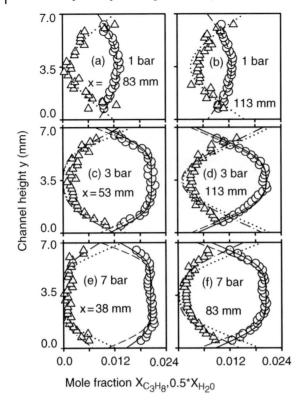

Figure 7.8 Measured (symbols) and predicted (lines) transverse profiles of propane and water mole fractions at three pressures: (a and b) $p = 1$ bar, $\varphi = 0.30$, (c and d) $p = 3$ bar, $\varphi = 0.27$, and (e and f) $p = 7$ bar, $\varphi = 0.26$. Measurements: C_3H_8 (circles), H_2O (triangles). Predictions: C_3H_8 (solid lines: pressure-corrected model in Eq. (7.2), dashed lines: original step of Garetto et al. [47]), H_2O (dashed-dotted lines: pressure-corrected model, dotted lines: step of Garetto et al. [47]). Adapted from Ref. [11].

range of $1 \leq p \leq 7$ bar, as seen in Figure 7.8. The overall pressure dependence of the catalytic reaction in Eq. (7.2) is thus 0.75. Simulations without the pressure correction term appreciably overpredicted the catalytic reactivity at pressures above 3 bar (see Figure 7.8e and f), thus signifying the care needed in extrapolating atmospheric pressure catalytic kinetic data to elevated pressures.

For the assessment of homogeneous ignition, Raman and OH-LIF measurements were employed, similar to the foregoing methane studies in Section 7.3.1.2. An optimized C_3 mechanism for homogeneous combustion by Qin et al. [48] (70 species, 14 irreversible and 449 reversible reactions) was used in the simulations. The combination of a global catalytic step with a detailed gaseous reaction mechanism in numerical models deserves some clarifications. It can be argued that a global catalytic step that captures, at least over a certain range of operating conditions, the heterogeneous fuel conversion can also reproduce the onset of homogeneous ignition when used in conjunction with a detailed gas-phase reaction scheme. Earlier numerical studies of methane (Section 7.3.1.2) and hydrogen (Section 7.3.3) with detailed hetero/homogeneous reaction schemes have attested the particularly weak chemistry coupling via radical adsorption/desorption reactions. Major hetero/homogeneous interactions arise from the near-wall catalytic fuel depletion, which in turn inhibits homogeneous ignition [42, 49]; a secondary chemical coupling originates from the heterogeneously produced major species (notably H_2O), which impact homogeneous ignition (promoting for methane as discussed in Section 7.3.1.2 and inhibiting for hydrogen [5]). The last two effects can be captured by a global catalytic step, however, at narrower pressure and temperature ranges compared to a detailed mechanism. Coupling via other intermediate species that participate in both reaction pathways, such as CO, can be important in determining the overall consumption of this species and its emission, but its impact on the homogeneous ignition of methane is secondary [7]. Methane and hydrogen studies (see Sections 7.3.1.2 and 7.3.3) point out that discrepancies in homogeneous ignition predictions stem primarily from inaccuracies in the low-temperature gaseous kinetics rather than in the heterogeneous kinetics. Moreover, analytical ignition criteria confirm that the homogeneous ignition position is much more sensitive to the gaseous rather than to the catalytic reactivity [42].

Comparisons between LIF-measured and numerically predicted distributions of the OH radical are illustrated in Figure 7.9. In all cases, the pressure-corrected heterogeneous reaction of Eq. (7.2) was used, coupled to the gas-phase mechanism of Qin et al. [48]. An overall good agreement is achieved: the ignition distance is underpredicted by 9% (1 bar), while pressures up to 5 bar consistently yield overpredictions by up to 10%. Increasing the pressure to 7 bar leads to greater overpredictions of x_{ig} (by 19%). These observations are in accordance with recent numerical studies showing increasing discrepancies of the Qin mechanism [48]

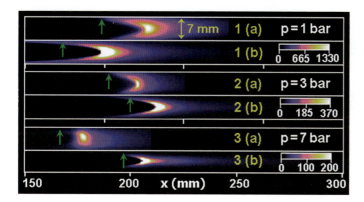

Figure 7.9 Measured and predicted OH maps in propane combustion over Pt at three pressures: OH-LIF measurements (a) and numerical predictions (b) with the Qin et al. [48] (gas-phase) and pressure-corrected (catalytic) reaction schemes. The arrows define the onset of homogeneous ignition and the color bars provide the predicted OH levels in ppmv. Adapted from Ref. [11].

when compared to measured ignition delay times at pressures greater than 4 bar [50]. In conclusion, the hetero/homogeneous models of this section can be used for reactor design over the provided pressure and pressure ranges. Nonetheless, a reduction of the gas-phase mechanism, in a fashion similar to methane hetero/homogeneous studies [8], will be needed for elaborate parametric reactor studies.

7.3.3
Fuel-Lean Combustion of Hydrogen/Air on Platinum

The fuel-lean ($\varphi \leq 0.32$) combustion of hydrogen/air mixtures over polycrystalline platinum was investigated at pressures of up to 15 bar and temperatures 950 K $\leq T_w \leq$ 1220 K [5, 12–14]. Here, a special water cooling arrangement was applied at the channel entry in order to suppress the superadiabatic surface temperatures [5] that would otherwise appear in the catalytic combustion of the highly diffusionally imbalanced hydrogen fuel (Lewis number, $Le_{H_2} \approx 0.3$). Homogeneous mechanisms have been assessed and the interplay of laminar transport and chemistry was clarified. On the other hand, catalytic kinetic studies could not be performed, since the high reactivity of hydrogen on platinum yielded mass transport-limited conversion, a fact further attested by the accompanying Raman measurements in Refs [5, 12–14].

Figure 7.10 compares LIF-measured and predicted 2D maps of the OH radical at 1 bar [5]. Model predictions were carried out using the catalytic scheme of Deutschmann et al. [37] and three different gas-phase schemes: Warnatz et al. [38], Mueller et al. [51], and Miller and Bowman [52]. The former gaseous scheme mildly underpredicts the measured homogeneous ignition distance, while the latter two yield significantly shorter ignition distances. The origin of the significant x_{ig} differences between the various schemes is largely attributed to the presence of catalytically

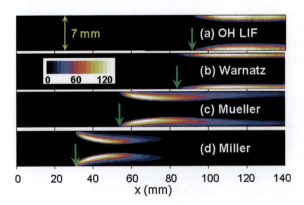

Figure 7.10 LIF-measured (a) and numerically predicted (b–d) distributions of the OH radical in hydrogen/air combustion over Pt: $p = 1$ bar, $\varphi = 0.28$. Predictions with the catalytic scheme of Deutschmann et al. [37] and the gas-phase schemes of (b) Warnatz et al. [38], (c) Mueller et al. [51], and (d) Miller and Bowman [52]. The arrows denote the onset of homogeneous ignition. The color bar provides the OH in ppmv and refers to (a) and (b).

formed H_2O, which is a very efficient third body in the chain-terminating step $H + O_2 + M = HO_2 + M$. The heterogeneously produced water inhibits chemically the onset of homogeneous ignition in fuel-lean combustion of hydrogen over platinum, an effect opposite to that observed in the earlier fuel-lean methane studies. It is emphasized that the corresponding differences in ignition delay times between the different gaseous schemes shown in Figure 7.10 are as large as 20 ms, a particularly long time for practical systems. The correct prediction of homogeneous ignition is crucial in the design of hydrogen-based catalytic reactors as will be clarified next.

High-pressure experimental and numerical studies have been reported more recently [12–14]; here, the gas-phase scheme of Li et al. [53] was used, which has been especially developed for elevated pressures. Measured and predicted OH distributions (ppmv) are illustrated in Figure 7.11(1, 2) for $\varphi = 0.28$, practically nonpreheated H_2/air mixtures (inlet temperatures $T_{IN} = 310$ K) and pressures of 1 and 4 bar. It is seen that gaseous combustion for nonpreheated reactants is largely suppressed at 4 bar, and this is also the case for pressures higher than 4 bar [12]. To understand the origin of this suppression, the pure homogeneous ignition characteristics of hydrogen are first investigated. Ignition delays of lean H_2/air mixtures ($\varphi = 0.10$ and 0.30) have been computed in a constant pressure ideal batch reactor and their inverses (quantities proportional to the gaseous reactivity) are plotted in Figure 7.12 as a function of pressure with parameter the temperature. At a moderate batch reactor temperature of 950 K (which for the inhomogeneous channel-flow reactor in Figure 7.11 represents an average between the inlet temperature and the wall temperature, with added weight on the latter), the gaseous reactivity initially decreases rapidly with rising pressure, while above 2 bar it only changes modestly. At $T \geq 1000$ K, the reactivity at first increases with rising pressure and then drops, with the turning point shifted to higher pressures for higher temperatures.

The rich behavior of hydrogen gaseous ignition characteristics shown in Figure 7.12 is caused by the competition between the chain-branching step $H + O_2 \Leftrightarrow O + OH$, the chain-terminating step $H + O_2 + M \Leftrightarrow HO_2 + M$ (the latter is favored at low temperatures and high pressures), and the chain-branching sequences $HO_2 + H_2 \Leftrightarrow H_2O_2 + H$ and $H_2O_2 + M \Leftrightarrow 2OH + M$ that overtake the stability of HO_2 in the termination step [54]. The implication for a hetero/homogeneous combustion system is that the catalytic pathway has the opportunity to consume significant amounts of hydrogen during the elongated gas-phase induction zones at high pressures, thus depriving fuel from the gaseous pathway and inhibiting homogeneous ignition. It is emphasized that the catalytic pathway is very effective in converting hydrogen due to the large molecular diffusivity of this species and its very high reactivity on platinum even at moderate temperatures. The suppression of gaseous combustion at high pressures is further aided by the presence of catalytically produced H_2O due to its impact on the chain-terminating step $H + O_2 + M \Leftrightarrow HO_2 + M$, which is accentuated at elevated pressures [12].

A modest increase of inlet temperature to about 400 K suffices to restore vigorous combustion at pressures up to 8 bar (see Figure 7.11(3, 4)). This is because by increasing the inlet temperature, the turning points in Figure 7.12 shift to higher

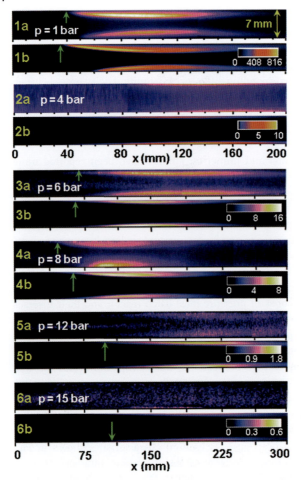

Figure 7.11 LIF-measured (a) and numerically predicted (b) OH distribution (ppmv) in combustion of fuel-lean H_2/air mixtures over Pt. Conditions: (1) $p=1$ bar, $T_{IN}=310$ K, $\varphi=0.28$; (2) $p=4$ bar, $T_{IN}=310$ K, $\varphi=0.28$; (3) $p=6$ bar, $T_{IN}=398$ K, $\varphi=0.25$; (4) $p=8$ bar, $T_{IN}=406$ K, $\varphi=0.25$; (5) $p=12$ bar, $T_{IN}=780$ K, $\varphi=0.10$; (6) $p=15$ bar, $T_{IN}=758$ K, $\varphi=0.10$. Color bars provide the predicted OH in ppmv and arrows define the onset of homogeneous ignition. Adapted from Refs [12–14].

pressures, thus leading to sufficiently high gas-phase reactivities that can in turn favorably compete against the catalytic pathway for hydrogen consumption. On the other hand, at the highest investigated pressures of 12 and 15 bar (Figure 7.11(5, 6)), gas-phase combustion could not be sustained even at inlet temperatures as high as 780 K [14]. This may appear surprising, since for such high preheats the corresponding behavior in Figure 7.12 (temperatures of 1200–1300 K) indicates for $p > 10$ bar either a modest drop or a small increase in reactivity with rising pressure. It has been recently shown [14] that the controlling parameter in this higher pressure range is not ignition chemistry but flame propagation characteristics.

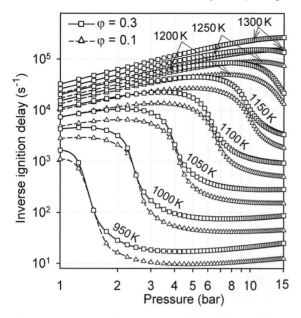

Figure 7.12 Computed inverse ignition delays of H_2/air mixtures in a batch reactor at different pressures and temperatures. Equivalence ratios $\varphi = 0.30$ (squares) and $\varphi = 0.10$ (triangles).

Laminar burning rates computed for 1D freely propagating flames are plotted as a function of pressure in Figure 7.13 for two H_2/air equivalence ratios, $\varphi = 0.30$ and 0.20, and two mixture preheats, $T_0 = 673$ and 773 K; this is to mimic the increase in mixture preheat and drop in hydrogen content with increasing axial distance along the catalytic channel. For $T_0 = 673$ K and $\varphi = 0.30$, the burning rate at 15 bar is 34% lower than the corresponding peak value at ~5 bar; while for $T_0 = 673$ K and $\varphi = 0.20$, the burning rate at 15 bar is 73% lower than the peak value at 1 bar. Similar trends are apparent for the higher preheat of $T_0 = 773$ K. The drop in laminar burning rates at the higher examined pressures leads to a push of the gaseous reaction zone closer to the wall, then to an increased leakage of hydrogen through the gaseous reaction zone, and finally to catalytic conversion of the escaped fuel at the wall [14]. The resulting significant suppression of gaseous combustion at $p > 10$ bar and high preheats in Figure 7.11(5, 6) reflects the combined effects of the reduced laminar burning rates at these pressures and also of the high diffusivity of hydrogen that decreases the availability of fuel in the near-wall hot ignitable region [14]. This picture is also consistent with the earlier hetero/homogeneous combustion studies of methane, where gas-phase combustion intensifies with rising pressure from 1 to 16 bar (see Figure 7.5): methane is nearly diffusionally neutral ($Le_{CH_4} \approx 1$), while its burning rate increases with rising pressure as $p^{0.5}$.

The aforementioned lack of appreciable gas-phase hydrogen combustion at certain regimes of pressure and preheat can be crucial in reactor design. It has been shown [5, 55] that contrary to many premises, the presence of a flame in hydrogen catalytic combustion is advantageous. The reason is that gaseous combustion

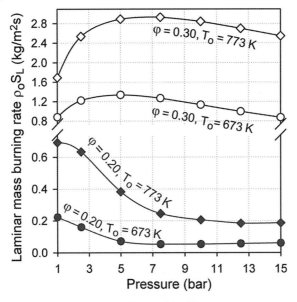

Figure 7.13 Computed 1D laminar burning rates versus pressure for $\varphi = 0.30$ and 0.20 H_2/air mixtures with upstream temperatures $T_0 = 673$ and 773 K.

moderates the reactor temperature by shielding the catalytic surfaces from the hydrogen-rich channel core, thus reducing the heterogeneous conversion that causes superadiabatic surface temperatures. The aforementioned surface temperature moderation can be up to 150 K for typical catalytic reactor confinements [55]. The absence of this moderation can be of concern, particularly for reactor thermal management at elevated pressures, thus exemplifying the need for correct gaseous chemistry description in catalytic combustion of either hydrogen or hydrogen-rich fuels, for example, syngases [55, 56].

7.3.4
Fuel-Rich Combustion of Methane/Air on Rhodium

The catalytic partial oxidation (CPO) of methane to syngas is of interest not only for chemical synthesis but also for new power generation combustion technologies [3, 57, 58]. While in the fuel-lean total oxidation methane studies of Section 7.3.1.1 the Raman data of the deficient fuel were adequate to deduce the catalytic reactivity, in CPO there is a need for Raman measurements of all major species due to the presence of multiple and competing heterogeneous pathways (e.g., complete and partial oxidation, steam reforming, water gas shift, and dry reforming). Raman measurements are used to assess key issues in CPO, such as the methane conversion and synthesis gas (CO and H_2) yields. CPO investigations in the channel-flow reactor of Figure 7.1 have been performed for fuel-rich ($2.5 \leq \varphi \leq 4.0$) methane/air mixtures with or without exhaust gas recycle (EGR) over supported rhodium at pressures up to 10 bar [15, 16].

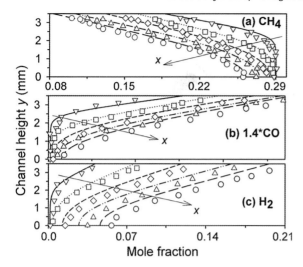

Figure 7.14 Predicted (lines) and measured (symbols) transverse profiles of species mole fractions over half the channel gap (the wall is located at $y=3.5$ mm and the symmetry plane at $y=0$). Catalytic partial oxidation of methane/oxygen over Rh, $p=6$ bar, $\varphi=4.0$, and 38% water dilution: $x=14$ mm (solid lines, lower triangles), $x=48$ mm (dotted lines, squares), $x=88$ mm (short-dashed lines, diamonds), $x=128$ mm (double-dotted dashed lines, upper triangles), $x=168$ mm (long-dashed lines, circles). Adapted from Ref. [16].

Raman-measured and predicted transverse species profiles in CPO of methane/air diluted with exhaust gas recycle over Rh/ZrO$_2$ technical catalysts are provided in Figure 7.14 at 6 bar [16]. Data in Figure 7.14 and earlier measurements in CPO without EGR [15] indicate that over the reactor extent where oxygen is still available, the elementary catalytic scheme of Deutschmann (Schwiedernoch et al. [59]) provides good agreement to the measured major species mole fractions. In the oxygen-depleted zones of the reactor and under operation with EGR, however, the heterogeneous scheme underpredicts the impact of steam reforming and water gas shift reactions, resulting in somewhat lower computed hydrogen yields. Kinetic corrections have been proposed in Ref. [15]; but for power generation systems where oxygen is not fully consumed, the original catalytic scheme suffices for engineering calculations, as will be shown in Section 7.3.5.

Homogeneous ignition at pressures of up to 10 bar has been investigated with LIF of formaldehyde, due to the resulting sub-ppm OH levels at fuel-rich stoichiometries. Figure 7.15 provides comparisons between LIF-measured and predicted 2D distributions of CH$_2$O [16] using the aforementioned catalytic scheme [59] and the C$_2$ gas-phase scheme of Warnatz et al. [38]. The employed gaseous reaction scheme reproduces well the onset of homogeneous ignition. The broadening of the LIF-measured formaldehyde distribution is due to well-known experimental factors [60], while the overprediction of the flame length in Figure 7.15(2b) points to model deficiencies in flame propagation at high pressures. However, as discussed in

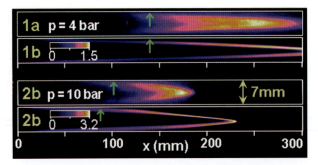

Figure 7.15 (a) LIF-measured and (b) numerically predicted distributions of formaldehyde in catalytic partial oxidation of methane over Rh: (1) $\varphi = 2.5$, $p = 4$ bar, and (2) $\varphi = 4.0$, $p = 10$ bar. The color bars provide the CH_2O in 10^3 ppmv. Adapted from Ref. [16].

Section 7.3.1.2, the propagation characteristics are probably not central in models used for reactor design.

7.3.5
Application of Kinetic Schemes in Models for Technical Systems

The kinetic models evaluated in the preceding sections have been implemented in multidimensional codes in order to assess design issues of practical reactors (e.g., modeling of subscale gas turbine catalytic reactors in Refs [39, 58, 61, 62]). An example is provided next on the ignition and extinction of CPO-based gas turbine honeycomb reactors coated with Rh/ZrO_2 catalysts [61]. In the simulation of technical catalysts and under conditions of negligible intraphase transport (in case, for example, of very thin catalyst coatings in Ref. [61]), the proper interfacial boundary condition for a gaseous species is

$$\left(\varrho Y_k \vec{V}_k\right)_w \cdot \vec{n} + F \dot{s}_k W_k = 0, \tag{7.3}$$

where $\vec{n}$ is the unit outward-pointing vector normal to the surface. The factor F in Eq. (7.3) is the ratio of the active to the geometrical surface area and can be determined with BET (based on physisorption of N_2 or Kr) and chemisorption (of CO or H_2) measurements [62]. A commonly used modeling method is to fit the F factor so as to match experimental results. Although this approach can be realistic in fuel-lean studies where the total oxidation pathway dominates, it may falsify the heterogeneous kinetics in CPO due to the presence of multiple and competing reaction pathways.

Transient CPO studies are performed in a FeCr alloy honeycomb reactor coated with Rh/ZrO_2. The reactor has a length of 75 mm (with only the central 55 mm coated), diameter of 35 mm, and individual channel hydraulic diameter of 1.2 mm (see inset in Figure 7.16). Operating conditions are $p = 5$ bar, $\varphi = 4.0$, exhaust gas dilution of 46.3% H_2O and 23.1% CO_2 per volume, inlet temperature $T_{IN} = 680$ K, and inlet velocity $U_{IN} = 5.6$ m/s. Temperatures within and at the exit of the reactor are monitored with thermocouples (see Figure 7.16), while gas chromatography provides

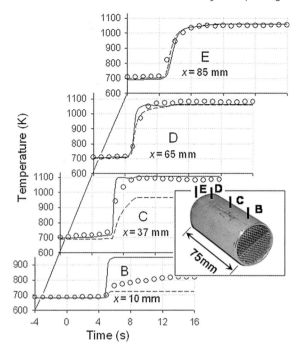

Figure 7.16 Catalytic partial oxidation of methane over Rh/ZrO$_2$, $p = 5$ bar, $T_{IN} = 680$ K, $\varphi = 4.0$, with 46.3% H$_2$O and 23.1% CO$_2$ volume dilution. Predicted (lines) and measured (symbols) temperatures at axial positions B through E in a honeycomb reactor. The reactor (shown in the inset) is 75 mm long, with catalyst applied at 10 mm $\leq x \leq$ 65 mm. The predictions refer to the wall temperature (solid lines) and the mean gas temperature (dashed lines). The origin of the timescale is arbitrary. Adapted from Ref. [61].

the exhaust gas composition. A 2D transient CFD model is used to simulate a single channel of the reactor. The model accounts for heat conduction in the solid, thermal radiation heat transfer between the inner channel surfaces, detailed transport, and elliptic flow description [61]. The experimentally deduced model factor F in Eq. (7.3) is 4.5. At the moderate pressure of 5 bar, gas-phase chemistry is altogether negligible [58, 61].

Measured and predicted temporal profiles of temperature are illustrated in Figure 7.16. Predictions are shown for both the surface and the radially averaged gas temperatures. The predictions capture, at all four positions, the measured elapsed times for the onset of abrupt temperature rise, the temporal extent of the main transient event, and the steady-state temperatures. The measured temperatures at late times ($t \geq 10$ s), where steady state has been practically achieved, are in good agreement with the computed surface temperatures. An exception is location B, with the measurements being in-between the predicted surface and mean gas temperatures. However, this can be attributed to the very steep spatial temperature gradients at B, which is located at the beginning of the catalytically coated section. Additional computations with $T_{IN} \leq 670$ K do not produce light-off (in the sense of a vigorous

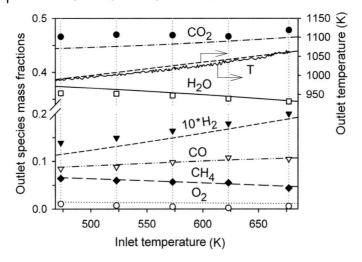

Figure 7.17 Predicted (lines) and measured (symbols) outlet species mass fractions in catalytic partial oxidation of methane, as a function of T_{IN}; the other parameters are as in Figure 7.16. On the same graph, the predicted (dashed line) and measured (solid line) outlet gas temperatures are also given. Adapted from Ref. [61].

burning solution). This is in good agreement with the experimentally assessed minimum inlet temperature for ignition ($T_{IN} = 680$ K).

Extinction was investigated by lowering the inlet temperature and monitoring the exhaust gas composition and the reactor temperature. Measurements and predictions in Figure 7.17 are in good agreement with each other, clearly showing that combustion can be sustained at inlet temperatures at least as low as 473 K in CPO with EGR. The model captures the increase in total oxidation products and the decrease in syngas yields with decreasing inlet temperature, despite some growing deviations at the lowest examined inlet temperatures. Such predictive capacity is important in defining the stable operating window for CPO-based gas turbine reactors. It is finally pointed out that contrary to gaseous reaction mechanisms, catalytic mechanisms may not always be thermodynamically and entropically consistent [63]. Care must be exercised in the theoretical construction of such mechanisms, since experiments in short contact time reactors (with operation far from equilibrium) may not be adequate in revealing such deficiencies.

7.4
Evaluation of Transport

Laminar interphase transport is encountered in most applications of heterogeneous catalysis. The use of mixture-average diffusion models, accounting for thermal diffusion when needed [5], suffices for most 2D simulations. However, this is not the case in 1D models with lumped heat and mass transport coefficients. Although Nusselt and Sherwood number correlations have been developed for catalytic laminar channel flows [64], their universal applicability is not warranted. Moreover, 1D models fail to describe homogeneous combustion due to the strong dependence of the gaseous processes on the boundary layer profiles of the species and temperature. Maintaining the focus of this chapter on 2D approaches, turbulence modeling is the main challenge in the interphase transport and is examined in Section 7.4.1. Intraphase transport is finally briefly outlined in Section 7.4.2.

7.4.1
Turbulent Transport in Catalytic Systems

In catalytic reactors of large turbines operated at full load, the inlet Reynolds numbers in each individual honeycomb channel can be up to 40 000. In entry channel flows with wall heat transfer, which is relevant to catalytic systems due to the heat release at the surface, the controlling parameters are the magnitude of the incoming turbulence and the laminarization of the flow [65]. Direct numerical simulation (DNS) in the entry region of a heated pipe airflow at a moderate Re_{IN} of 4300 [66], supplemented with heat transfer experiments, has led to the development of advanced low Reynolds (LR) number models with damping functions suitable for strongly heated and developing channel flows [65]. Appropriate turbulent models for CST flows were developed in Refs [33, 34, 36]. A moment closure approach was adopted [34, 36], where the Favre-averaged transport equation of any gas-phase variable φ is given by

$$\frac{\partial(\bar{\varrho}\tilde{u}\tilde{\varphi})}{\partial x} + \frac{\partial(\bar{\varrho}\tilde{v}\tilde{\varphi})}{\partial y} - \frac{\partial}{\partial x}\left(\Gamma_{\text{eff}} \frac{\partial\tilde{\varphi}}{\partial x}\right) - \frac{\partial}{\partial y}\left(\Gamma_{\text{eff}} \frac{\partial\tilde{\varphi}}{\partial y}\right) = \tilde{S}_{\varphi}, \quad (7.4)$$

$$\Gamma_{\text{eff}} = \Gamma_1 + \mu_t/\sigma_{\varphi} \quad \text{and} \quad \mu_t = c_\mu f_\mu \bar{\varrho} \tilde{k}^2/\tilde{\varepsilon}. \quad (7.5)$$

For the catalytic reaction rates, the large thermal inertia of the solid wall suppresses the surface temperature fluctuations and eliminates the major reaction nonlinearity [34], thereby greatly simplifying the surface treatment under turbulent flow conditions.

Three different LR turbulence models have been tested in Ref. [34]. The first is the two-layer model of Chen and Patel [67] developed for nonheated flows. The other approaches are the LR model of Ezato *et al.* [65] that has been validated in strongly laminarizing heated channel flows, and the LR model of Hwang and Lin [68] that was developed for channel flows with and without heat transfer. Raman and OH-LIF measurements were carried out in hydrogen/air turbulent combustion over platinum at $p = 1$ bar, in a manner similar to the laminar studies of Section 7.3.3. In addition, particle image velocimetry provided the 2D velocity field. The PIV laser sheet had the same orientation as the OH-LIF sheet in Figure 7.1. The gas-phase scheme or Warnatz *et al.* [38] (validated in the laminar studies of Section 7.3.3) and the catalytic of Deutschmann *et al.* [37] were used in the simulations.

Predicted and measured transverse profiles of the mean velocity $\tilde{U}$ and the turbulent kinetic energy $\tilde{k}$ are shown in Figure 7.18 for two cases with comparable Re_{IN}, one reacting and another nonreacting (the flow consists of air). In the latter, the

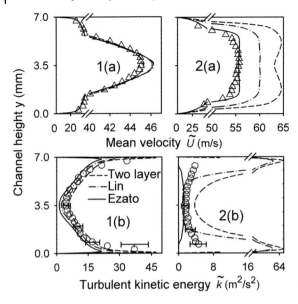

Figure 7.18 Transverse profiles of (a) mean axial velocities $\tilde{U}$ and (b) turbulent kinetic energies $\tilde{k}$ for the nonreacting case 1 (300 K airflow with $Re_{IN} = 35\,240$) and the reacting case 2 (wall temperatures up to 1220 K, $Re_{IN} = 30\,150$). The symbols are PIV measurements and the lines are predictions with different LR turbulence models. Adapted from Ref. [34].

two-layer model [67] provides very good agreement with the measurements. The heat transfer models of Hwang and Lin [68], and particularly Ezato et al. [65], appreciably underpredict (overpredict) $\tilde{k}$ ($\tilde{U}$) (Figure 7.18(1a, 1b)). These models are thus over-dissipative, clearly demonstrating the aptness of the two-layer model and the inapplicability of the heat transfer LR models in isothermal channel flows. On the other hand, the LR model of Ezato provides good agreement to the measured $\tilde{U}$ for the reacting case in Figure 7.18(2a); here, the Hwang and Lin model [68], and to a greater extent the two-layer model [67], overpredicts $\tilde{U}$ significantly. The last two models yield considerably higher near-wall $\tilde{k}$ (Figure 7.18(2b)) that in turn enhances the transport of heat from the hot wall to the channel core, leading to an unrealistic acceleration of the mean flow (Figure 7.18(2a)). In conclusion, from the tested heat transfer LR models, only the Ezato model provides a realistically strong turbulence damping for catalytic combustion applications.

Measured and predicted 2D maps of the OH radical have revealed [34] that only the model of Ezato [65] reproduces the measured location of homogeneous ignition and the flame shape in all cases. This is because the two-layer and Hwang and Lin models, with their enhanced turbulent transport, lead to increased upstream catalytic conversion that deprives hydrogen from the homogeneous pathway, thus inhibiting homogeneous ignition. Using the validated LR model of Ezato, additional computations of hydrogen/air catalytic combustion over platinum have been carried out in commercial channel geometries [34]. Uniform turbulent kinetic energies and turbulent dissipation rates are used as inlet boundary conditions, according to

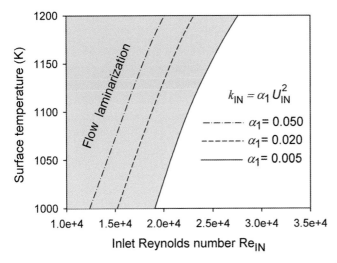

Figure 7.19 Domains of flow laminarization in catalytic combustion. The lines correspond to different inlet turbulent kinetic energies according to Eq. (7.6). Adapted from Ref. [34]

$$\tilde{k}_{IN} = a_1 \tilde{U}_{IN}^2 \quad \text{and} \quad \tilde{\varepsilon}_{IN} = (c_\mu/a_2) k_{IN}^{3/2}/r_h, \tag{7.6}$$

where r_h is the channel hydraulic radius and $\alpha_2 = 0.03$. The computed flow laminarization domains are depicted in Figure 7.19. In the shaded zones, the flow laminarizes and the turbulent hydrogen conversion differs by less than 10% from the corresponding conversion of a simpler laminar model. As seen in Figure 7.19, the inlet turbulence (determined by the parameter α_1 in Eq. (7.6)) is crucial in determining the extent of the flow laminarization domain.

7.4.2
Modeling Directions in Intraphase Transport

Modeling of intraphase transport is important when diffusion limitations arise in the porous catalyst layer. Present models define an effective diffusivity by combining Knudsen and molecular diffusion and describe the porous transport with macroscopic approaches such as the dusty gas model [69]. A comparative study of different macroscopic models for intraphase transport (effectiveness factor and multidimensional reaction–diffusion approaches) has been carried out in Mladenov et al. [70] for automotive applications. More fundamental approaches suited for mesoscale flows (Knudsen numbers up to 0.1) where the continuous Navier Stokes equations break down include the lattice Boltzmann models. Lattice Boltzmann techniques involve solution of discrete velocity distributions, a major simplification to the kinetic theory continuous Boltzmann velocity distribution [71]. Moreover, LB models can be particularly efficient in the simulation of complex geometries and multiphase flows.

Multispecies description in the LB formulation has been proposed in Refs [72–76]; in particular, Arcidiacono et al. [73] included separate velocity distributions for each individual chemical species. Thermal models allowing an LB-consistent treatment of the energy equation have also advanced [77]. The inclusion of catalytic reactions in LB models has followed suit [78–81]. LB models can capture the physicochemical processes in microchannels and porous media with catalytic surfaces and can either be incorporated as submodels in larger multiscale CFD codes or used to refine existing macroscopic models. The validation of LB models against experiments is a demanding task and requires further advancement of measuring techniques at the microscale. Validation is in many cases possible by comparing with results of direct simulation Monte Carlo. This approach was used to demonstrate that LB multicomponent models can capture the dependence of the velocity slip coefficient on the species concentrations [73].

7.5
Conclusions

In situ techniques for the measurement of gas-phase thermoscalars over the catalyst boundary layer have fostered the fundamental investigation of hetero/homogeneous chemistry coupling at realistic operating conditions. The methodology for validating reaction mechanisms was presented for the fuel-lean combustion of methane, propane, and hydrogen over polycrystalline platinum and for the fuel-rich combustion of methane over supported rhodium. It has been shown that even for simple fuels, the underlying hetero/homogeneous kinetics and their coupling were not formerly well understood. The impact of pressure on the catalytic reactivity of hydrocarbon fuels has been clarified. Key to the aptness of any heterogeneous reaction scheme to industrially relevant pressures is the capture of the reduction of the free surface sites with rising pressure, which in turn moderates the rate of increase of the catalytic reactivity with increasing pressure. Moreover, while the gaseous reactivity of hydrocarbon fuels increases with rising pressure, for the particular temperature ranges of catalytic systems the gaseous reactivity of hydrogen may either decrease or increase with rising pressure, for pressures up to 15 bar. The inclusion of validated kinetic schemes in advanced CFD codes has provided a powerful tool for the design of modern catalytic reactors, as shown by experiments and simulations in industrial geometries.

In addition to kinetics, fluid transport deserves special attention in models used for heterogeneous catalysis. Emphasis was placed on interphase transport, from the bulk of the gas to the catalyst surface. It was shown that the turbulent transport in heterogeneous combustion systems is controlled by the flow laminarization due to heat transfer from the hot catalytic walls. Near-wall turbulence models adapted from heat transfer studies describe adequately the degree of relaminarization under the intense wall heating relevant to catalytic systems. In high-temperature catalytic combustion applications, the domains of laminarization have been mapped out in terms of the inlet turbulent flow properties and the wall temperature. Finally, intraphase transport in the catalyst porous structure was briefly addressed with new lattice Boltzmann modeling methods. These techniques are suited for mesoscale flows relevant to many catalytic systems. The validation of LB models, however, would require advancement in measuring techniques at the microscale and also in *ab initio* simulations.

Acknowledgments

Support was provided by the Swiss Competence Center of Energy and Mobility (CCEM), Federal Office of Energy (BFE), European Union under the H_2-IGCC project, and Alstom Power of Switzerland. Material in certain figures, as denoted in the corresponding captions, has been reproduced with permission from Elsevier and Taylor & Francis.

Notation

c_μ	turbulence constant
f_μ	damping function
k	turbulent kinetic energy
S_φ	source term for variable φ
$\dot{s}_k$	catalytic molar reaction rate of species k
u, v	streamwise and transverse velocity components
$\vec{V}_k$	diffusion velocity vector of species k
W_k	species molecular weight

Greek Symbols

$\Gamma_l, \Gamma_{\text{eff}}$	laminar and effective turbulent transport coefficients
ε	dissipation rate of turbulent kinetic energy
μ_t	turbulent viscosity
ϱ	density
σ_φ	turbulent Prandtl or Schmidt number
φ	scalar variable or fuel-to-air equivalence ratio

Superscripts

$\sim, -$	Favre and Reynolds averages

Subscripts

ag, ig	appreciable gaseous conversion, homogeneous ignition
w, IN	wall, inlet

Abbreviations

CPO catalytic partial oxidation
CST catalytically stabilized thermal combustion
GHSV gas hourly space velocity
LIF laser induced fluorescence

References

1 Jannasch, A.K., Silversand, F., Berger, M., Dupuis, D., and Tena, E. (2006) *Catal. Today*, **117**, 433–437.

2 Karagiannidis, S., Mantzaras, J., Jackson, G., and Boulouchos, K. (2007) *Proc. Combust. Inst.*, **31**, 3309–3317.

3 Griffin, T., Winkler, D., Wolf, M., Appel, C., and Mantzaras, J. (2004) ASME Turbo Expo (2004), Paper No. 2004–54101.

4 Griffin, T.A., Calabrese, M., Pfefferle, L.D., Sappey, A., Copeland, R., and Crosley, D.R. (1992) *Combust. Flame*, **90**, 11–33.

5 Appel, C., Mantzaras, J., Schaeren, R., Bombach, R., Inauen, A., Kaeppeli, B., Hemmerling, B., and Stampanoni, A. (2002) *Combust. Flame*, **128**, 340–368.

6 Reinke, M., Mantzaras, J., Schaeren, R., Bombach, R., Inauen, A., and Schenker, S. (2005) *Proc. Combust. Inst.*, **30**, 2519–2527.

7 Reinke, M., Mantzaras, J., Schaeren, R., Bombach, R., Inauen, A., and Schenker, S. (2004) *Combust. Flame*, **136**, 217–240.

8 Reinke, M., Mantzaras, J., Bombach, R., Schenker, S., and Inauen, A. (2005) *Combust. Flame*, **141**, 448–468.

9 Dogwiler, U., Mantzaras, J., Benz, P., Kaeppeli, B., Bombach, R., and Arnold, A. (1998) *Proc. Combust. Inst.*, **27**, 2275–2282.

10 Reinke, M., Mantzaras, J., Schaeren, R., Bombach, R., Kreutner, W., and Inauen, A. (2002) *Proc. Combust. Inst.*, **29**, 1021–1029.

11 Karagiannidis, S., Mantzaras, J., Schenker, S., and Boulouchos, K. (2009) *Proc. Combust. Inst.*, **32**, 1947–1955.

12 Mantzaras, J., Bombach, R., and Schaeren, R. (2009) *Proc. Combust. Inst.*, **32**, 1937–1945.

13 Ghermay, Y., Mantzaras, J., and Bombach, R. (2010) *Combust. Flame*, **157**, 1942–1958.

14 Ghermay, Y., Mantzaras, J., Bombach, R., and Boulouchos, K. (2011) *Combust. Flame*, **158**, 1491–1506.

15 Appel, C., Mantzaras, J., Schaeren, R., Bombach, R., Inauen, A., Tylli, N., Wolf, M., Griffin, T., Winkler, D., and Carroni, R. (2005) *Proc. Combust. Inst.*, **30**, 2509–2517.

16 Schneider, A., Mantzaras, J., Bombach, R., Schenker, S., Tylli, N., and Jansohn, P. (2007) *Proc. Combust. Inst.*, **31**, 1973–1981.

17 Carroni, R., Schmidt, V., and Griffin, T. (2002) *Catal. Today*, **75**, 287–295.

18 Christmann, K., Freund, H.J., Kim, J., Koel, B., Kuhlenbeck, H., Morgenstern, M., Panja, C., Pirug, G., Rupprechter, G., Samano, E., and Somorjai, G.A. (2006) *Adsorption of Molecules on Metal, Semiconductor and Oxide Surfaces*, Springer, New York.

19 Somorjai, G.A. and Rupprechter, G. (1999) *J. Phys. Chem. B*, **103**, 1623–1638.

20 Frenken, J. and Hendriksen, B. (2007) *Mater. Res. Soc.*, **32**, 1015–1021.

21 McMillan, N., Snively, C., and Lauterbach, J. (2007) *Surf. Sci.*, **601**, 772–780.

22 Kissel-Osterrieder, R., Behrendt, F., Warnatz, J., Metka, U., Volpp, H.R., and Wolfrum, J. (2000) *Proc. Combust. Inst.*, **28**, 1341–1348.

23 McCarty, J.G. (1995) *Catal. Today*, **26**, 283–293.

24 Groppi, G., Ibashi, W., Tronconi, E., and Forzatti, P. (2001) *Catal. Today*, **69**, 399–408.

25 Lyubovsky, M. and Pfefferle, L. (1999) *Catal. Today*, **47**, 29–44.

26 Song, X., Williams, W.R., Schmidt, L.D., and Aris, R. (1991) *Combust. Flame*, **84**, 292–311.

27 Ikeda, H., Sato, J., and Williams, F.A. (1995) *Surf. Sci.*, **326**, 11–26.

28 Deutschmann, O., Schmidt, R., Behrendt, F., and Warnatz, J. (1996) *Proc. Combust. Inst.*, **26**, 1747–1754.
29 Aghalayam, P., Park, Y.K., and Vlachos, D.G. (2000) *Proc. Combust. Inst.*, **28**, 1331–1339.
30 Sidwell, R.W., Zhu, H.Y., Kee, R.J., and Wickham, D.T. (2003) *Combust. Flame*, **134**, 55–66.
31 Coltrin, M.E., Kee, R.J., Evans, G.H., Meeks, E., Rupley, F.M., and Grcar, J.M. (1996). SPIN: a Fortran program for modeling one-dimensional rotating-disk/stagnation-flow chemical vapor deposition reactors, report No. SAND 91-8003, Sandia National Laboratories, report No. SAND 91-8003, Livermore, CA.
32 Taylor, J.D., Allendorf, M.D., McDaniel, A.H., and Rice, S.F. (2003) *Ind. Eng. Chem. Res.*, **42**, 6559–6566.
33 Appel, C., Mantzaras, J., Schaeren, R., Bombach, R., Kaeppeli, B., and Inauen, A. (2002) *Proc. Combust. Inst.*, **29**, 1031–1038.
34 Appel, C., Mantzaras, J., Schaeren, R., Bombach, R., and Inauen, A. (2005) *Combust. Flame*, **140**, 70–92.
35 Dogwiler, U., Benz, P., and Mantzaras, J. (1999) *Combust. Flame*, **116**, 243–258.
36 Mantzaras, J., Appel, C., Benz, P., and Dogwiler, U. (2000) *Catal. Today*, **59**, 3–17.
37 Deutschmann, O., Maier, L.I., Riedel, U., Stroemman, A.H., and Dibble, R.W. (2000) *Catal. Today*, **59**, 141–150.
38 Warnatz, J., Dibble, R.W., and Maas, U. (1996) *Combustion, Physical and Chemical Fundamentals, Modeling and Simulation*, Springer, New York.
39 Carroni, R., Griffin, T., Mantzaras, J., and Reinke, M. (2003) *Catal. Today*, **83**, 157–170.
40 Westbrook, C.K. and Dryer, F.L. (1981) *Combust. Sci. Technol.*, **27**, 31–43.
41 Moffat, H.K., Glaborg, P., Kee, R.J., Grcar, J.F., and Miller, J.A. (1993). Surface PSR: a Fortran program for modeling well-stirred reactors with gas and surface reactions, report No. SAND91-8001, Sandia National Laboratories, Livermore, CA.
42 Mantzaras, J. and Appel, C. (2002) *Combust. Flame*, **130**, 336–351.
43 Reinke, M., Mantzaras, J., Bombach, R., Schenker, S., Tylli, N., and Boulouchos, K. (2006) *Combust. Sci. Technol.*, **179**, 553–600.
44 Smith, G.P., Golden, D.M., Frenklach, M., Moriarty, N.W., Eiteneer, B., Goldenberg, M., Bowman, C.T., Hanson, R.K., Song, S., Gardiner, W.C., Lissianski, V., and Qin, Z. (2000) BT An Optimized Detailed Chemical Reaction Mechanism for Methane Combustion, Gas Research Institute, http://www.me.berkeley.edu/gri_mech.
45 Hughes, K.J., Turanyi, T., Clague, A., and Pilling, M.J. (2001) *Int. J. Chem. Kinet.*, **33**, 513–538, http://www.chem.leeds.ac.uk/Combustion/methane.htm.
46 Friedrichs, G., Herbon, J.T., Davidson, D.F., and Hanson, R.K. (2002) *Phys. Chem. Chem. Phys.*, **4**, 5778–5788.
47 Garetto, T.F., Rincon, E., and Apesteguia, C.R. (2004) *Appl. Catal. B*, **48**, 167–174.
48 Qin, Z., Lissianski, V.V., Yang, H., Gardiner, W.C., Davis, S.G., and Wang, H. (2000) *Proc. Combust. Inst.*, **28**, 1663–1669.
49 Mantzaras, J. and Benz, P. (1999) *Combust. Flame*, **119**, 455–472.
50 Jomass, G., Zheng, X.L., Zhu, D.L., and Law, C.K. (2005) *Proc. Combust. Inst.*, **30**, 193–200.
51 Mueller, M.A., Kim, T.J., Yetter, R.A., and Dryer, F.L. (1999) *Int. J. Chem. Kinet.*, **31**, 113–125.
52 Miller, J.A. and Bowman, C.T. (1989) *Prog. Energ. Combust. Sci.*, **15**, 273–338.
53 Li, J., Zhao, Z., Kazakov, A., and Dryer, F.L. (2004) *Int. J. Chem. Kinet.*, **36**, 566–575.
54 Glassman, I. (1996) *Combustion*, Academic Press, London.
55 Mantzaras, J. (2008) *Combust. Sci. Technol.*, **180**, 1137–1168.
56 Ghermay, Y., Mantzaras, J., and Bombach, R. (2011) *Proc. Combust. Inst.*, **33**, 1827–1835.
57 Castaldi, M.J., Etemed, S., Pfefferle, W.C., Khanna, V., and Smith, K.O. (2005) *J. Eng. Gas Turb. Power Trans. ASME*, **127**, 27–35.
58 Schneider, A., Mantzaras, J., and Jansohn, P. (2006) *Chem. Eng. Sci.*, **61**, 4634–4646.
59 Schwiedernoch, R., Tischer, S., Correa, C., and Deutschmann, O. (2003) *Chem. Eng. Sci.*, **58**, 633–642.

60 Shin, D.I., Dreier, T., and Wolfrum, J. (2001) *Appl. Phys. B*, **72**, 257–261.
61 Schneider, A., Mantzaras, J., and Eriksson, S. (2008) *Combust. Sci. Technol.*, **180**, 89–126.
62 Eriksson, S., Schneider, A., Mantzaras, J., Wolf, M., and Jaras, S. (2007) *Chem. Eng. Sci.*, **62**, 3991–4011.
63 Mhadeshwar, A.B., Wang, H., and Vlachos, D.G. (2003) *J. Phys. Chem. B*, **107**, 12721–12733.
64 Hayes, R.E. and Kolaczkowski, S.T. (1999) *Catal. Today*, **47**, 295–303.
65 Ezato, K., Shehata, A.M., Kunugi, T., and McEligot, D.M. (1999) *J. Heat Transf. Trans. ASME*, **121**, 546–555.
66 Satake, S., Kunugi, T., Shehata, A.M., and McEligot, D.M. (2000) *Int. J. Heat Fluid Flow*, **21**, 526–534.
67 Chen, H.C. and Patel, V.C. (1988) *AIAA*, **26**, 641–648.
68 Hwang, C.B. and Lin, C.A. (1998) *AIAA*, **36**, 38–43.
69 Kaviany, M. (1999) *Principles of Heat Transfer in Porous Media*, Springer, New York.
70 Mladenov, N., Koop, J., Tischer, S., and Deutschmann, O. (2010) *Chem. Eng. Sci.*, **65**, 812–826.
71 Succi, S. (2001) *The Lattice Boltzmann Equation*, Oxford University Press, New York.
72 Arcidiacono, S., Mantzaras, J., Ansumali, S., Karlin, I.V., Frouzakis, C., and Boulouchos, K. (2006) *Phys. Rev. E*, **74**, 056707.
73 Arcidiacono, S., Karlin, I., Mantzaras, J., and Frouzakis, C. (2007) *Phys. Rev. E*, **76**, 046703.
74 Shan, X.W. and Doolen, G. (1996) *Phys. Rev. E*, **54**, 3614.
75 Asinari, P. (2006) *Phys. Rev. E*, **73**, 056705.
76 Luo, L.S. and Girimaji, S.S. (2003) *Phys. Rev. E*, **67**, 036302.
77 Prasianakis, N. and Karlin, I. (2007) *Phys. Rev. E*, **76**, 016702.
78 Succi, S. (2002) *Phys. Rev. Lett.*, **89**, 064502.
79 Yamamoto, K., Takada, N., and Misawa, M. (2005) *Proc. Combust. Inst.*, **30**, 1509–1515.
80 Arcidiacono, S., Mantzaras, J., and Karlin, I. (2008) *Phys. Rev. E*, **78**, 046711.
81 Succi, S., Smith, G.P., and Kaxiras, E. (2002) *J. Stat. Phys.*, **107**, 343–366.

8
Computational Fluid Dynamics of Catalytic Reactors
Vinod M. Janardhanan and Olaf Deutschmann

8.1
Introduction

Catalytic reactors are generally characterized by the complex interaction of various physical and chemical processes. Monolithic reactors can serve as example, in which partial oxidation and reforming of hydrocarbons, combustion of natural gas, and reduction of pollutant emissions from automobiles are frequently carried out. Figure 8.1 illustrates the physics and chemistry in a catalytic combustion monolith that glows at a temperature of about 1300 K due to the exothermic oxidation reactions. In each channel of the monolith, the transport of momentum, energy, and chemical species occurs not only in flow (axial) direction but also in radial direction. The reactants diffuse to the inner channel wall, which is coated with the catalytic material, where the gaseous species adsorb and react on the surface. The products and intermediates desorb and diffuse back into the bulk flow. Due to the high temperatures, the chemical species may also react homogeneously in the gas phase. In catalytic reactors, the catalyst material is often dispersed in porous structures such as washcoats or pellets. Mass transport in the fluid phase and chemical reactions are then superimposed by diffusion of the species to the active catalytic centers in the pores. The temperature distribution depends on the interaction of heat convection and conduction in the fluid, heat release due to chemical reactions, heat transport in the solid material, and thermal radiation. If the feed conditions vary in time and space and/or heat transfer occurs between the reactor and the ambience, a nonuniform temperature distribution over the entire monolith will result, and the behavior will differ from channel to channel.

Today, the challenge in catalysis is not only the development of new catalysts to synthesize a desired product but also the understanding of the interaction of the catalyst with the surrounding reactive flow field. Sometimes, the exploitation of these interactions can lead to the desired product selectivity and yield. Hence, a better understanding of gas–solid flows in chemical reactors is understood as a critical need in chemical technology calling for the development of reliable simulation tools that

---

Modeling and Simulation of Heterogeneous Catalytic Reactions: From the Molecular Process to the Technical System,
First Edition. Edited by Olaf Deutschmann.
© 2012 Wiley-VCH Verlag GmbH & Co. KGaA. Published 2012 by Wiley-VCH Verlag GmbH & Co. KGaA.

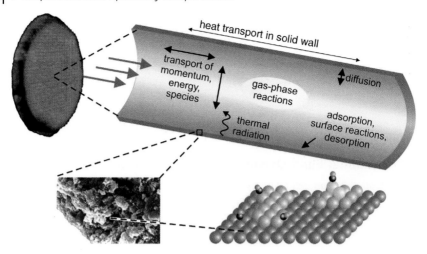

Figure 8.1 Catalytic combustion monolith and physical and chemical processes occurring in the single monolith channel.

integrate detailed models of reaction chemistry and computational fluid dynamics (CFD) modeling of macroscale flow structures.

Computational fluid dynamics is able to predict very complex flow fields, even combined with heat transport, due to the recently developed numerical algorithms and the availability of faster and bigger (memory) computer hardware. The consideration of detailed models for chemical reactions, in particular for heterogeneous reactions, however, is still very challenging due to the large number of species mass conservation equations, their highly nonlinear coupling, and the wide range of timescales introduced by the complex reaction networks.

This chapter introduces the application of CFD simulations to obtain a better understanding of the interactions between mass and heat transport and chemical reactions in catalytic reactors. Concepts for modeling and numerical simulation of catalytic reactors are presented, which describe the coupling of the physical and chemical processes in detail. The elementary kinetics and dynamics as well as ways for modeling the intrinsic chemical reaction rates (microkinetics) by various approaches such as Monte Carlo (MC), mean field approximation (MF), and lumped kinetics are discussed in the earlier chapters of this book. In this chapter, it is assumed that models exist that can compute not only the local heterogeneous but also the homogeneous reaction rate as function of the local conditions such as temperature and species concentrations in the gas phase and of the local and temporal state of the catalyst. These chemical source terms are here coupled with the fluid flow and used to numerically simulate the catalytic reactor.

The ultimate objective of CFD simulations of catalytic reactors is (1) to understand the interactions of physics (mass and heat transport) and chemistry in the reactor, (2) to support reactor design and engineering, and (3) eventually, to find optimized operating conditions for maximization of the desired product's yield and minimization of undesired side products and/or pollutants. Though computational fluid dynamics covers a wide range of problems, ranging from simulation of the flow around airplanes to laminarization of turbulent flows entering a microchannel, this chapter focuses on the principal ideas and the potential applications of CFD in heterogeneous catalysis; textbooks [1, 2] and specific literature are frequently referenced for more details. Specific examples taken from literature and our own work will be used for illustration of the state-of-the-art CFD simulation of chemical reactors with heterogeneously catalyzed reactions. The next chapters of the book will cover some specific topics of numerical simulation of catalytic reactors in more detail.

8.2
Modeling of Reactive Flows

8.2.1
Governing Equations of Multicomponent Flows

As long as a fluid can be treated as a continuum, the most accurate description of the flow field of multicomponent mixtures is given by the transient three-dimensional (3D) Navier–Stokes equations coupled with the energy and species governing equations, which will be summarized in this section. More detailed introduction to fluid dynamics and transport phenomena can be found in a number of textbooks [1–5]. Other alternative concepts such as Lattice-Boltzmann models have also been discussed for simulation of catalytic reactors as introduced in Section 8.4.1.

Governing equations, which are based on conservation principles, can be derived by consideration of the flow within a certain spatial region, which is called the *control volume*.

The principle of mass conservation leads to the mass continuity equation

$$\frac{\partial \varrho}{\partial t} + \frac{\partial (\varrho v_i)}{\partial x_i} = S_m, \tag{8.1}$$

with ϱ being the mass density, t the time, x_i ($i = 1,2,3$) are the Cartesian coordinates, and v_i the velocity components. The source term S_m vanishes unless mass is either deposited on or ablated from the solid surfaces. The Einstein convention is used here, that is, whenever the same index appears twice in any term, summation over that index is implied, except when the index refers to a chemical species. The principle of momentum conservation for Newtonian fluids leads to three scalar equations for the momentum components ϱv_i

$$\frac{\partial (\varrho v_i)}{\partial t} + \frac{\partial (\varrho v_i v_j)}{\partial x_j} + \frac{\partial p}{\partial x_i} + \frac{\partial \tau_{ij}}{\partial x_j} = \varrho g_i, \tag{8.2}$$

where p is the static pressure, τ_{ij} is the stress tensor, and g_i are the components of the gravitational acceleration. The above equation is written for Cartesian coordinates. Gravity, the only body force taken into account, can often be neglected when modeling catalytic reactors. The stress tensor is given as

$$\tau_{ij} = -\mu\left(\frac{\partial v_i}{\partial x_j} + \frac{\partial v_j}{\partial x_i}\right) + \left(\frac{2}{3}\mu - \kappa\right)\delta_{ij}\frac{\partial v_k}{\partial x_k}. \quad (8.3)$$

Here, κ and μ are the bulk viscosity and mixture viscosity, respectively, and δ_{ij} is the Kronecker delta, which is unity for $i=j$, else zero. The bulk viscosity vanishes for low-density monoatomic gases and is also commonly neglected for dense gases and liquids [1]. The coupled mass continuity and momentum governing equations have to be solved for the description of the flow field.

In multicomponent mixtures, not only the flow field is of interest but also the mixing of the chemical species and reactions among them, which can be described by an additional set of partial differential equations. Here, the mass m_i of each of the N_g gas-phase species obeys a conservation law that leads to

$$\frac{\partial(\varrho Y_i)}{\partial t} + \frac{\partial(\varrho v_j Y_i)}{\partial x_j} + \frac{\partial(j_{i,j})}{\partial x_j} = R_i^{\text{hom}}, \quad (8.4)$$

with Y_i is the mass fraction of species i in the mixture ($Y_i = m_i/m$) with m as total mass, R_i^{hom} is the net rate of production due to homogeneous chemical reactions. The components $j_{i,j}$ of the diffusion mass flux caused by concentration and temperature gradients are often modeled by the mixture average formulation [6]:

$$j_{i,j} = -\varrho \frac{Y_i}{X_i} D_i^M \frac{\partial X_i}{\partial x_j} - \frac{D_i^T}{T}\frac{\partial T}{\partial x_j}. \quad (8.5)$$

D_i^M is the effective diffusion coefficient of species i in the mixture, D_i^T is the thermal diffusion coefficient, which is significant only for light species, and T is the temperature. The molar fraction X_i is related to the mass fraction Y_i using the species molar masses M_i by

$$X_i = \frac{1}{\sum_{j=1}^{N_g} Y_j/M_j}\frac{Y_i}{M_i}. \quad (8.6)$$

Heat transport and heat release due to chemical reactions lead to spatial and temporal temperature distributions in catalytic reactors. The corresponding governing equation for energy conservation is commonly expressed in terms of the specific enthalpy h:

$$\frac{\partial(\varrho h)}{\partial t} + \frac{\partial(\varrho v_j h)}{\partial x_j} + \frac{\partial j_{q,j}}{\partial x_j} = \frac{\partial p}{\partial t} + v_j\frac{\partial p}{\partial x_j} - \tau_{jk}\frac{\partial v_j}{\partial x_k} + S_h, \quad (8.7)$$

with S_h being the heat source, for instance, due to thermal radiation. In multicomponent mixtures, diffusive heat transport is significant due to heat conduction and mass diffusion, hence

$$j_{q,j} = -\lambda\frac{\partial T}{\partial x_j} + \sum_{i=1}^{N_g} h_i j_{i,j}. \quad (8.8)$$

Here, λ is the thermal conductivity of the mixture. The temperature is then related to the enthalpy by the definition of the mixture-specific enthalpy

$$h = \sum_{i=1}^{Ng} Y_i h_i(T), \tag{8.9}$$

with h_i being the specific enthalpy of species i, which is a monotonic increasing function of temperature. The temperature is then commonly calculated from Eq. (8.9) for known h and Y_i by employing a root finding algorithm.

Heat transport in solids such as reactor walls and catalyst materials can also be modeled by an enthalpy equation, for instance, in the form of

$$\frac{\partial(\varrho h)}{\partial t} - \frac{\partial}{\partial x_j}\left(\lambda \frac{\partial T}{\partial x_j}\right) = S_h, \tag{8.10}$$

where h is the specific enthalpy and λ the thermal conductivity of the solid material. S_h accounts for heat sources, for instance, due to heat release by chemical reactions and electric or radiative heating of the solid.

This system of governing equations is closed by the equation of state to relate the thermodynamic variable density ϱ, pressure p, and temperature T. The simplest model of this relation for gaseous flows is the ideal gas equation

$$p = \frac{\varrho RT}{\sum_{i=1}^{N_g} X_i M_i}, \tag{8.11}$$

with the universal gas constant $R = 8.314 \text{ J}/(\text{mol K})$.

The transport coefficients μ, D_i^M, D_i^T, and λ appearing in Eqs. (8.3), (8.5), and (8.8) depend on temperature and mixture composition. They are derived from the transport coefficients of the individual species and the mixture composition by applying empirical approximations [1, 2, 4], which eventually lead to two physical parameters for each species, a characteristic diameter (the Lennard–Jones collision diameter), σ_i, and a characteristic energy (the Lennard–Jones potential well depth), ε_i, which can be taken from databases [7].

The specific enthalpy h_i is a function of temperature and can be expressed in terms of the heat capacity

$$h_i = h_i(T_{\text{ref}}) + \int_{T_{\text{ref}}}^{T} c_{p,i}(T') dT', \tag{8.12}$$

where $c_{p,i}$ is the specific heat capacity at constant pressure. The specific standard enthalpy of formation $\Delta h_{f,298,i}^0$ can be used as integration constant $h_i(T_{\text{ref}} = 298.15 \text{ K}, p_0 = 1 \text{ bar})$. Experimentally determined and estimated standard enthalpies of formation, standard entropies, and temperature-dependent heat capacities can be found in databases [8–10] or estimated by Benson's additivity rules [11].

8.2.2
Turbulent Flows

Turbulent flows are characterized by continuous fluctuations of velocity, which can lead to fluctuations in scalars such as density, temperature, and mixture composition. Turbulence can be desired in catalytic reactors to enhance mixing and reduce mass transfer limitations, but can be unwanted due to the increased pressure drop and energy dissipation. An adequate understanding of all facets of turbulent flows is still lacking [4, 12, 13]. In the area of catalytic systems, some progress has recently been made in turbulent flow modeling, for example, in catalytically stabilized combustion [14, 15]. The Navier–Stokes equations as presented above are, in principle, able to model turbulent flows (direct numerical simulation). However, in practice, the solutions of the Navier–Stokes equations for turbulent flows in technical reactors demand a prohibitive amount of computational time due to the huge number of grid points needed to resolve the small scales of turbulence. Therefore, several concepts were developed to model turbulent flows by the solution of averaged governing equations. However, the equation system is not closed, meaning that a model has to be set up to describe the so-called Reynold stresses that are the correlations between the velocity fluctuations and the fluctuations of all the quantities of the flow (velocity, enthalpy, and mass fractions). The $k - \varepsilon$ model [16] is one of the most widely used concept for modeling the Reynold stresses at high Reynolds numbers, which adds two additional partial differential equations for the description of the turbulent kinetic energy, k, and the dissipation rate, ε, to the governing equations. Although the model has well-known deficiencies, it is today implemented in most commercial CFD codes and also widely used for the simulation of catalytic reactors. Recently, turbulent flow field simulations are often based on large-eddy simulation (LES), which combines DNS for the larger scales with a turbulence model, for example, $k - \varepsilon$ model, for the unresolved smaller scales.

Aside form this closure problem, one still has to specify the averaged chemical reaction rates [4, 12]. Because of the strong nonlinearity of the rate coefficients due to the exponential dependence on temperature and the power law dependence on partial pressure, the source terms of chemical reactions in turbulent flows cannot be computed using average concentrations and temperature. Here, probability density functions (PDFs) [4], either derived by transport equations [13] or empirically constructed [17], are used to take the turbulent fluctuations into account when calculating the chemical source terms. For the simulation of reactions on catalysts, it is important to use appropriate models for the flow laminarization at the solid surface.

8.2.3
Three-Phase Flow

Three-phase flows involve the participation of solid, liquid, and gaseous phases. In certain cases, the solid phase will be a porous medium, and the fluids will flow though the pore networks. In certain other cases, all phases will be mobile and these flows are usually characterized by various regimes such as particle-laden flow, fluidized bed flow, slug flow, bubbly flow, and so on. Examples for three-phase flow devices with chemical reactions are fluidized bed reactors. They are one of the most important classes of multiphase reactors used in chemical, petrochemical, and biochemical processing. Simulating multiphase reactors is a challenge due to the numerous physicochemical processes occurring in the reactor. For example, one has to account for interactions between and among various phases, lift, buoyancy, virtual mass forces, particle agglomeration, and bubble coalescence [18].

Either the Euler–Lagrange model or Euler–Euler model can be used to solve the three-phase flow problem. The former adopts a continuum description for the liquid phase and tracks the discrete phases using Lagrangian particle trajectory analysis. The Euler–Euler model is based on the concept of interpenetrating continua. Here, all the phases are treated as continua with properties analogous to those of a fluid. That is, conservation equations are derived for each of the phases and constitutive relations that are empirical in nature closes the equation set. Therefore, the accuracy of this method heavily relies on the empirical constitutive relations used. Furthermore, this approach has limitation in predicting certain characteristics of discrete flow. For instance, the method cannot account for particle size effects, particle agglomeration, bubble coalescence, and bubble breakage. On the other hand, the Euler–Lagrange model has empirical equations and can provide detailed information of discrete phases. However, it is computationally more expensive. A detailed description of three-phase flow modeling is beyond the scope of this chapter and interested readers can refer to textbooks [19–21].

8.2.4
Momentum and Energy Equations for Porous Media

Porous media are present everywhere in catalytic reactors [22, 23], for instance, fixed bed reactors, catalytic filters, washcoat layers, perforated plates, flow distributors, tube banks, membranes, electrodes, fiber materials, and so on. Modeling the transport and reactions in the actual tortuous structure on the microscopic level is a rather formidable task [23, 24]. Chapter 5 of this book deals in detail with this topic. Due to this complexity, it is often necessary to work with small representative volume elements where the porous medium and other properties are assumed to be homogenized. Several methods have been developed to include porous media and reactions in CFD simulations.

Most porous media models in CFD codes incorporate an empirically determined flow resistance accounting for the pressure drop, which is a sink in the governing momentum equation (8.2). In case of simple homogeneous porous media, a source term is added to the right side of Eq. (8.2),

$$S_i = -\left(\frac{\mu}{\alpha} v_i + \frac{C}{2} \varrho |\vec{v}| v_i \right), \tag{8.13}$$

where α is the permeability (Darcy's law) and C is the inertial resistance, which can be viewed as a loss per unit length along the flow direction. Concerning the temperature profile in porous media, the enthalpy equations (8.7) and (8.10) have to be adapted.

The total enthalpy is now a sum of the enthalpies of the fluid and the solid. Their partition is defined by porosity. An effective thermal conductivity is used based on the porosity and the thermal conductivities of the fluid and the solid. This continuum approach has to be used carefully; for instance, the effect of the porous medium on turbulent flows can be barely approximated within this concept. The approach, which assumes constant unidirectional flow, also breaks down for fixed bed reactors with reactor diameter being less than ten times the particle size. Thus, the model cannot predict the velocity maximum in the vicinity of the wall observed experimentally for those reactors [25]. An averaged velocity with a radial varying axial component can be provided by further modification of the momentum balance [25–27] as improvement of the classical model.

8.3
Coupling of the Flow Field with Heterogeneous Chemical Reactions

Depending on the spatial resolution of the different catalyst structures such as flat surface, gauzes, single pellets, and in porous media, the species mass fluxes due to catalytic reactions at these structures are differently coupled with the flow field.

8.3.1
Given Spatial Resolution of Catalyst Structure

In the first case considered, the catalytic layer is resolved in space, that is, the surface of the catalyst is directly exposed to the fluid flow. Examples are thin catalytically coated walls in honeycomb structures, disks, plates, and well-defined porous media (fixed bed reactors, foams, and washcoats), in which the shape of the individual pellet or channel is spatially resolved in the CFD simulation. The chemical processes at the surface are then coupled with the surrounding flow field by boundary conditions for the species continuity equation (8.4) at the gas–surface interface [2, 28]:

$$\vec{n}(\vec{j}_i + \varrho \vec{v}_{\text{Stef}} Y_i) = R_i^{\text{het}}. \tag{8.14}$$

Here, $\vec{n}$ is the outward-pointing unit vector normal to the surface, $\vec{j}_i$ is the diffusion mass flux of species i as discussed in Eq. (8.4), and R_i^{het} is the heterogeneous surface reaction rate, which is given per unit geometric surface area, corresponding to the reactor geometry, in kg/(m² s). Approaches to model the heterogeneous reaction rates R_i^{het} are discussed in Chapter 4 of this book.

The Stefan velocity $\vec{v}_{\text{Stef}}$ occurs at the surface if there is a net mass flux between the surface and the gas phase:

$$\vec{n}\vec{v}_{\text{Stef}} = \frac{1}{\varrho} \sum_{i=1}^{N_g} R_i^{\text{het}}. \tag{8.15}$$

Under steady-state conditions, this mass flux vanishes unless mass is deposited on the surface, for example, chemical vapor deposition, or ablated, for example, material etching. Equation (8.14) basically means that for $\vec{v}_{\text{Stef}} = 0$ the amount of gas-phase molecules of species i, which are consumed/produced at the catalyst by adsorption/desorption, have to diffuse to/from the catalytic wall (Eq. (8.5)). Only for fast transient ($<10^{-4}$ s) adsorption/desorption processes, for example, during ignition of catalytic oxidation, does Eq. (8.14) break down and special treatment of the coupling is needed [29, 30]. In that case, accumulation of species in the near-catalyst zone has to be considered, for example, through [29]

$$\int \varrho \frac{\partial Y_i}{\partial t} dV = -\int (\vec{j}_i + \varrho \vec{v}_{\text{Stef}} Y_i) \vec{n} dA + \int R_i^{\text{het}} dA. \tag{8.16}$$

In that case, special care has to be taken in the spatial discretization procedure [2]. Furthermore, those fast transient processes may lead to heat accumulation terms [29] and to additional convective transport and associated pressure gradients in the fluid phase above the catalyst [30].

Calculation of R_i^{het} is straightforward if the catalytic surface corresponds to the geometrical surface of the fluid–solid interphase of the flow field simulation, for example, wires and flat plates without any porosity. In that case, R_i^{het} is the production rate of species i per *catalyst surface area* due to catalytic reactions (Chapter 4). It should be noted that the catalyst surface area is the surface area (layer on which we find adsorbed species) of the catalytic particle exposed to the ambient gas (fluid) phase, which can be measured, for example, by chemisorptions with sample molecules such as CO and hydrogen. The catalyst surface area should not be confused with the BET surface area.

8.3.2
Simple Approach for Modeling the Catalyst Structure

Most catalystic systems, however, exhibit a certain structure, for instance, they may occur as dispersed particles on a flat or in a porous substrate. The simplest way to account for that structure and the active catalytic surface area consists in scaling the intrinsic reaction rate at the fluid–solid interphase by two parameters. The first parameter represents the amount of catalytically active surface area in relation to the geometric surface area of the fluid–solid interphase, here denoted by $F_{\text{cat/geo}}$:

$$R_i^{\text{het}} = \eta F_{\text{cat/geo}} M_i \dot{s}_i. \tag{8.17}$$

Here, $\dot{s}_i$ is the molar net production rate of gas-phase species i, given in mol/(m^2 s); the area now refers to the actual catalytically active surface area. $F_{\text{cat/geo}}$ can be consequently determined experimentally, for example, by chemisorption measurements. Recently, it was shown that this ratio ($F_{\text{cat/geo}}$) can also serve as parameter to describe the dependence of the overall reaction rate of catalyst loadings and effects of hydrothermal aging for structure-insensitive catalysts [31]. This concept was even applied to model the variation in performance of on-road aged three-way catalysts [32].

The simplest model to include the effect of internal mass transfer resistance for catalysts dispersed in a porous media is the effectiveness factor η based on the Thiele modulus [5, 33]. The effectiveness factor of species i, η_i, is defined as

$$\eta_i = \frac{\dot{s}_{i,\text{mean}}}{\dot{s}_i}, \qquad (8.18)$$

with $\dot{s}_{i,\text{mean}}$ as mean surface reaction rate in the porous structure. Assuming a homogeneous porous medium, time-independent concentration profiles, and a rate law of first order, the effectiveness factor can be analytically calculated in terms of

$$\eta_i = \frac{\tanh(\Phi_i)}{\Phi_i}, \qquad (8.19)$$

with Φ_i as Thiele module defined as

$$\Phi_i = L\sqrt{\frac{\dot{s}_i \gamma}{D_{\text{eff},i} c_{i,0}}}. \qquad (8.20)$$

Here, L is the thickness of the porous medium (washcoat), γ is the ratio of catalytic active surface area to washcoat volume, and $c_{i,0}$ are the species concentrations at the fluid/porous media interface. The Thiele module is a dimensionless number. The value in the root term of Eq. (8.20) represents the ratio of intrinsic reaction rate to diffusive mass transport in the porous structure. Since mass conservation has to be obeyed (Eq. (8.17)), the same effectiveness factor has to be applied for all chemical species. Therefore, this simple model can be applied only under conditions at which the reaction rate of one species determines overall reactivity. Furthermore, this model then implies that mass diffusion inside the porous media can be described by the same diffusion coefficient for all species.

In most fixed bed reactors with large numbers of catalytic pellets, both for nontrivial shapes of the catalysts and for catalyst dispersed in porous media, the structure of the catalyst cannot be resolved geometrically. In these cases, the catalytic reaction rate is expressed per volumetric unit, which means R_i^{het} is now given in kg/(m³ s); the volume here refers to the volume of a computational cell in the geometrical domain of fluid flow. Then R_i^{het} simply represents an additional source term on the right side of the species continuity equation(8.4) and is computed by

$$R_i^{\text{het}} = \eta S_V M_i \dot{s}_i, \qquad (8.21)$$

where S_V is the active catalytic surface area per volumetric unit, given in m^{-1}, determined experimentally or estimated. Both $F_{\text{cat/geo}}$ and S_V can be expressed as function of the reactor position and time to account for inhomogeneously distributed catalysts and loss of activity, respectively. In reactors with more than one catalytic material, a different value for $F_{\text{cat/geo}}$ or S_V can be given for every individual active material or phase, respectively.

8.3.3
Reaction Diffusion Equations

The dispersion of the catalyst material in porous layers or pellets easily leads to a reduced overall reaction rate due to finite diffusion of the reactants to and products from the active sites. The simplest model to account for this mass transport limitation is the effectiveness factor η as introduced above. However, this model fails under conditions in which the reaction rate and diffusion coefficient of more than a single species determine overall reactivity. Like in this case, the interaction of diffusion and reaction demands more detailed models if mass transport in the porous media is dominated rather by diffusion than by convection.

Concentration gradients inside the porous media result in spatial variations in the surface reaction rates $\dot{s}_i$. In thin catalyst layers (washcoats), these are primarily significant in normal direction to the fluid/washcoat boundary. Therefore, one-dimensional reaction diffusion equations are applied with their spatial coordinate in that direction. Each chemical species leads to one reaction diffusion equation, which is written in steady state as

$$\frac{\partial}{\partial r}\left(-D_i^{\text{eff}} \frac{\partial c_i^W}{\partial r}\right) - S_V \dot{s}_i = 0. \tag{8.22}$$

Here, c_i^W denotes the species concentration in the washcoat in normal direction to the boundary fluid/washcoat. D_i^{eff} is the effective diffusion coefficient, which can account for the different diffusion processes in macro- and micropores and can be derived from the binary diffusion coefficients [23, 34]. In addition to Eq. (8.22), the surface coverages can be calculated, assuming a *microkinetics* model is available, according to

$$\frac{d\theta_i}{dt} = \frac{\dot{s}_i \sigma_i}{\Gamma}. \tag{8.23}$$

A heat balance, in which Eqs. (8.7) and (8.10) are combined, may be added to the model to account for temperature variations in the porous media. Since Eq. (8.22) is applicable only for thin catalytic layers or small pellets without net mass fluxes (ablation, deposition, etc.) and internal pressure-driven flows, temperature variations can generally be neglected. Equation (8.22) is coupled with the surrounding flow field, Eq. (8.5), at the interface between open fluid and catalytic layer/pellet, where the diffusion fluxes normal to this interface must compensate. In this model, the species concentrations, catalytic reaction rates, and surface coverages do not only depend on the position of the catalytic layer/pellet in the reactor but also vary inside the catalyst layer/particle leading to CPU time-consuming computations.

8.3.4
Dusty Gas Model

Fluxes within porous media that are driven by gradients in concentration and pressure, that is, diffusion and convection, can be described by the dusty gas model (DGM) [23, 34]. This model, which is also applicable for three-dimensional and larger porous media, not only is superior to the ones discussed in the previous two sections but also leads to more sophisticated computational efforts. The conservation equation (8.4) for reactive porous media species transport at steady state is now written as

$$\frac{\partial(j_{i,j})}{\partial x_j} = R_i^{\text{hom}} + R_i^{\text{het}} = R_i^{\text{hom}} + S_V M_i \dot{s}_i. \tag{8.24}$$

The components j of the gas-phase mass fluxes, $j_{i,j}$, of species i are evaluated by an implicit relationship among the molar concentrations, concentration gradients, and pressure gradients [23, 34]:

$$j_{i,j} = \left[-\sum_{l=1}^{N_g} D_{il}^{\text{DGM}} \frac{\partial c_l}{\partial x_j} - \left(\sum_{l=1}^{N_g} D_{il}^{\text{DGM}} \frac{c_l}{D_{l,\text{Kn}}^{\text{eff}}} \right) \frac{\alpha}{\mu} \frac{\partial p}{\partial x_j} \right] M_i. \tag{8.25}$$

Here, D_{il}^{DGM} are the DGM diffusion coefficients and $D_{i,\text{Kn}}^{\text{eff}}$ are the effective Knudsen diffusion coefficients. The first term on the right-hand side of Eq. (8.25) represents the diffusive flux and the second the viscous flux. The DGM diffusion coefficients can be represented as a matrix inverse $D_{il}^{\text{DGM}} = H^{-1}$, where the elements of the H matrix are given by

$$h_{il} = \left[\frac{1}{D_{i,\text{Kn}}^{\text{eff}}} + \sum_{j \neq i} \frac{X_j}{D_{ij}^{\text{eff}}} \right] \delta_{il} + (\delta_{il} - 1) \frac{X_j}{D_{il}^{\text{eff}}}. \tag{8.26}$$

The effective binary diffusion coefficients D_{il}^{eff} in the porous media are related to the ordinary binary diffusion coefficient D_{il} by

$$D_{il}^{\text{eff}} = \frac{\Phi_g}{\tau_g} D_{il}, \tag{8.27}$$

with $\Phi_g =$ porosity and $\tau_g =$ tortuosity. The effective Knudsen diffusion coefficient can be expressed as

$$D_{i,\text{Kn}}^{\text{eff}} = \frac{2}{3} \frac{\Phi_g}{\tau_g} r_p \sqrt{\frac{8RT}{\pi M_i}}, \tag{8.28}$$

where r_p is the average pore radius.

A critical evaluation of transport models including DGM and the development of a more general concept have been proposed by Kerkhof [35, 36]. For more on transport in porous media in interaction with catalytic reactions, the reader may refer to Chapters 5 and 6.

8.4
Numerical Methods and Computational Tools

There are a variety of methods to solve the coupled system of partial differential equations (PDE) and algebraic equations, which were presented in the previous sections for modeling catalytic reactors. Very often, the transient three-dimensional governing equations are simplified (no time dependence, symmetry, preferential flow direction, infinite diffusion, etc.) as much as possible, but still taking care of all significant processes in the reactor. Simplifications often are not straightforward and need to be conducted with care. Special algorithms were developed for special types of reactors to achieve a converged solution or to speed up the computation.

8.4.1
Numerical Methods for the Solution of the Governing Equations

An analytical solution of the PDE system is possible only in very limited special cases; for all practical cases, a numerical solution is needed. Numerical solution means that algebraic equations are derived that approximate the solution of the PDE system at discrete points of the geometrical space of the reactor. The way of selection of these grid points and the derivation of algebraic equations, which are finally solved by the computer, is called discretization. Since the solution of the discretized equations is only an approximation of the solution of the PDE system, an error analysis is an essential feature of the interpretation of every CFD simulation.

The three major methods of discretization [37] are the methods of finite differences (FDM), finite volumes (FVM), and finite elements (FEM). The simplest method is FDM, which is based on a Taylor series expansion of the solution vector between neighboring grid points and applied for well-structured grids. The chosen number of terms of the Taylor series determines the accuracy. In contrast to FDM, the finite volume method can be applied for unstructured grids so that for regions with larger gradients more grid points can be chosen, well adapted to the reactor behavior. FVM calculates the dependent variables not for certain points but for certain volumes. Source terms within cells and fluxes through the boundaries of these cells are considered to derive the local values, which not only makes this method very physically descriptive but also allows simple error estimation.

The most universal method from a mathematical point of view is FEM [38, 39]; FDM and FVM can be considered as special cases of FEM. FEM originates from structural mechanics and has meanwhile found increased use in CFD. FEM generates the computational grid in a very adaptive way and is therefore ideal for complex geometries. Furthermore, FEM-based codes are suitable for the application of parallel computing. The great flexibility of FEM regarding the description of the solution and its convergence comes at the cost of a higher complexity of the computer program. Today, all commercial CFD codes are based on either FVM or FEM.

Very different from these three methods are the lattice–Boltzmann methods (LBMs) [40], which have become popular in particular for the simulation of complex flow structures found in fixed beds [41–43]. The LBM may be considered as a finite difference method for a discrete Boltzmann equation. The method simulates hydrodynamic or mass transport phenomena by tracking the time evolution of particle distribution functions confined to a lattice moving with discrete velocity during discrete advances in time. Each time step is subdivided into separate streaming and collision steps. It could be shown that correctly chosen particle distribution functions recover the Navier–Stokes equations. LBM for reaction engineering applications is still under development; in particular, the implementation of heat transport and complex reaction schemes seems to be difficult. There is no commercial code based on LBM available yet.

8.4.2
CFD Software

Available multipurpose commercial CFD codes can simulate very complex flow configurations including turbulence and multicomponent transport based on FVM and FEM. However, CFD codes still have difficulties to implement complex models for the chemical processes. One problem is the insufficient number of reactions and species the codes can handle. An area of recent development is the implementation of detailed models for heterogeneous reactions.

Several software packages have been developed for modeling complex reaction kinetics in CFD such as CHEMKIN [44], CANTERA [45], DETCHEM [46], which also offer CFD codes for special reactor configurations such as channel flows and monolithic reactors. These kinetic packages and also a variety of user written subroutines for modeling complex reaction kinetics have meanwhile been coupled to several commercial CFD codes. Aside from the commercially widespread multipurpose CFD software packages such as ANSYS FLUENT [47], STAR-CD [48], FIRE [49], CFD-ACE + [50], CFX [51], a variety of multipurpose and specialized CFD codes have been developed in academia and at research facilities such as MP-SALSA [52]. The latter ones are often customized for special reactor types and therefore more efficient. Another tool for the solution of PDE systems based on the finite element method is the FEMLAB software package [53], which has been applied for CFD simulations of catalytic reactors as well. Recently, the free, open source CFD software package OpenFOAM (OpenCFD Ltd) gained popoularity.

8.4.3
Solvers for Stiff ODE and DAE Systems

Model simplification and numerical algorithms make it possible to convert the PDE system of the governing equations to an ordinary differential equation (ODE) system or a coupled system of ODEs and algebraic equations called differential algebraic equation (DAE) system. In these equation systems, time or one spatial component is the independent variable. Several computer codes have been developed to solve ODE and DAE systems. In particular, suitable for reactive flows are DASSL [54], LSODE [55], LIMEX [56, 57], SUNDIALS from LLNL, and VODE [58], which are written in FORTRAN language. The Trilinos Project [59] offers ODE and DAE solvers written in C programming language. For the underlying theory of the numerical solution of DAE systems and software implementation, one can refer to the textbook by Ascher and Petzold [60].

8.5
Reactor Simulations

In the remainder of this chapter, recent and challenging CFD simulations of catalytic reactors will be discussed according to the type of reactor.

8.5.1
Flow through Channels

There is a wide variety of chemical reactors, in which the reactive mixtures flow trough channel-like devices such as tubular chemical reactors, automotive catalytic converters, and catalytic combustion monoliths.

Pipes with diameters ranging from a few centimeters to meters are one class of those reactors. The flow field here is in most cases turbulent, guaranteeing good mixing of the reactants. A fine resolution of flow field details is rarely of interest, and aside from that, such a task exceeds today's computer capacities. Therefore, averaged equations and turbulence models are applied as discussed above.

Mantzaras et al. [15] applied the k–ε model, a presumed (Gaussian) probability density function for gaseous reactions, and a laminar-like closure for surface reactions to study turbulent catalytically stabilized combustion of lean hydrogen–air mixtures in plane platinum-coated channels. They also examined different low-Reynolds number near-wall turbulence models and compared the numerically predicted results with data derived from planar laser-induced fluorescence measurements of OH radicals, Raman measurements of major species, and laser doppler velocimetry measurements of local velocities and turbulence [61]. They found that discrepancies between predictions and measurements are ascribed to the capacity of the various turbulence models to capture the strong flow laminarization induced by heat transfer from the hot catalytic surfaces. A more detailed discussion on laminar and turbulent flows in catalytically coated channels can be found in Chapter 7.

Another class of tube-like reactors is the monolith or honeycomb structure, which consists of numerous passageways with diameters reaching from one-tenth of a millimeter to few millimeters. The flow field in the thin channels of this reactor type is usually laminar. The catalytic material is mostly dispersed in a washcoat on the inner channel wall. Monolith channels are manufactured with various cross-sectional shapes, for example, circular, hexagonal, square, or sinusoidal. Several recent CFD studies were conducted to understand the impact of the real washcoat shape on transport and overall reaction rate [33, 62, 63]. Hayes et al. [63] recently showed for a catalytic structure used for exhaust gas cleanup in automobiles that the internal diffusion resistance, expressed in terms of an effectiveness factor, cannot be represented in terms of a unique curve using the generalized Thiele modulus approach to model diffusion and reaction in the washcoat of a catalytic monolith reactor. The most significant deviation occurs when the washcoat has the greatest variation in thickness. As shown in Figure 8.2, only a thin layer of the washcoat in the corners of the channel is needed for conversion, which implies that the corners can be coated with a catalyst-free layer to reduce the amount of expensive noble metals. Mladenov et al. recently coupled the three-dimensional Navier–Stokes equations with washcoat diffusion models and an elementary-step-like heterogeneous reaction mechanism consisting of 74 reactions among 11 gas-phase and 22 adsorbed surface species to study mass transfer in single channels of a honeycomb-type automotive catalytic converter operated under direct oxidation conditions [64]. The resulting concentration profiles (Figure 8.3) at constant temperature were compared with 17

266 | 8 Computational Fluid Dynamics of Catalytic Reactors

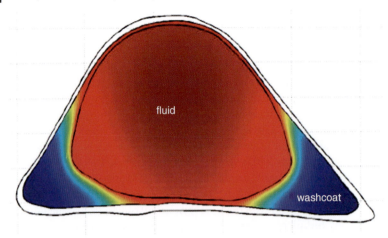

Figure 8.2 Concentrations distribution in the washcoat for a "sinusoidal" channel with highly nonuniform washcoat at 700 K. Adapted from Hayes et al. [63].

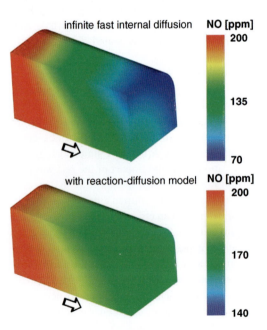

Figure 8.3 Comparison of the NO concentration profiles in a single channel of a honeycomb-type automotive catalytic converter operated under direct oxidation conditions at 250 °C. 3D Navier–Stokes simulation with infinite fast diffusion in the washcoat (*top*) and detailed reaction-diffusion model to describe the internal mass transport. The arrows at the symmetry axes show the main flow direction. Adapted from Mladenov et al. [64].

different numerical models with increasingly simplifying models concerning channel cross section and external and internal diffusion. Again, internal diffusion in the washcoat determines overall activity. Even heat balances also accounting for heat conducting channel walls and external heat loss were coupled with the three-dimensional Navier–Stokes equations including detailed reaction mechanisms for the simulation of partial oxidation of methane to synthesis gas on rhodium-coated monoliths with a rectangular channel cross section [65]. On this level, computing time easily increases to several hours.

Since those 3D simulations require long computing times, the single channel is often approximated by a perfect cylindrical geometry, even for noncircular cross sections. Furthermore, the inlet flow pattern is assumed to follow this geometry. Hence, the flow through the single channel can be treated as the flow through a tubular reactor that means two dimensional (2D) with the axial and radial position as independent variables. The resulting 2D Navier–Stokes equations still describe an elliptic flow; that means information in the channel may travel not only downstream but also upstream, which makes the numerical solution still expensive. As the flow rate in the channel increases (i.e., high Reynolds number but still laminar), the axial diffusive transport is diminished in comparison to the radial diffusion and the convective transport. Hence, all the second derivatives in axial direction can be eliminated in Eqs. (8.2), (8.4) and (8.7) [2, 66, 67]. Mathematically, the character of the equations is changed from elliptic to parabolic – a huge simplification, leading to a much more efficient computational solution. This well-known simplification is generally known as the boundary layer approximation, which is widely used in fluid mechanics. The boundary layer equations form a DAE system, with the time-like direction being the axial coordinate. These simplifications now permit the coupling of the flow field simulation with even very large reaction mechanisms. In terms of these assumptions, the catalytic partial oxidation of the gasoline surrogate *iso*-octane over a Rh-coated monolith was studied in terms of complex reaction mechanisms consisting of 7193 homogeneous and 58 heterogeneous reactions among 857 gas phase and 17 surface species [68]. This detailed description led to the explanation of the experimentally observed coke formation in the downstream section of the catalyst. As Figure 8.4 reveals, the formation of hydrogen is mass transfer limited in the first section of the catalyst, the diffusion of oxygen being the rate-limiting process. The very low oxygen concentration at the catalytic channel wall leads to some formation of hydrogen in a region where the oxygen concentration in the gaseous bulk phase is still sufficiently high to promote total oxidation. In general, the reaction sequence is very similar to the behavior observed for light hydrocarbons by many groups [69–73]: after a short initial total oxidation zone leading to steam and CO_2, the oxygen deficiency at the catalytic surface leads to the formation of hydrogen by steam reforming and partial oxidation. Due to the high temperature of approximately 1000 °C, some remaining fuel is pyrolyzed by gas-phase reactions to form the coke precursors ethylene and propylene (Figure 8.4), a relatively slow process that is kinetically controlled but presenting a threat to any downstream system such as fuel cell devices [74–76].

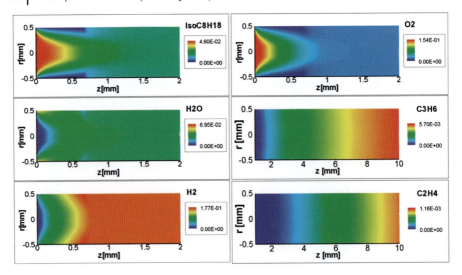

Figure 8.4 Catalytic partial oxidation of iso-octane in Rh-coated monolithic channels at C/O = 1.2 and 800 °C. Numerically predicted molar fractions of reactants, hydrogen, water (all for the initial section of 2 mm), and the coke precursors propylene and ethylene (along the entire catalyst of 1 cm). Flow direction is from left to right. Adapted from Hartmann et al. [68].

A further simplification of modeling channel flows consists of the assumption of infinite radial mass transport or at least very fast radial mass transport, leading to vanishing gradients of radial concentration and temperature. There is a large amount of literature discussing this so-called plug flow reactor (PFR) model [5], which has been the model of choice until recently, including a variety of extensions such as mass transfer limitations [77] or two-phase approaches [78]. The application of the PFR model becomes unreliable for systems in which fast catalytic reactions [67] and/or homogeneous gas-phase reactions occur [79].

Further detailed simulations were carried out, for instance, by Hayes et al. [80], who developed a 2D finite-element model for simulation of a single channel of honeycomb-type monolith catalytic reactor; and Wanker et al. [81] conducted transient two-dimensional simulations of a single channel of a catalytic combustor, taking into account the effects occurring in the gas phase, in the washcoat layer, and in the substrate. They also applied their model to simulate a wood-fired domestic boiler [82].

8.5.2
Monolithic Reactors

The simplest way to model honeycomb-like structures, as shown in Figure 8.1, is based on the assumption that all channels behave essentially alike and therefore only one channel needs to be analyzed. If upstream heat conduction does not matter, parabolic approaches as the boundary layer approximation may be used [67], otherwise an elliptic ansatz is needed [65, 79]; both approaches are discussed above. Heat transfer at the outer boundary of the monolith, spatially varying inlet conditions at the front face of the monolith, and different catalyst coatings will demand models that consider the entire monolithic structure. Since the detailed simulation of every individual channel is usually not tractable, simplifying algorithms are needed [83]. Catalytic monoliths, for instance, have been treated as porous media [84], which can save computational time but can yield unreliable results if the interaction of transport and reactions in the individual channels matters.

Another approach combines the simulation of a representative number of channels with the simulation of the temperature profiles of the solid structure treating the latter one as continuum [85–88]. This approach also is the basis for the computer code DETCHEM^{MONOLITH} [46], which has been applied to model the transient behavior of catalytic monoliths. The code combines a transient three-dimensional simulation of a catalytic monolith with a 2D model of the single-channel flow field based on the boundary layer approximation. It uses detailed models for homogeneous gas-phase chemistry, heterogeneous surface chemistry, and contains models for the description of pore diffusion in washcoats. The numerical structure of the code as sketched in Figure 8.5 is based on the following idea: The residence time of the reactive gas in the monolith channels is much smaller than the unsteadiness of the inlet conditions and the thermal response of the solid monolith structure. Under these assumptions, the timescales of the channel flow are decoupled from the temporal temperature variations of the solid, and the following procedure can be applied: A transient multidimensional heat balance is solved for the monolithic structure including the thermal insulation and reactor walls, which are treated as porous continuum. This simulation of the heat balance provides temperature profiles along the channel walls. At each time step, the reactive flow through a representative number of single

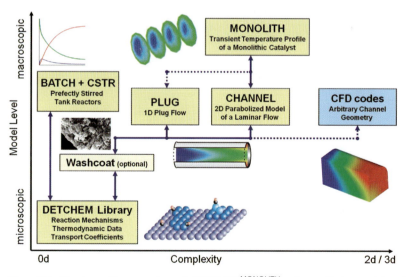

Figure 8.5 Structure of the computer code DETCHEM^{MONOLITH} and some further modules of the software package DETCHEM™ [46].

channels is simulated including detailed transport and chemistry models. These single-channel simulations also calculate the heat flux from the fluid flow to the channel wall due to convective and conductive heat transport in the gaseous flow and heat released by chemical reactions. Thus, at each time step, the single-channel simulations provide the source terms for the heat balance of the monolith structure while the simulation of the heat balance provides the boundary condition (wall temperature) for the single-channel simulations. At each time step, the inlet conditions may vary. This very efficient iterative procedure enables a transient simulation of the entire monolith without sacrificing the details of the transport and chemistry models, as long as the prerequisites for the timescales remain valid. Furthermore, reactors with alternating channel properties such as flow directions, catalyst materials, and loadings can be treated. The code has been applied to model transient behavior of automotive catalytic converters, catalytic combustion monoliths for gas turbine applications, and high-temperature catalysis. Exemplarily, two recently discussed cases are presented as follows:

- In Figure 8.6, the impact of flow rate on the temperature distribution in the monolithic sections of a short-contact time reactor for reforming *iso*-octane to hydrogen-rich synthesis gas reveals that higher flow rates lead to an increase in temperature, conversion, and consequently higher hydrogen yields [89]. This counterintuitive increase in fuel conversion with decreasing residence time

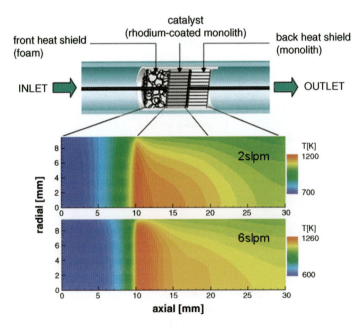

Figure 8.6 Sketch of the catalyst section of a reformer for logistic fuels (*iso*-octane as surrogate) with two heat shields (*top*) and numerically predicted steady-state monolith temperature at C/O = 1.0 and at flow rates of 2 slpm (*top*) and 6 slpm (*bottom*). The symmetry axis of the monolith is at radial dimension of zero. Reproduced from Maier et al. [89].

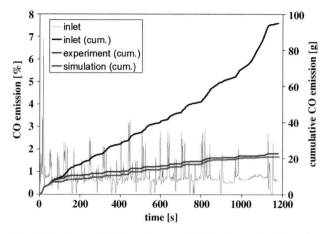

Figure 8.7 Cumulated CO emissions in MVEG driving cycle of an automotive catalytic converter, simulation experiment. The continuously varying raw emissions (inlet, gray color) shown in the background serve as inlet conditions for the simulation. Reproduced from Tischer et al. [93].

(increasing flow rate) can be explained by analyzing the ratio of chemical heat release to heat loss in the reactor [90].

- The second example is the simulation of driving cycles to be used by legislation to test automotive catalytic converters. These cycles last about 20 min and cover a very wide range of conditions. Due to spatially varying inlet conditions and radial temperature profiles over the monolithic catalyst structure, a rather large number of channels need to be considered in the simulations [91]. Furthermore, the continuously temporarily varying inlet temperature, exhaust gas composition, and mass flows make such a simulation a formidable task. The code DTECEM^{MONOLITH} can handle this challenge quite well due to the approach discussed above [92–94]. Figure 8.7 presents a comparison of the experimental and computed time-resolved CO emission in a realistic automobile driving cycle.

8.5.3
Fixed Bed Reactors

The understanding of fluid dynamics and their impact on conversion and selectivity in fixed bed reactors is still very challenging [95, 96]. For large ratios of reactor width to pellet diameter, simple porous media models are usually applicable [97]. This simple approach becomes questionable as this ratio decreases [25, 98]. At small ratios, the individual local arrangement of the particles and the corresponding flow field are significant for mass and heat transfer and, hence, the overall product yields. Therefore, several attempts have recently been made to resolve the flow field in the actual configuration, that is, by a direct numerical simulation (DNS). Even though the governing equations are relatively simple for laminar flows, this approach can be applied usually for small and periodic regions of the reactor only, which is caused by the huge number of computational cells needed to resolve all existing boundary layers [99–105]. Ideally, for simulation of fixed bed reactors, one should account for the transport of chemical species from the bulk of the gas phase to the pellet surface, and then the diffusion and reaction of the species within the catalyst pellets, which may be made up of microporous materials.

Exemplarily, two DNS studies of the group of Dixon will be presented, in which the actual structure of the catalytic fixed bed reactors was taken into account [106–110]. Having spheres as catalyst particles, the modeled turbulent flow and heat transport in a periodic test cell with a tube-to-particle diameter ratio of 4 was simulated [111]. The turbulence was modeled by the renormalization group (RNG) k–ε model [112], and two different wall functions (standard [113] and nonequilibrium) were applied to model the flow field near solid surfaces. Attempts to correlate the local wall heat flux with local properties of the flow field, such as velocity components, velocity gradients, and components of vorticity, led to the conclusion that local heat transfer rates do not correlate statistically with the local flow field. Instead, a conceptual analysis was used to suggest that local patterns of wall heat flux are related to larger-scale flow structures in the bed. Recently, the same group studied the interplay of 3D transport and reaction occurring inside cylindrical pellets and in the gas flow around the pellets used for propane dehydrogenation to better understand catalyst deactivation by carbon deposition (Figure 8.8) [114].

Lattice Boltzmann methods have also been applied for a better understanding of fluid flow in complex reactor configurations [42, 43, 115]. The packing of spheres in cylindrical columns can be created either from experimental observations, such as magnetic resonance imaging (MRI), or by computer simulations. The created topology is then divided into a Cartesian grid, where individual elements are labeled

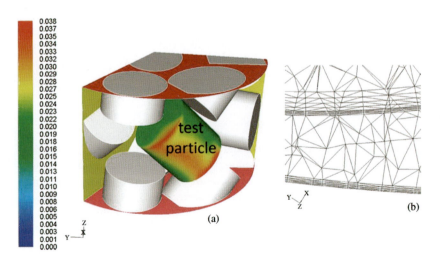

Figure 8.8 Example of a direct numerical simulation of a fixed bed reactor with cylindrical packings showing contours of propane dehydrogenation rate on fresh catalyst (kmol/m^3(solid) s) (a) and details of mesh (b). Reproduced from Dixon et al. [114].

as solid or fluid regions. A high resolution of the grid leads to accurate flow profiles. Zeiser et al. [43] generated the geometrical structures of the fixed bed with a Monte Carlo method. This allowed to efficiently simulate the placement of randomly packed spheres in a cylinder and to obtain detailed information of statistical properties, such as the distribution of the void fraction. This geometrical information was the basis for subsequent numerical flow simulation using LBM. This approach allowed the prediction of the local fluid velocity distribution in the bed, as well as the transport and rate of simple chemical reactions. Yuen et al. [115] studied correlations between local conversion and hydrodynamics in a 3D fixed bed esterification process by applying an LBM and comparing its results with data from *in situ* magnetic resonance visualization techniques.

8.5.4
Wire Gauzes

Wire gauze reactors have been applied for high-temperature catalytic reactions in industry for quite a long time. For example, ammonia is oxidized over Pt/Rh wire gauzes to produce NO (Ostwald process), and similarly, HCN is synthesized by ammoxidation of methane (Andrussov process). Due to the complex 3D geometry, wire gauze reactors have been frequently treated by simpler two-dimensional simulations [116, 117]. However, since mass and heat transport are the dominating processes in wire gauze reactors, simplification of the flow field is risky. Therefore, CFD studies were performed using 3D simulations of the flow field. The 3D flow field through knitted gauzes applied for ammonia oxidation was simulated by Neumann et al. [118]. Catalytic partial oxidation (CPOX) of light alkanes was also studied in wire gauze reactors. De Smet et al. [119] studied CPOX of methane with oxygen at atmospheric pressure in a continuous flow reactor containing a single Pt metal gauze. They used 3D computations of simultaneous heat and mass transfer in case of a simple surface reaction on the gauze catalyst to derive intrinsic kinetics. This experiment was later simulated using even detailed surface and gas-phase reaction schemes [120]. Figure 8.9 exemplarily shows the computed temperature profile around a Pt/Rh wire gauze used for ammonia oxidation, which was carried out with the commercial CFD code FLUENT [121] coupled with a multistep surface reaction mechanism.

8.5.5
Catalytic Reactors with Multiphase Fluids

CFD simulations have recently been applied to quite a number of catalytic reactor types with multiphase flow fields such as fluidized bed reactors with and without circulation, slurry reactors, trickle bed reactors, membrane reactors, electrocatalytic devices (e.g., fuel cells), and reactive distillation devices. These multiphase reactors are of multiscale structures, that is, single particles, particle clusters/bubbles, and reactor vessel, and of multiple physics, that is, hydrodynamics, heat and mass transfer, and reaction kinetics. The formation of complex structures/patterns in each regime is a result of a compromise between dominant mechanisms at multiple

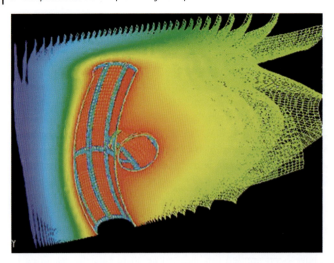

Figure 8.9 Computed temperature profile (max. 950 °C) around a Pt/Rh wire gauze used for ammonia oxidation (Ostwald process).

scales. Coupling of hydrodynamics, heat and mass transfer, and reaction kinetics takes place at molecular and particle levels where conductive and convective transfer and diffusion within the internal pores of the catalyst are accompanied by the adsorption, surface reaction, and desorption of reactant and product on the surface. Even though this complexity is challenging for CFD simulations, computations are a promising tool to achieve a better understanding of multiphase reactors.

A detailed description of the fundamentals and modeling attempts of these multiphase reactors is beyond the scope of this chapter; here, it is referred to general textbooks [19–21]. Instead, few examples may serve as illustration of the potential of CFD simulations of reactors with multiphase flow fields.

Heterogeneously catalyzed gas–liquid reactions, such as hydrogenations, oxidations, hydroformylations, and Fischer–Tropsch synthesis are frequently carried out in slurry reactors. The catalysts of typical diameter of $1-100\,\mu m$ are suspended in the liquid phase, the injecting gas providing the mixing and catalyst suspension. The advantages of bubble slurry reactors over fixed bed reactors are better temperature control and high reaction rates using small catalyst particles. The Khinast group studied heterogeneously catalyzed reactions close to bubbles using advanced CFD tools [122]. Their study revealed the influence of bubble size and shape and particle properties on selectivity of fast heterogeneously catalyzed gas–liquid reactions. The reaction was shown to occur primarily in the wake of the bubble for fast gas–liquid–solid reactions (Figure 8.10), and is thus dependent only on mixing in this region.

Recently, Fischer–Tropsch synthesis slurry bubble column reactors have been the objective of several modeling studies [123–125]. Troshko and Zdravistch recently conducted a CFD study based on a Eulerian multifluid formulation with both the liquid–catalyst slurry and the syngas bubbles phases [125]. The model includes

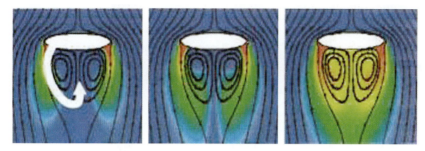

Figure 8.10 Snapshots of the computed developing concentration of the product in the wake of a bubble in a gas–liquid–solid reaction. Reproduced from Raffensberger et al. [122]

variable gas bubble size, effects of the catalyst suspended in the liquid phase and chemical reactions. Findings of this study are that the highly localized FT reaction rate appears next to the gas injection region leading to heat release maxima in that region.

The mass and heat transfer in three-phase flow is determined not only by the transport through the interphases but also by convective transport within the bubbles and the liquid structure (Figure 8.11) [126]. Based on a volume-of-fluid method with interface reconstruction computation, bubble shapes were recently computed and experimentally validated in Taylor flows in a viscous liquid within a square channel of 1 mm hydraulic diameter (Figure 8.12) by Wörner and coworkers [127]. These simulations are of great interest for gas–liquid–solid reactions in microstructures, in which the catalyst is dispersed on the solid wall.

8.5.6
Material Synthesis

Chemical reactors for material synthesis are often characterized by interactions between more than one phase, for example, gas phase and solid phase. Chemical vapor deposition (CVD) and chemical vapor infiltration (CVI) are two commonly used methods for material synthesis. The methodology to treat those systems is very close to the one discussed in this book for heterogeneous catalytic reactions and, therefore, a few remarks shall be made.

CVD is widely used for manufacturing thin solid films in semiconductor industry. The complex interactions of a large number of chemical species with flow and heat transport prohibited early CFD models to include detailed chemistry into the reactor models. Nevertheless, for simple configurations such as stagnation flow configuration, codes were developed in the early stage of modeling heterogeneous reactive flows, when Kee and coworkers developed the first tools for CFD modeling of heterogeneous reactions [28, 128, 129]. These tools were applied for catalysis and material synthesis [28, 128, 130, 131]. With increasing computer performance, it became possible to include more detailed descriptions of transport phenomena and reaction chemistry into CFD models. Kleijn [132], for instance, carried out a full 2D

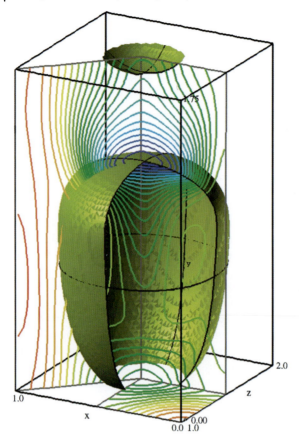

Figure 8.11 Numerical simulation of gas–liquid mass transfer in cocurrent downward Taylor flow; reproduced from Kececi et al. [126].

simulation of rotating disk/stagnation flow CVD reactor using CFD models with detailed surface chemistry and could show that properties at the outer edges of the reactor can vary significantly from that of the centerline, especially in the case of stagnation flow.

Modeling CVI is more challenging compared to CVD due to the temporal densification of the porous substrate and hence the changing surface area. These changes significantly influence the interaction between the gas-phase and the surface kinetics. Therefore, one has to incorporate additional model equations that describe the temporal changes in porosity and surface area into the CFD models. Li et al. modeled the chemical vapor infiltration (CVI) of hydrocarbons for synthesis of carbon–carbon composites. They coupled the CVI model with COMSOL to simulate the densification of the porous substrate as a function of time [133, 134] and studied the densification of a porous carbon felt using CH_4 precursor. CFD simulations have also found their way into modeling of the synthesis of catalytic particles by flame

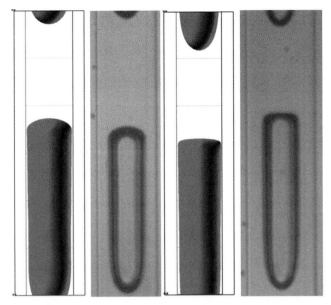

Figure 8.12 Comparison of bubble shape in experiment (*left*) and simulation (*right*) for viscous cocurrent downward Taylor flow in a square minichannel. Reproduced from Keskin et al. [127]

synthesis [135], carbon nanotubes [136], and fibrous active materials [137, 138], to name a few more examples related to catalysis.

8.5.7
Electrocatalytic Devices

CFD simulations using heterogeneous reactions are extensively applied to fundamental research pertaining to electrochemical systems, as discussed in more detail for SOFC modeling in Chapter 6. Also, battery dynamics studied using CFD techniques give insightful understanding of its discharge characteristics. Several one-dimensional models are reported in literature that simulate the electrolyte transport and discharge characteristics [139, 140]. Although the one-dimensional models are simple and efficient in predicting the discharge characteristics of battery systems, multidimensional models without any *ad hoc* approximations can be very valuable in the fundamental understanding of processes that occur in battery systems. The ability to visualize flow patterns during operation of a device is a uniqueness of CFD, which is very difficult if not impossible to realize in pure experimentation. Gu *et al.*, for instance, developed a CFD model to predict the transient behavior of electric-vehicle lead acid batteries during charge and discharge processes [141]. The growing interest in lithium ion batteries also led to first CFD applications in this field [142]. CFD simulation has also been used to understand experimentally observed phenomena in electrocatalytic flow cells, in which mass transport interacts with the electric and chemical processes leading to a complex dependence of the Faradaic current on the potential [143].

8.6
Summary and Outlook

From a reaction engineering perspective, computational fluid dynamics simulations have matured into a powerful tool for understanding mass and heat transport in catalytic reactors. Initially, CFD calculations focused on a better understanding of mixing, and mass transfer to enhance reaction rates, diffusion in porous media, and heat transfer. Over the past decade, the flow field and heat transport models have also been coupled with models for heterogeneous chemical reactions. So far, most of these models are based on the mean field approximation, in which the local state of the surface is described by its coverage with adsorbed species averaged on a microscopic scale. The increasing research activities on surface reactions under practical conditions will certainly boost the application of CFD codes that combine fluid flow and chemistry. New insights into the complexity of heterogeneous catalysis, however, will also reveal the demand for more sophisticated chemistry models. Their implementation into CFD simulations will then require even more sophisticated numerical algorithms and computer hardware. Hence, CFD simulations of reactive systems will remain a very active field and the implementation of more adequate and complex models will continue.

The simulation results will always remain a reflection of the models and physical parameters applied. The careful choice of the submodels (geometry, turbulence, diffusion, species, reactions involved, etc.) and the physical parameters (inlet and boundary conditions, conductivity, permeability, viscosity, etc.) is a precondition for reliable simulation results. Therefore, only the use of appropriate models and parameters, which describe all significant processes in the reactor, can lead to reliable results. Furthermore, numerical algorithms never give an accurate solution of the model equations but only an approximated solution. Hence, error estimation is needed. Bearing these crucial issues in mind, CFD can really serve as a powerful tool in understanding the behavior in catalytic reactors and in supporting the design and optimization of reactors and processes.

Acknowledgments

The authors would like to thank R.J. Kee (Colorado School of Mines), S. Tischer, M. Wörner, and L. Maier (all Karlsruhe Institute of Technology) for very stimulating discussions on modeling and simulation of chemical reactors and Y. Dedecek (Karlsruhe Institute of Technology) for editorial corrections of the manuscript. Financial support by the German Research Foundation (DFG) and the Helmholtz Association is gratefully acknowledged.

References

1 Bird, R.B., Stewart, W.E., and Lightfoot, E.N. (2001) *Transport Phenomena*, 2nd edn, John Wiley & Sons, Inc., New York.
2 Kee, R.J., Coltrin, M.E., and Glarborg, P. (2003) *Chemically Reacting Flow*, Wiley-Interscience.
3 Patankar, S.V. (1980) *Numerical Heat Transfer and Fluid Flow*, McGraw-Hill, New York.
4 Warnatz, J., Dibble, R.W., and Maas, U. (1996) *Combustion, Physical and Chemical Fundamentals, Modeling and Simulation, Experiments, Pollutant Formation*, Springer, New York.
5 Hayes, R.E. and Kolaczkowski, S.T. (1997) *Introduction to Catalytic Combustion*, Gordon and Breach Science Publ., Amsterdam.
6 Hirschfelder, J.O., Curtiss, C.F., and Bird, R.B. (1964) *Molecular Theory of Gases and Liquids*, rev. edn, John Wiley & Sons, Inc., New York.
7 Kee, R.J., Dixon-Lewis, G., Warnatz, J., Coltrin, M.E., and Miller, J.A. (1986) A Fortran Computer Code Package for the Evaluation of Gas-Phase Multicomponent Transport Properties, SAND86-8246, Sandia National Laboratories.
8 Chase, M.W., Jr., Davis, C.A., Downey, J.R., Jr., Frurip, D.J., McDonald, R.A., and Syverud, A.N. (1985) *J. Phys. Chem. Ref. Data*, **14**, 1.
9 Kee, R.J., Rupley, F.M., and Miller, J.A. (1987) The Chemkin Thermodynamic Database, SAND87-8215, Sandia National Laboratories, Livermore.
10 Burcat, A. (1984) in *Combustion Chemistry* (ed. W.C. Gardiner), Springer, New York, p. 455.
11 Benson, S.W. (1976) *Thermochemical Kinetics*, John Wiley & Sons, Inc., New York.
12 Libby, P.A. and Williams, F.A. (1993) *Turbulent Reactive Flow*, Academic Press, London.
13 Pope, S.B. (1985) *Prog. Energ. Combust.*, **11**, 119.
14 Appel, C., Mantzaras, J., Schaeren, R., Bombach, R., and Inauen, A. (2005) *Combust. Flame*, **140**, 70.
15 Mantzaras, J., Appel, C., Benz, P., and Dogwiler, U. (2000) *Catal. Today*, **59**, 3.
16 Laudner, B.E. and Spalding, D.B. (1972) *Mathematical Models of Turbulence*, Academic Press, London.
17 Gutheil, E. and Bockhorn, H. (1987) *Physicochem. Hydrodyn.*, **9**, 525.
18 Zhang, X.Y. and Ahmadi, G. (2005) *Chem. Eng. Sci.*, **60**, 5089.
19 Jakobsen, H.A. (2008) *Chemical Reactor Modeling: Multiphase Reactive Flows*, Springer, Heidelberg.
20 Crowe, C.T., Schwarzkopf, J.D., Sommerfeld, M., and Tsuji, Y. (1998) *Multiphase Flows with Droplets and Particles*, CRC Press.
21 Ishii, M. and Hibiki, T. (2006) *Thermo-Fluid Dynamics of Two-Phase Flow*, Springer.
22 Aris, R. (1975) *The Mathematical Theory of Diffusion and Reaction in Permeable Catalysts*, Clarendon Press, Oxford.
23 Keil, F. (1999) *Diffusion und Chemische Reaktionen in der Gas-Feststoff-Katalyse*, Springer, Berlin.
24 Keil, F.J. (2000) *Catal. Today*, **53**, 245.
25 Bey, O. and Eigenberger, G. (1997) *Chem. Eng. Sci.*, **52**, 1365.
26 Giese, M., Rottschafer, K., and Vortmeyer, D. (1998) *Am. Inst. Chem. Eng. J.*, **44**, 484.
27 Winterberg, M., Tsotsas, E., Krischke, A., and Vortmeyer, D. (2000) *Chem. Eng. Sci.*, **55**, 967.
28 Coltrin, M.E., Kee, R.J., and Rupley, F.M. (1991) SURFACE CHEMKIN (Version 4.0): A Fortran Package for Analyzing Heterogeneous Chemical Kinetics at a Solid-Surface–Gas-Phase Interface, SAND91-8003B, Sandia National Laboratories.
29 Deutschmann, O., Schmidt, R., Behrendt, F., and Warnatz, J. (1996) *Proc. Comb. Inst.*, **26** 1747.
30 Raja, L.L., Kee, R.J., and Petzold, L.R. (1998) *Proc. Comb. Inst*, **27** 2249.
31 Boll, W., Tischer, S., and Deutschmann, O. (2010) *Ind. Eng. Chem. Res.*, **49**, 10303.
32 Kang, S.B., Kwon, H.J., Nam, I.S., Song, Y.I., and Oh, S. (2011) *Ind. Eng. Chem. Res.*, **50** 5499.

33 Papadias, D., Edsberg, L., and Björnbom, P.H. (2000) *Catal. Today*, **60**, 11.
34 Mason, E. and Malinauskas, A. (1983) *Gas Transport in Porous Media: The Dusty-Gas Model*, Elsevier, New York.
35 Kerkhof, P. and Geboers, M.A.M. (2005) *Am. Inst. Chem. Eng. J.*, **51**, 79.
36 Kerkhof, P. (1996) *Chem. Eng. J.*, **64**, 319.
37 Ranade, V.V. (2002) *Computational Flow Modeling for Chemical Reactor Engineering*, Acadamic Press.
38 Hayes, R.E., Kolaczkowski, S.T., and Thomas, W.J. (1997) *Comput. Chem. Eng.*, **16**, 654.
39 Burnett, D.S. (1987) *Finite Elemnet Analysis*, Addison-Wesley Publ. Co., Reading.
40 Succi, S. (2001) *The Lattice Boltzmann Equation for Fluid Dynamics and Beyond*, Oxford University Press.
41 Sullivan, S.P., Sani, F.M., Johns, M.L., and Gladden, L.F. (2005) *Chem. Eng. Sci.*, **60**, 3405.
42 Freund, H., Zeiser, T., Huber, F., Klemm, E., Brenner, G., Durst, F., and Emig, G. (2003) *Chem. Eng. Sci.*, **58**, 903.
43 Zeiser, T., Lammers, P., Klemm, E., Li, Y.W., Bernsdorf, J., and Brenner, G. (2001) *Chem. Eng. Sci.*, **56**, 1697.
44 Kee, R.J., Rupley, F.M., Miller, J.A., Coltrin, M.E., Grcar, J.F., Meeks, E., Moffat, H.K., Lutz, A.E., Dixon-Lewis, G., Smooke, M.D., Warnatz, J., Evans, G.H., Larson, R.S., Mitchell, R.E., Petzold, L.R., Reynolds, W.C., Caracotsios, M., Stewart, W.E., Glarborg, P., Wang, C., and Adigun, O. (2000.) *CHEMKIN*, 3.6 edn, Reaction Design, Inc., San Diego, www.chemkin.com.
45 Goodwin, D.G. (2003) CANTERA. An open-source, extensible software suite for CVD process simulation, www.cantera.org.
46 Deutschmann, O., Tischer, S., Correa, C., Chatterjee, D., Kleditzsch, S., and Janardhanan, V.M. (2004) Detchem Software Package, 2.0 edn, Karlsruhe, www.detchem.com.
47 (2005). FLUENT, Fluent Incorporated, Lebanon, www.fluent.com.
48 in CD-adapco, London Office, 200 Shepherds Bush Road, London, W6 7NY, United Kingdom, www.cd-adapco.com.
49 (2005) FIRE, AVL LIST GmbH, Graz, Austria, www.avl.com.
50 (2005) CFD-AC+, CFD Research Corporation, Huntsville, AL, www.cfdrc.com.
51 (2005). CFX, www-waterloo.ansys.com.
52 Shadid, J., Hutchinson, S., Hennigan, G., Moffat, H., Devine, K., and Salinger, A.G. (1997) *Parallel Comput.*, **23**, 1307.
53 (2005) FEMLAB, www.comsol.com.
54 Brenan, K.E., Campbell, S.L., and Petzold, L.R. (1996) *Numerical Solution of Initial-Value Problems in Differential-Algebraic Equations*, 2nd edn, SIAM, Philadelphia, PA.
55 Hindmarsh, A.C. (1983) *Scientific Computing* (ed. R.S. Stepleman), North Holland Publishing Co., Amsterdam, p. 55.
56 Deuflhardt, P., Hairer, E., and Zugk, J. (1987) *Numer. Math.*, **51**, 501.
57 Deuflhard, P. and Nowak, U. (1987) *Progr. Sci. Comput.*, **7**, 37.
58 Brown, P.N., Byrne, G.D., and Hindmarsh, A.C. (1989) *SIAM J. Sci. Statist. Comput.*, **10**, 1038.
59 (2011) The Trilinos Project, Sandia National Laboratories, http://trilinos.sandia.gov/.
60 Ascher, U.M. and Petzold, L.R. (1998) *Computer Methods for Ordinary Differential Equations and Differential-Algebraic Equations*, SIAM, Philadelphia, PA.
61 Appel, C., Mantzaras, J., Schaeren, R., Bombach, R., Kaeppeli, B., and Inauen, A. (2003) *Proc. Comb. Inst.*, **29**, 1031.
62 Hayes, R.E., Liu, B., and Votsmeier, M. (2005) *Chem. Eng. Sci.*, **60**, 2037.
63 Hayes, R.E., Liu, B., Moxom, R., and Votsmeier, M. (2004) *Chem. Eng. Sci.*, **59**, 3169.
64 Mladenov, N., Koop, J., Tischer, S., and Deutschmann, O. (2010) *Chem. Eng. Sci.*, **65**, 812.
65 Deutschmann, O., Schwiedernoch, R., Maier, L., and Chatterjee, D. (2001) *Natural Gas Conversion VI, Studies in Surface Science and Catalysis*, vol. 136

(eds E. Iglesia, J.J. Spivey, and T.H. Fleisch), Elsevier, Alaska, p. 251.
66 Schlichting, H. and Gersten, K. (1999) *Boundary-Layer Theory*, 8th edn, Springer, Heidelberg.
67 Raja, L.L., Kee, R.J., Deutschmann, O., Warnatz, J., and Schmidt, L.D. (2000) *Catal. Today*, **59**, 47.
68 Hartmann, M., Maier, L., Minh, H.D., and Deutschmann, O. (2010) *Combust. Flame*, **157**, 1771.
69 Schwiedernoch, R., Tischer, S., Correa, C., and Deutschmann, O. (2003) *Chem. Eng. Sci.*, **58**, 633.
70 Karagiannidis, S., Mantzaras, J., Jackson, G., and Boulouchos, K. (2007) *Proc. Comb. Inst.*, **31**, 3309.
71 Maestri, M., Vlachos, D.G., Beretta, A., Forzatti, P., Groppi, G., and Tronconi, E. (2009) *Top. Catal.*, **52**, 1983.
72 Beretta, A., Groppi, G., Lualdi, M., Tavazzi, I., and Forzatti, P. (2009) *Ind. Eng. Chem. Res.*, **48**, 3825.
73 Horn, R., Degenstein, N.J., Williams, K.A., and Schmidt, L.D. (2006) *Catal. Lett.*, **110**, 169.
74 Hartmann, M., Kaltschmitt, T., and Deutschmann, O. (2009) *Catal. Today*, **147**, S204.
75 Hebben, N., Diehm, C., and Deutschmann, O. (2010) *Appl. Catal. A Gen.* **2010**, *388*, 225
76 Kaltschmitt, T., Maier, L., and Deutschmann, O. (2010) *Proc. Comb. Inst.*, **33**, 3177.
77 Hayes, R.E. and Kolaczkowski, S.T. (1999) *Catal. Today*, **47**, 295.
78 Veser, G. and Frauhammer, J. (2000) *Chem. Eng. Sci.*, **55**, 2271.
79 Zerkle, D.K., Allendorf, M.D., Wolf, M., and Deutschmann, O. (2000) *J. Catal.*, **196**, 18.
80 Hayes, R.E., Kolaczkowski, S.T., and Thomas, W.J. (1992) *Comput. Chem. Eng.*, **16**, 645.
81 Wanker, R., Raupenstrauch, H., and Staudinger, G. (2000) *Chem. Eng. Sci.*, **55**, 4709.
82 Wanker, R., Berg, M., Raupenstrauch, H., and Staudinger, G. (2000) *Chem. Eng. Technol.*, **23**, 535.
83 Kolaczkowski, S.T. (1999) *Catal. Today*, **47**, 209.
84 Mazumder, S. and Sengupta, D. (2002) *Combust. Flame*, **131**, 85.
85 Jahn, R., Snita, D., Kubicek, M., and Marek, M. (1997) *Catal. Today*, **38**, 39.
86 Koltsakis, G.C., Konstantinidis, P.A., and Stamatelos, A.M. (1997) *Appl. Catal. B Environ.*, **12**, 161.
87 Tischer, S., Correa, C., and Deutschmann, O. (2001) *Catal. Today*, **69**, 57.
88 Tischer, S. and Deutschmann, O. (2005) *Catal. Today*, **105**, 407.
89 Maier, L., Hartmann, M., Tischer, S., and Deutschmann, O. (2011) *Combust. Flame*, **158**, 796.
90 Hartmann, M., Maier, L., and Deutschmann, O. (2010) *Appl. Catal. A Gen.* **2011**, *391*, 144.
91 Windmann, J., Braun, J., Zacke, P., Tischer, S., Deutschmann, O., and Warnatz, J. (2003) SAE Technical Paper, 2003-01-0937.
92 Braun, J., Hauber, T., Többen, H., Windmann, J., Zacke, P., Chatterjee, D., Correa, C., Deutschmann, O., Maier, L., Tischer, S., and Warnatz, J. (2002) SAE Technical Paper, 2002-01-0065.
93 Tischer, S., Jiang, Y., Hughes, K.W., Patil, M.D., and Murtagh, M. (2007) SAE Technical paper, 2007-01-1071.
94 Koop, J. and Deutschmann, O. (2009) *Appl. Catal. B Environ.*, **91**, 47.
95 Elnashaie, S.S.E.H. and Elnashaie, S.S. (1995) *Modelling, Simulation and Optimization of Industrial Fixed Bed Catalytic Reactors*, G + B Gordon and Breach.
96 Dixon, A.G. and Nijemeisland, M. (2001) *Ind. Eng. Chem. Res.*, **40**, 5246.
97 Bizzi, M., Saracco, G., Schwiedernoch, R., and Deutschmann, O. (2004) *AICHE J.*, **50**, 1289.
98 Bauer, M. and Adler, R. (2003) *Chem. Eng. Technol.*, **26**, 545.
99 Sorensen, J.P. and Stewart, W.E. (1974) *Chem. Eng. Sci.*, **29**, 827.
100 Dalman, M.T., Merkin, J.H., and McGreavy, C. (1986) *Comput. Fluids*, **14**, 267.
101 Lloyd, B. and Boehm, R. (1994) *Numer. Heat Tr. A Appl.*, **26**, 237.
102 Debus, K., Nirschl, H., Delgado, A., and Denk, V. (1998) *Chem. Ing. Tech.*, **70**, 415.

103 Logtenberg, S.A., Nijemeisland, M., and Dixon, A.G. (1999) *Chem. Eng. Sci.*, **54**, 2433.
104 Calis, H.P.A., Nijenhuis, J., Paikert, B.C., Dautzenberg, F.M., and van den Bleek, C.M. (2001) *Chem. Eng. Sci.*, **56**, 1713.
105 Petre, C.F., Larachi, F., Iliuta, I., and Grandjean, B.P.A. (2003) *Chem. Eng. Sci.*, **58**, 163.
106 Dixon, A.G., Taskin, M.E., Nijemeisland, M., and Stitt, E.H. (2008) *Chem. Eng. Sci.*, **63**, 2219.
107 Taskin, M.E., Dixon, A.G., Nijemeisland, M., Stitt, E.H. (2008) *Ind. Eng. Chem. Res.*, **47**, 5966.
108 Dixon, A.G., Taskin, M.E., Stitt, E.H., and Nijemeisland, M. (2007) *Chem. Eng. Sci.*, **62**, 4963.
109 Dixon, A.G., Nijemeisland, M., and Stitt, E.H. (2005) *Ind. Eng. Chem. Res.*, **44**, 6342.
110 Behnam, M., Dixon, A.G., Nijemeisland, M., and Stitt, E.H. (2005) *Ind. Eng. Chem. Res.*, **49**, 10641.
111 Nijemeisland, M. and Dixon, A.G. (2004) *Am. Inst. Chem. Eng. J.*, **50**, 906.
112 Yakhot, V. and Orszag, S.A. (1986) *Phys. Rev. Lett.*, **57**, 1722.
113 Laudner, B.E., and Spalding, D.B. (1974) *Comput. Methods Appl. Math. Eng.*, **3**, 169.
114 Behnam, M., Dixon, A.G., Nijemeisland, M., and Stitt, E.H. (2010) *Ind. Eng. Chem. Res.*, **49**, 10641.
115 Yuen, E.H.L., Sederman, A.J., Sani, F., Alexander, P., and Gladden, L.F. (2003) *Chem. Eng. Sci.*, **58**, 613.
116 Zakharov, V.P., Zolotarskii, I.A., and Kuzmin, V.A. (2003) *Chem. Eng. J.*, **91**, 249.
117 O'Connor, R.P., Schmidt, L.D., and Deutschmann, O. (2002) *Am. Inst. Chem. Eng. J.*, **48**, 1241.
118 Neumann, J., Golitzer, H., Heywood, A., and Ticu, I. (2002) *Rev. Chim. Bucharest*, **53**, 721.
119 de Smet, C.R.H., de Croon, M.H.J.M., Berger, R.J., Marin, G.B., and Schouten, J.C. (1999) *Appl. Catal. A Gen.*, **187**, 33.
120 Quiceno, R., Perez-Ramirez, J., Warnatz, J., and Deutschmann, O. (2006) *Appl. Catal. A Gen.*, **303**, 166.
121 (1997) FLUENT, 4.4 edn, Fluent Inc., Lebanon.
122 Raffensberger, J.A., Glasser, B.J., and Khinast, J.G. (2005) *AICHE J.*, **51**, 1482.
123 Lozano-Blanco, G., Thybaut, J.W., Surla, K., Galtier, P., and Marin, G.B. (2009) *AICHE J.*, **55**, 2159.
124 van Baten, J.M. and Krishna, R. (2004) *Ind. Eng. Chem. Res.*, **43**, 4483.
125 Troshko, A.A. and Zdravistch, F. (2009) *Chem. Eng. Sci.*, **64**, 892.
126 Kececi, S., Worner, M., Onea, A., and Soyhan, H.S. (2009) *Catal. Today*, **147**, S125.
127 Keskin, O., Worner, M., Soyhan, H.S., Bauer, T., Deutschmann, O., and Lange, R. (2010) *AICHE J.*, **56**, 1693.
128 Coltrin, M.E., Kee, R.J., Evans, G.H., Meeks, E., Rupley, F.M., and Grcar, J.F. (1991) Sandia National Laboratories.
129 Coltrin, M.E., Kee, R.J., and Evans, G. (1989) *J. Electrochem. Soc.*, **136**, 819.
130 Meeks, E., Kee, R.J., Dandy, D.S., and Coltrin, M.E. (1993) *Combust. Flame*, **92**, 144.
131 Ruf, B., Behrendt, F., Deutschmann, O., and Warnatz, J. (1996) *Surf. Sci.*, **352**, 602.
132 Kleijn, C.R. (2000) *Thin Solid Films*, **365**, 294.
133 Li, A., Norinaga, K., Zhang, W.G., and Deutschmann, O. (2008) *Compos. Sci. Technol.*, **68**, 1097.
134 Li, A.J. and Deutschmann, O. (2007) *Chem. Eng. Sci.*, **62**, 4976.
135 Kammler, H.K., Madler, L., and Pratsinis, S.E. (2001) *Chem. Eng. Technol.*, **24**, 583.
136 Kim, H., Kim, K.S., Kang, J., Park, Y.C., Chun, K.Y., Boo, J.H., Kim, Y.J., Hong, B.H., and Choi, J.B. (2011) *Nanotechnology*, **22**, 5.
137 Chub, O.V., Borisova, E.S., Klenov, O.P., Noskov, A.S., Matveev, A., and Koptyug, I.V. (2005) *Catal. Today*, **105**, 680.
138 De Greef, J., Desmet, G., and Baron, G. (2005) *Catal. Today*, **105**, 331.
139 Tsaur, K.C. and Pollard, R. (1984) *J. Electrochem. Soc.*, **131**, 975.
140 Evans, T.I., Nguyen, T.V., and White, R.E. (1989) *J. Electrochem. Soc.*, **136**, 328.
141 Gu, W.B., Wang, C.Y., and Liaw, B.Y. (1997) *J. Electrochem. Soc.*, **144**, 2053.
142 Smith, K.A., Rahn, C.D., and Wang, C.Y. (2007) *Energ. Convers. Manage.*, **48**, 2565.
143 Zhang, D., Deutschmann, O., Seidel, Y.E., and Behm, R.J. (2011) *J. Phys. Chem. C*, **115**, 468.

9
Perspective of Industry on Modeling Catalysis
Jens R. Rostrup-Nielsen

9.1
The Industrial Challenge

Since the beginning of the 1990s, two new types of focused companies are emerging: the molecule suppliers and the problem solvers [1, 2]. The molecule suppliers include those delivering commodities and fine chemicals. The problem solvers supply functional chemicals such as additives and pharmaceuticals for which the client is interested in the effect rather than the chemistry. Each type has its own characteristics reflected by the role of the catalysts that themselves can be considered functional chemicals.

Production costs are essential for commodity suppliers. Catalyst life ("on-stream factor") is crucial for large commodity plants. Production may be stopped after a few days because of a catalyst failure crucial for the plant economy. It means that secondary phenomena such as catalyst deactivation are important issues.

In contrast, process optimization is of less importance for suppliers of fine chemicals. The key issue is flexibility of the process equipment and normally batch-wise manufacture is preferred. The most active catalysts may show a lifetime of a few hours.

Commodity plants have become larger to take advantage of the *economy of scale*. Today, ammonia plants are built at a capacity of 3000 MTPD and methanol plants are being considered at capacities of 10 000 MTPD. This corresponds to the size of synthetic fuel plants based on FT synthesis (35 000 bpd).

Plants have become more *integrated* to minimize energy consumption. As an example, the energy consumption of ammonia production has decreased over the past 50 years from about 40 to 29 GJ/t, corresponding to a thermal efficiency (LHV) of 65% or 73% of the theoretical minimum [3]. Commodity plants depend on steady improvement and sophistication of the technology. Even small improvements in the process scheme may show low payback times. On the other hand, the uncertainties associated with new technology may outbalance the economic advantage of a new process. Improvement of one process step might easily result in less favorable performance of another process step. The high degree of integration means that the

Modeling and Simulation of Heterogeneous Catalytic Reactions: From the Molecular Process to the Technical System, First Edition. Edited by Olaf Deutschmann.
© 2012 Wiley-VCH Verlag GmbH & Co. KGaA. Published 2012 by Wiley-VCH Verlag GmbH & Co. KGaA.

weakest part of the chain may determine the performance of the entire plant. As an example, there is a need for more coke-resistant catalysts and often deactivation phenomena determine the process layout and the optimum process conditions to be applied [4, 5]. Industrial R&D must deal with not only the basic principles, but also solutions to a series of secondary problems [5].

The composition of the ammonia synthesis catalyst has been known for almost 100 years. The optimum catalyst composition was established after a tedious combinatorial work by Mittasch [6]. However, it is not sufficient to have the catalyst at hand. It is the knowledge about the system that provides the strength in industrial catalysis. It is the knowledge of the characteristics of a catalyst that should be controlled during the manufacture, the activation, and the operation, as well as the knowledge of the impact of trace components in the feed on the aging properties of the catalyst.

In practice, there appears to be a relationship between activity and lifetime of commercial catalysts [7]. The most active catalysts may also exhibit a short life of a few hours. It is evident that catalyst life, that is, *on-stream factor*, is crucial for large-scale commodity plants in contrast to batch-wise manufacture of fine chemicals. For each case, there is a minimum *space time yield* (production rate) and corresponding stability that is commercially applicable. For large-scale operation, economic arguments will limit the minimum space time yield to about 0.1 t product/m^3 reactor/h [7] and with a typical catalyst life of 5 years. This corresponds to a catalyst consumption of less than 0.2 kg/t product. For the ammonia synthesis, a typical figure is 0.03 kg/t NH_3. On the other hand, the heat removal capacity of conventional space time yield reactors rarely permits a reactor productivity in excess of $\sim$10 t product/m^3 reactor/h [7].

When dealing with gradual improvements of existing technologies, it is important to assess what is the room for improvement of the technology—in other words, to know the position of the technology on the S-curve (Figure 9.1) [8]. Another equally important fraction of innovation deals with the second S-curve, that is, possibilities for breakthrough for a completely new approach. A new emerging technology may

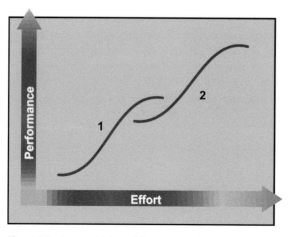

Figure 9.1 S-curves. Adapted from Foster [8].

appear inferior to present technologies, but again it is important to analyze the potential at an early stage. Many exciting schemes may turn out to represent no potential for better process economy.

Many well-established processes are approaching their theoretically achievable efficiency, selectivity, and so on, but new challenges have been introduced by the environmental objectives formulated by society. They have not only led to the introduction of new products, but also necessitated the development of new processes. One example is the refinery industry. Fuels are being specified by chemical composition rather than by performance, and a change of many processes is necessary to meet the requirements of legislation Therefore, the environmental challenges represent a major room for breakthroughs in the catalytic process industry [9].

The risk involved in large-scale operation of highly integrated plants means that it has become more expensive to develop new processes. It is required to a larger extent that the processes are demonstrated [5]. This means that the scale-up of catalytic processes must be well managed to ensure the use of minimum resources and time [9]. Modeling is a useful tool to direct this process.

9.2
The Dual Approach

The traditional bench-scale testing of catalysts simulates the process, but it cannot give results of fundamental character, nor provide data useful for scaling up to industrial conditions. Therefore, a more efficient approach is to take advantage of the progress in both surface science and computer methods to aim at more fundamental studies (*in situ* studies) and early process simulation and scale-up to pilot operation (Figure 9.2) [5]. The scale-up to semi-industrial operation gives the opportunity to

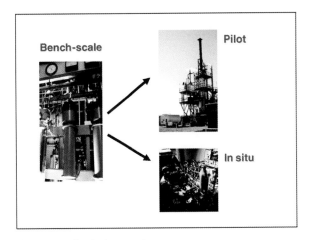

Figure 9.2 The dual approach.

Table 9.1 Integral tests.

Phenomena to be checked in integral tests	Requirements of integrated tests
• Product inhibition at high conversion • Validity of rate equation • Mass and heat transfer restrictions • Gas film related deactivation phenomena • Other secondary phenomena • Feedstock characteristics and reactor performance	• Full conversion • Industrial pellet size • Industrial pressure • Real feedstock • Industrial mass velocity • Pseudoadiabatic conditions

study the process at industrial mass velocities and heat transfer rates, meaning that results are relevant for engineering of industrial units. Practice has also shown that it is at this scale that secondary phenomena such as unforeseen deactivation are best recognized. On the other hand, the progress in surface science has provided a detailed understanding of poisoning phenomena. This knowledge can be used in characterizing samples of spent catalyst from the pilot plant [10]. It is true that the application of fundamental research and rapid scale-up to pilot has made R&D more expensive, but far more efficient.

It is important at an early stage to *analyze constraints* of the entire process and for the individual process steps to identify as early as possible what is the real problem–to focus on rapid understanding of the key phenomenon.

There is a need for *integral tests* to verify the rate expression at full conversion to check the impact of the phenomena listed in Table 9.1. The conditions must be selected carefully. It is important that the integral test can be carried out at industrial pellet size and pressure to avoid severe errors due to the complex relationship between the pore volume distribution of the catalyst and the effective diffusion coefficient. It is also important that scale-up is carried out with real feedstock to identify possible impact of impurities. The integral test should be carried out at sufficient scale to identify important secondary deactivation phenomena that may be related to film diffusion effects. The scale-up can lead to surprises for new processes because it may not be possible to observe the deactivation problem before approaching industrial mass velocities.

Whenever possible, it is advantageous to *split* the *chemical process* problems *from* problems related to *hydrodynamics* and solids handling. The study of the latter problems typically requires larger scale than do the "chemical" catalyst problems. This can be done more cheaply by cold flow experiments in "*mock-up*" units. It is in particular useful when studying fluid bed reactions.

The challenge for the scale-up of catalytic processes appears to identify the smallest scale required [5]. It is cheaper, safer, and quicker to carry out experiments at a smaller scale and it also means a larger flexibility in the approach. Progress in making more accurate pumps, instruments, and analytical equipment for small-scale units and the use of advanced control systems and process computers makes it easier to establish exact mass balances for small units.

Large-scale pilot units are justified for scientific reasons when the process is affected by flow-related heat and mass transport, involves handling of solid particles, and applies complex new reactor systems with recycle streams or involves the combination of catalysis and separation. Also, a large-scale unit may be required simply to provide sufficient amounts of the product for testing its application.

The installation of a semi-industrial plant or *process demonstration unit* (PDU) is costly and may delay the introduction of the process to industry. In other situations, it may be necessary to demonstrate the process at sufficient scale to convince the client irrespective of a solid design basis.

The scale-up (by Exxon) of one of the most complicated processes, the *fluid bed cracking* (FCC), took place in a 100 bbl/day pilot unit that was started in 1940 at the same time as the first full-scale plant (13 600 bbl/day) was designed [11]. The results from the pilot plant were immediately incorporated into the ongoing design. The pilot plant operated from April 1940 to June 1941 and the commercial plant was started up in May 1942. The speed of scale-up and the use of simultaneous engineering are remarkable. Although this happened under the pressure of World War II, it illustrates that even without advanced models and computer techniques, there were no scientific or engineering constraints for a quick scale-up.

Today, pilot-scale work provides a more precise knowledge for the design models, reducing the risks involved in scale-up.

9.3
The Role of Modeling

9.3.1
Reactor Models

The introduction of computer methods have allowed much more sophisticated models to be used in the reaction engineering. These models have been the key to process innovation.

The reactor models may be defined as follows: *Homogeneous* models do not consider gradients in and around catalyst pellets, which are handled by *heterogeneous* models. *One-dimensional* models consider only axial gradients in the reactor, whereas radial gradients are included in two-dimensional models. The main equations for a *two-dimensional model* [13, 14] are shown in Figure 9.3.

Most reactor models assume plug flow, which is not fulfilled in many complex systems. These situations should be modeled by combining *CFD* (computational fluid dynamics) analysis with the reactor model. Examples are monolith reactors, fuel cells, microchannel reactors, and reactors involving combustion.

At industrial conditions, the catalytic reaction is influenced by mass and heat transfer restrictions in the gas film surrounding the catalyst pellet as well as inside the pore system of the catalyst pellet. This is reflected by the *effectiveness factor*, which is defined as ratio of the rate of the pellet relative to the intrinsic rate of the catalyst. This is incorporated into heterogeneous models that may be one or two dimensional.

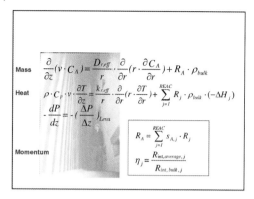

Figure 9.3 Two-dimensional reactor model [14].

Reaction kinetics is important for the scale-up of catalytic processes, not the least for making reliable computer simulations. This requires careful experimental work at well-controlled conditions.

Most kinetic studies deal with initial rates far from equilibrium that may reflect the situation in only a small part of the industrial reactor. Most of the catalyst volume will work close to equilibrium. Hence, there is a need for integral tests (Table 9.1).

The reaction engineering methods resulted in computer models capable of describing the complex interaction between intrinsic kinetics and mass and heat transfer inside the pellet and in the film surrounding the catalyst pellet. These models were included at an early stage in reactor engineering models for design [12, 13, 15]. In conclusion, kinetics and reaction engineering was developed almost into perfection 25 years ago, and reaction engineering was considered a mature field of science.

The success almost turned catalysis into the *"noble art of modeling"* with more and more complex differential equations with dimensionless parameters [10]. It brought many chemical engineers in front of the computer screen far from the reality of industrial problems. The models must be linked to the real world.

9.3.2
Surface Science and Breakdown of the Simplified Approach

Irrespective of high sophistication of the mathematical modeling of chemical reactors, most work was still based on semi-empirical *Langmuir–Hinshelwood kinetics*. However, the simplified Langmuir–Hinshelwood approach breaks down in a number of situations [16]. The "rds" as well as the "masi" may change through the reactor as a function of conversion or temperature. If it changes with conversion, it will also change inside the catalyst pellet, if the effectiveness factor is low.

The problems are aggravated for the so-called *"ppm reactions"* aiming at complete conversion to the ppm level [12, 16]. It involves large conversion in single reactors. The interaction between diffusion restrictions and intrinsic kinetics may become complex as the concentration of reactants approaches zero.

Table 9.2 Breakdown of simple approach [12].

- rds/masi and
- Required "ensemble" may change with conversion, temperature
- Catalyst structure may change with conversion, temperature
- "Mean field" assumption may not reflect situation

Surface science (LEED, STM, etc.) has demonstrated that the structure of the adsorbed layer does not reflect random adsorption. On the contrary, the typical picture is the formation of island structures. Hence, the *"mean field" assumption* behind rate equations stating that the rate of a surface reaction between two adsorbed species is proportional to their surface coverage may reflect a simplified picture of the real situation. It may be more likely that reactions take place at the edges of surface islands. In conclusion, there are several reasons that the simplified Langmuir–Hinshelwood approach may break down, as listed in Table 9.2 [12].

One may speculate about the *meaning of* ∗ in the kinetic sequence [16, 17]. The size of the ensemble required may change with conversion and temperature. Also, the surface may be reconstructed under reaction conditions [18, 19], depending on the composition of the gas at various positions in the reactor. Step sites have been demonstrated to play a decisive role in a number of important catalytic reactions [17]. The supported liquid-phase catalyst for SO_2 oxidation of sulfuric acid may be considered as an extreme. The composition of the vanadium potassium salt melt changes not only through the reactor, but also inside the catalyst pellet [20].

The classical way of overcoming the situation described above is to use steady-state kinetics with no assumption of a rate-determining step. This approach easily leads to complicated equations with a number of constants difficult to determine in catalytic reactions [13, 15]. However, the progress in surface science has provided strong tools for estimating the kinetic constants of various reaction steps.

The success of surface science almost turned catalysis into the *"noble art of characterization"* [10]. There is a need to study reactions rather than characterizing sites. This is not the least the success of the so-called *in situ* methods by which the catalytic reaction is studied at the same time as changes of the structure of the catalyst or surface intermediates can be identified.

The input from surface science to *microkinetics* has made it meaningful to solve the steady-state rate equations, which is possible with modern computer techniques [21, 22]. Microkinetics analysis of various reaction sequences is a useful tool to combine the kinetic data with information from related kinetic studies (TPD, tracer studies, etc.) with data from surface science (sticking coefficients, surface bond energies, etc.). The pre-exponential factors for the elementary steps in the reaction sequence may be estimated by using transition state and collision theory and thermodynamic consistency may be achieved by statistical mechanics for estimating equilibrium constants [21, 22].

The microkinetic analysis should be considered a tool for providing a consistent description, which may then form the basis for formulation of more predictive rate equations, for reformulation of the catalyst, or for exploration of new reaction paths.

The consolidation of spectroscopic kinetic and surface chemical data may provide a basis for new catalytic cycles and new catalyst systems to study. This approach is facilitated by theoretical models.

9.3.3
Theoretical Methods

The progress in theoretical methods has led to a detailed understanding of many catalytic systems. The input from "*ab initio*"*density functional theory* (DFT) based calculations has become well established [23, 24]. The DFT approach often gives precise explanations of catalysis on bimetallic systems or promoted metal catalysts. The DFT method can be expanded to nonmetallic systems [25].

The discussions of catalytic phenomena have often been based on whether the observations should be related to geometric or electronic effects. Purely geometrical effects have been identified for dissociation (and association) for diatomic molecules (N_2, CO, NO, O_2) [26]. The reason is that the transition state complexes for these molecules are quite extended and at the step it is possible to involve more metal atoms to stabilize them.

There are, however, also electronic effects distinguishing steps from close-packed surfaces. It is generally found that steps with undercoordinated metal atoms have high-lying d states that are able to bond stronger to adsorbates (and transition state complexes) than metal atoms with a higher coordination number. This "electronic" effect can be quantified by DFT calculations [27], as shown for methane activation on nickel in Figure 9.4.

The energy barrier for the first step is plotted against the center of the d bands projected onto the metal atoms involved with the process. It is seen that the d band center rationalizes not only the effect of steps, but also the effect of alloying with Au, the effect of preadsorbed C and S atoms, and the effect of strain.

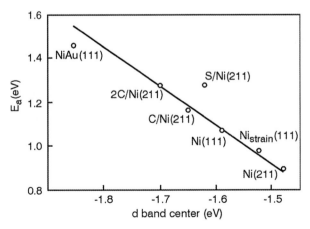

Figure 9.4 Activation energy for CH_3-H bond breaking and local d band center for nickel atoms. Adapted from Ref. [27].

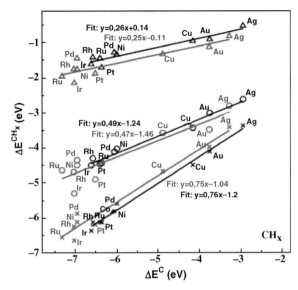

Figure 9.5 Adsorption energies of CH_x intermediates and adsorption energy of C. $\Delta E(AH_x) = \gamma \Delta E(A) + \xi$. Adapted from Ref. [28].

The Nørskov group was able to push this further by the so-called *scaling model* introduced by Abild-Pedersen et al. [28]. The aim was to provide a simple tool to estimate the bonding energy assumed to be the key quantity describing the surface reaction. It was found that the adsorption energy of a molecule AH_x was linearly correlated with the adsorption energy of the atom A and that the slope was related to the "valency" of the absorbate, as illustrated for CH_x in Figure 9.5.

It was further shown that the constant ξ (Figure 9.5) can be obtained from any transition metal as it is related to the sp states, being essentially the same for the transition metals. This means that it is possible to estimate adsorption energy for all other transition metals. As stated by Abild-Pedersen et al. [28]: "By combining this simple scaling model with the Brøndsted-Evans-Polyani type correlations, it should be possible to estimate the full potential energy diagram for a surface reaction on any transition metal on basis of the S, N, O and S chemisorption energies and a calculation for a single metal. This may lead to *computer aided design of catalysts* etc."

9.4
Examples of Modeling and Scale-Up of Industrial Processes

9.4.1
Ammonia Synthesis

The ammonia synthesis is a rare example where the catalyst activity is the most important parameter for reactor design. It involves only one reaction and the catalyst

S-300

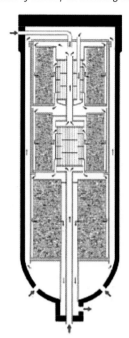

Figure 9.6 Radial flow S-300 converter for ammonia synthesis. Haldor Topsøe A/S.

is well protected at the back-end of the plant after several catalytic reactors that may retain potential poisons.

Well-defined laboratory experiments [29] formed the basis for a one-dimensional heterogeneous reactor model [15] that included the concentration and temperature gradients inside and outside the catalyst particles. The computer analysis showed that the effectiveness factor was below 1. This meant smaller catalyst particles that, however, would result in higher pressure drop. The solution was the *radial flow converter* [30, 31], as shown in Figure 9.6. The development of the radial flow converters for ammonia synthesis is an example of a *deductive* approach [9].

These data formed the basis for an advanced computer model [15] simulating the operation of the synthesis loop. It was this modeling work that allowed the jump to the complex radial converter [9]. It was not possible to test the concept at pilot site. A systematic feedback from industrial ammonia plants provided a big amount of data to check the model [32]. The industrial data replaced the requirement for integral testing at industrial pilot scale on real feed.

It has been possible to *"bridge the gap"* between measurements of adsorption and desorption rates at ultrahigh vacuum on well-defined surfaces and the activity measurements for ammonia synthesis at industrial conditions (Figure 9.7). This was done by *microkinetic analysis* by Stoltze and Nørskov [22]. Later, a more detailed DFT analysis led to a similar correlation for a ruthenium catalyst [33].

Christensen *et al.* [34] were able to analyze activity trends for various metals in volcano plots by combining results from a microkinetic for the ammonia

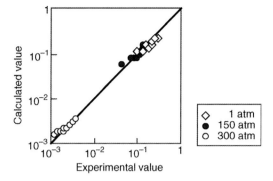

Figure 9.7 NH₃ synthesis. Calculated and measured conversions. Adapted from Ref. [22].

synthesis model with DFT calculations, leading to prediction of optimum catalyst filling depending on reaction conditions. The catalyst activity was assumed to depend only on the binding energy of nitrogen. For each set of reaction conditions, the nitrogen binding energy corresponding to the optimum in the *volcano plot* was calculated. One example [35] is shown in Figure 9.8. It illustrates the well-known observations that iron catalyst is the best for low ammonia concentrations (reactor inlet), whereas ruthenium is the preferred choice for high partial pressures of ammonia (reactor exit).

Several attempts have been made to find catalyst for the ammonia synthesis closer to the optimum [35, 36]. However, it is not sufficient to find the optimum catalyst composition. It is essential to achieve high active surface area and stability as well. Still it seems difficult to beat the promoted iron-based catalyst discovered 100 years ago.

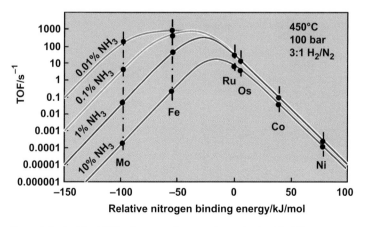

Figure 9.8 Activity (TOF) of promoted ammonia synthesis with different ammonia catalysts (conversion levels). 450 °C, 100 bar, H₂/N₂ = 3. Adapted from Ref. [35].

9.4.2
Syngas Manufacture

9.4.2.1 Steam Reforming

Steam reforming for syngas is carried out in a tubular reactor in a furnace [37, 38]. The heat is transferred to the catalyst for driving the endothermic reaction and to heat up the process gas to the exit temperatures to ensure high conversion. In principle, steam reforming may appear straightforward as composition and heat balances are determined by simple thermodynamics, but in reality it is a *complex coupling* of catalysis, heat transfer, and mechanical design.

The strong *radial temperature gradients* in a tubular steam reformer were simulated by a two-dimensional homogeneous reformer model [14, 37] (see Figure 9.3), but the model was no better than the accuracy of the many constants.

It was decided to test the model of the process in the *full-size monotube* reformer [14] shown in Figure 9.9. In principle, tests were carried out to determine the radial Peclet number at very special conditions.

For steam reforming, the "*ab initio*" methods have given a detailed understanding of reaction mechanism [39], but the industrial value has been the input to understanding the mechanism of carbon formation and to rationalize the methods for catalyst promotion. This is related to the role of step sites [17]. The role of highly coordinated sites was discussed in the 1960s in terms of the so-called B_5 sites. The activity of nickel catalysts for steam reforming correlated with the presence of B_5 sites [39, 40].

Figure 9.9 Monotube steam reformer. The radial diffusivity (D_{er} in Figure 9.3) depends on the radial Peclet number: $Pe_{m,r} = G_M d_p / \varrho D_r$, mr

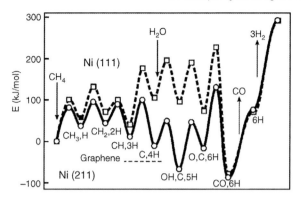

Figure 9.10 Energy diagram for steam reforming of methane. DFT calculations. Adapted from Ref. [41].

It was also shown that promotion of the catalyst with alkali removed the B_5 sites, causing a drastic decrease in activity [40]. At the same time, the presence of alkali inhibited the formation of carbon.

The role of the step site was confirmed and quantified by *DFT calculations* by Bengaard et al. [41] for steam reforming of methane on the Ni(111) surface and the Ni (211) surface representing a stepped surface. The energetics of the full reaction is shown in Figure 9.10. The step sites are the most reactive. They have lower activation barriers and the intermediates are more strongly bound than on the close-packed Ni(111).

This analysis was expanded [42] by applying the scaling principles for activation of methane (Figure 9.4) and carbon monoxide and at the same time using values for the free energy and not the total energy for the reaction scheme. The *scaling method* led to prediction of the activity for the reforming reaction as function of the adsorption energies for carbon and oxygen as shown in Figure 9.11.

Although this approach has led to prediction of catalysts with improved reforming activity [42], this is not the main problem for the reaction in conventional reformers working with a high surplus of activity [39]. The main industrial impact of the DFT approach has been a better understanding of the *mechanism of carbon formation* [17, 39].

Figure 9.8 shows that carbon is particularly strongly bound at the step sites. It means that the step sites may also be the sites for nucleation of carbon [39]. The energy of a graphene layer is lower than that of the adsorbed carbon atoms, thus creating a driving force for the nucleation of carbon. *In situ* HREM [43] confirmed the theoretical conclusions. It was possible to observe how graphene layers grow from the surface steps that are created during the nucleation and move as the graphene layers grow. This is illustrated in Figure 9.12.

DFT calculations [44] led to the conclusion that surface diffusion of carbon (or perhaps through the subsurface layer of nickel) was involved in the nucleation of the carbon whisker, whereas the previous view [37] involving diffusion of carbon through the bulk nickel was unlikely. This understanding formed the basis for a new approach for catalyst *promotion by blockage* of step sites [17, 38].

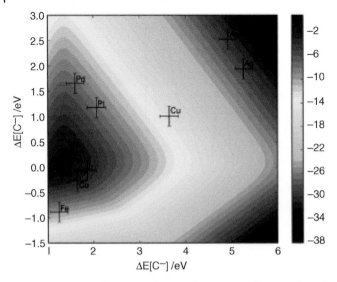

Figure 9.11 Activity for steam reforming ($\log_{10}$ TOF) as function of C and O adsorption energies. 500 °C, 1 bar absolute. Adapted from Ref. [42].

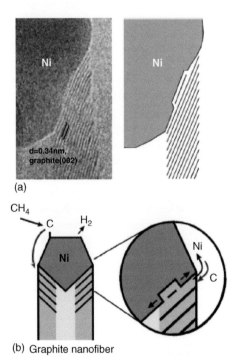

Figure 9.12 Snapshot of a whisker growth, HREM. The step sites move as the graphite plane grows. Adapted from Ref. [43].

The blockage of the step sites by alkali was indicated above. Gold is also preferentially located on step sites and it was demonstrated on a model catalyst that addition of gold to a nickel could eliminate carbon formation [45] – again in agreement with DFT calculations showing a weaker carbon bonding in the vicinity of gold and a strong tendency for gold to decorate the steps. Similar effects may be expected from other bimetallic catalysts.

The blockage of sites for carbon nucleation was the idea of the *SPARG process* [46], with chemisorbed sulfur "passivating the ensembles" for nucleation of carbon. Sulfur is also found to be chemisorbed more strongly on step sites than on the flat surface as demonstrated by DFT calculations [41] and STM observations [47].

When the SPARG process was developed in the 1980s, these details were not known. However, LEED/Auger studies and chemisorption studies had provided a quantitative description of Ni/H_2S system, allowing prediction of sulfur coverages [37]. Electron microscopic investigations and TGA measurements had also given quantitative description of the formation of whisker carbon [37]. These inputs formed the scientific basis for explaining the effect of *sulfur passivation* at partial sulfur coverage, resulting in elimination of carbon formation at conditions for which thermodynamics would predict potential for carbon formation [46]. This fundamental understanding was essential to develop a design model and to convince process engineers as sulfur is a well-known poison for steam reformers, often resulting in carbon formation (by cracking of higher hydrocarbons). This basis and demonstration in a full-size monotube reformer (Figure 9.9) led to the introduction of the SPARG process to industry [48].

9.4.2.2 Autothermal Reforming

Oxygen-blown autothermal reforming (ATR) developed by Topsoe is the preferred technology for large-scale Fischer–Tropsch plants for converting natural gas into liquid (GTL) plants (Figure 9.13) [49, 50]. It becomes cheaper than steam (and CO_2)

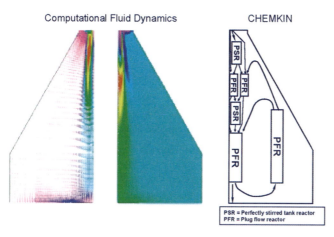

Figure 9.13 ATR reactor. Fluid mechanics simulation [52].

Table 9.3 Scale-up of the Topsoe ATR process for FT synthesis.

	Nm³ natural gas/h	Year
Pilot operation	100	1995
Industrial demo	31.000	1999
FT plant RSA	2 × 75000	2004
GTL plant/Qatar	2 × 160.000	2007

reforming because of a more favorable economy of scale for oxygen plants than for the tubular reformer. The feedstock is reacted with a mixture of oxygen and steam by the use of a special burner and a fixed catalyst bed for the equilibration of the gas and removal of soot precursors. Study of the autothermal reforming process was hardly possible at laboratory scale. Most of the pioneering work was done on some of the first industrial units. Further optimization was carried out at pilot scale during the 1990s [50]. The pilot work was supplemented by modeling using cold flow simulations and CFD calculations [51, 52].

Fluid mechanics simulation by *CFD calculations* was essential in the development of the burner for the autothermal reformer. By means of a 3D programme, it was possible to map temperature and flow gradients for different burner designs. The reaction patterns depend on the residence time/temperature distribution. Hence, it is important to couple the kinetic models with CFD simulations [53]. The reactions may also involve the formation of soot. The CFD simulation was supplemented by cold flow experiments. The mixing in the burner was simulated by two water schemes, one alkaline and the other containing phenolphthalein. The momentum of the two streams corresponded to those in the reactor.

The modeling was essential to minimize the number of expensive experiments to be carried out in the pilot plant [50]. On the basis of the pilot tests and modeling work, the scale-up was fast as shown in Table 9.3.

9.5
Conclusions

The dual approach with *scale-down* to well-controlled experiments using so-called "*in situ*" techniques and advanced characterization and *scale-up* to pilot plants operating at industrial mass velocity and industrial heat transfer rates provide information allowing a better description of the mechanism as well as providing data for fast scale-up. Modeling is the tool linking the two. However, modeling must not be the aim for its own sake. Models should be used to formulate new experiments to expand our knowledge.

Laboratory and pilot studies should be carried out in parallel and the loop is closed by characterizing samples of spent catalyst from the pilot plants (and later the

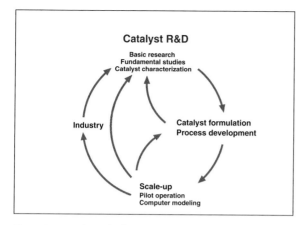

Figure 9.14 Industrial R&D in catalysis. A learning process [10].

industrial plants) by means of the advanced surface science techniques. This gives a feedback to the formulation of new catalysts and to better understanding of catalyst poisoning, coking, sintering, and so on. This makes catalyst development a *learning process* (Figure 9.14) and not an empirical screening process.

The *linear model for scale-up* with sequential phases for basic research, applied research, development, and engineering is not effective. It is necessary to work with a simultaneous approach from research, explorative reaction engineering, and process analysis. It is essential to evaluate both the industrial potential and the feasibility of the scale-up process on an ongoing basis. It means working with overlapping phases: in all these steps, modeling tools will accelerate the speed of development.

It is essential for process innovation to continuously analyze what are the reasons for the *limits to the technology*, rather than optimizing to the immediate flow sheet needs. It is not dangerous to move close to a cliff if you know where the cliff is.

Industrial catalysis needs the interaction with *basic research* because the knowledge of catalysis allows a more systematic approach to catalyst development than the tedious trial and error method applied by Mittasch. Nevertheless, catalysis remains a field for surprises. It is important to *leave room for explorative research* driven by curiosity and ambition to find new solutions. That is the only way to create a basis for the second S-curve. The modeling work may not give the precise answer, but may function as a useful compass.

The industrial development projects should be supplemented by *research to understand* the phenomena that give strength to cope with problems and that of course is also important in the marketing. Part of this research can be published in scientific journals. This type of research strongly depends on fundamental methods to characterize the nanostructure of the catalyst material and the development of tools for modeling. Needless to say that this is an ideal theme for collaboration with universities [17, 19, 22, 26, 33, 34, 36, 39, 41–45, 47, 53].

References

1. Rostrup-Nielsen, J.R. (2004) *Catal. Rev. Sci. Eng.*, **46**, 246.
2. Felcht, U.-H. (2003) *Chemical Engineering: Visions of the World* (eds A.C. Darton, R.G.H. Prince, and D.G. Wood), Elsevier, Amsterdam, p. 41.
3. Dybkjær, I. (1995) *Ammonia* (eds A. Nielsen and I. Dybkjær), Springer, Berlin, p. 199.
4. Rostrup-Nielsen, J.R. (1997) *Catal. Today*, **137**, 225.
5. Rostrup-Nielsen, J.R. (2000) *Combinatorial Catalysis and High Throughput Catalyst Design and Testing* (ed. E.G. Derouane et al..), Kluwer Academic Publishers, Dordrecht, NL, p. 337.
6. Mittasch, A. (1951) *Geschichte der Ammoniaksynthese*, Verlag Chemie, Weinheim, p. 113.
7. Lange, J.P. (2001) *Cattech*, **5** (2), 85.
8. Foster, R. (1986) *Innovation*, Summit Books, New York.
9. Rostrup-Nielsen, J.R. (1995) *Chem. Eng. Sci.*, **50**, 4061.
10. Rostrup-Nielsen, J.R. (1993) *Catal. Today*, **18**, 125.
11. Reichle, A.D. (1992) *Oil Gas J.*, **90** (5), 41.
12. Rostrup-Nielsen, J.R. (2000) *J. Mol. Catal. A*, **163**, 157.
13. Froment, G.F. and Bischoff, K.B. (1979) *Chemical Reactor Design and Analysis*, John Wiley & Sons, Inc., New York.
14. Rostrup-Nielsen, J.R., Christiansen, L.J., and Bak Hansen, J.-H. (1988) *Appl. Catal.*, **43**, 287.
15. Kjaer, J. (1972) *Computer Methods in Catalytic Reactor Calculations*, Topsoe, Vedbaek.
16. Rostrup-Nielsen, J.R. (1994) *Catal. Today*, **22**, 295.
17. Rostrup-Nielsen, J.R. and Nørskov, J.K. (2006) *Top. Catal.*, **40**, 45.
18. Ruan, L., Besenbacher, F., Steensgaard, I., and Laegsgaard, E. (1992) *Phys. Rev. Lett.*, **69**, 3523.
19. Hansen, P.L., Wagner, J.B., Helweg, S., Rostrup-Nielsen, J.R., Clausen, B.S., and Topsøe, H. (2002) *Science*, **295**, 2053.
20. Villadsen, J. and Livbjerg, H. (1978) *Catal. Rev. Sci. Eng.*, **17** (2), 203.
21. Dumesic, J.A., Rudd, D.R., Aparicio, L.M., Rekoske, J.E., and Trevito, A.A. (1993) *The Microkinetics of Heterogeneous Catalysis, ACS Professional Reference Book*, American Chemical Society, Washington, DC.
22. Stoltze, P. and Nørskov, J.K. (1985) *Phys. Rev. Lett.*, **55**, 2502.
23. Chorkendorff, I. and Niemantsverdriet, J.W. (2003) *Concepts of Modern Catalysis and Kinetics*, Wiley-VCH Verlag GmbH, Weinheim.
24. Hammer, B. and Nørskov, J.K. (2000) *Adv. Catal.*, **45**, 71.
25. Fernandez, E.M., Moses, P.G., Toftelund, A., Hansen, H.A., Martinez, J.I., Abild-Pedersen, F., Kleib, J., Hinnemann, B., Rossmeisl, J., Bligaard, T., and Nørskov, J.K. (2008) *Angew. Chem., Int. Ed.*, **47**, 4683.
26. Bligaard, T., Nørskov, J.K., Dahl, S., Mathiesen, J., Christensen, C.H., and Sehested, J. (2004) *J. Catal.*, **224**, 206.
27. Abild-Pedersen, F., Greeley, J., and Nørskov, J.K. (2005) *Catal. Lett.*, **105**, 9.
28. Abild-Pedersen, F., Greeley, J., Studt, F., Rossmeisl, J., Munter, T.R., Moses, P.G., Skulason, E., Bligaard, T., and Nørskov, J.K. (2007) *Phys. Rev. Lett.*, **99**, 16105.
29. Nielsen, A. (1910) *An Investigation on Promoted Iron Catalysts for the Synthesis of Ammonia*, 3rd edn, Julius Gjellerups Forlag, Copenhagen.
30. Dybkjær, I. (1995) *Ammonia* (ed. A. Nielsen), Springer, Berlin, p. 199.
31. Nielsen, A. (1972) *Ammonia Plant Saf.*, **14**, 46.
32. Jarvan, J.E. (1970) *Ber. Bunsensges. Phys. Chem.*, **74**, 142.
33. Honkala, K., Hellman, A., Remediakis, I.N., Logadottir, A., Carlson, A., Dahl, S., Christensen, C.H., and Nørskov, J.K. (2005) *Science*, **307**, 555.
34. Jacobsen, C.J.H., Dahl, S., Boisen, A., Clausen, B.S., Topsøe, H., Logadottir, A., and Nørskov, J.K. (2002) *J. Catal.*, **205**, 382.

35 Jacobsen, C.J.H. and Nielsen, S.E. (2002) *Ammonia Tech. Manual*, 212.
36 Jacobsen, C.H., Dahl, S., Clausen, B.G.S., Bahn, S., Logadottir, A., and Nørskov, J.K. (2001) *J. Am. Chem. Soc.*, **123**, 8404.
37 Rostrup-Nielsen, J.R. (1984) Chapter 1, in *Catalysis, Science and Technology*, vol. **5** (eds J.R. Anderson and M. Boudart), Springer, Berlin.
38 Rostrup-Nielsen, J.R. (2008) Chapter 13.11, in *Handbook of Heterogeneous Catalysis* (eds G. Ertl, H. Knözinger, F. Schüth, and J. Weitkamp), Wiley-VCH Verlag GmbH, pp. 1–24.
39 Rostrup-Nielsen, J.R., Sehested, J., and Nørskov, J.K. (2002) *Adv. Catal.*, **47**, 65.
40 Rostrup-Nielsen, J.R. (1973) *J. Catal.*, **31**, 173.
41 Bengaard, H.S., Nørskov, J.K., Sehested, J.S., Clausen, B.S., Nielsen, L.P., Molenbroek, A.M., and Rostrup-Nielsen, J.R. (2002) *J. Catal.*, **209**, 365.
42 Jones, G., Kleis, J., Anderson, M.P., Rossmeisl, J., Abild-Pederesen, F., Bligaard, T., Nørskov, J.K., Chorkendorff, I., Jakobsen, J.C., Shim, S.S., Helweg, S., Hinnemann, B., Rostrup-Nielsen, J.R., and Sehested, J., (2008) *J. Catal*, **259**, 147.
43 Helveg, S., López-Cartes, C., Sehested, J., Hansen, P.L., Clausen, B.S., Rostrup-Nielsen, J.R., Abild-Pedersen, F., and Nørskov, J.K. (2004) *Nature*, **427**, 426.
44 Abild-Pedseren, F., Nørskov, J.K., Rostrup-Nielsen, J.R., Sehested, J., and Helweg, S. (2006) *Phys. Rev. Lett. B.*, **73**, 115449.
45 Besenbacher, F., Chorkendorff, I., Clausen, B.S., Hammer, B., Molenbroek, A.M., Nørskov, J.K., and Steensgaard, I. (1998) *Science*, **279**, 1913.
46 Rostrup-Nielsen, J.R. (1984) *J. Catal*, **85**, 31.
47 Vang, R.T.K., Honkala, S., Dahl, S., Vestgergaard, E.K., Schnadt, J., Laegsgaard, E., Clausen, B.S., Nørskov, J.K., and Besenbacher, F. (2005) *Nat. Mater.*, **4**, 160.
48 Udengaard, N.R., Bak Hansen, J.-H., Hanson, D.C., and Stal, J.A. (1992) *Oil Gas J.*, **90** (10), 62.
49 Christensen, T.S. and Primdahl, I.I. (1994) *Hydrocarbon Process.*, **73** (3), 39.
50 Christensen, T.S., Christensen, P.S., Dybkjær, I., Bak Hansen, J.-H., and Primdahl, I.I. (1998) *Stud. Surf. Sci. Catal.*, **119**, 883.
51 Christensen, T.S., Dybkjær, I., Hansen, L., and Primdahl, I.I. (1995) *Ammonia Plant Saf.*, **35**, 205.
52 Christensen, T.S. and Østberg, M. (2001) AIChE Annual Meeting, Reno, paper 348b.
53 Skjøtt Rasmussen, M.S., Glarborg, P., Østberg, M., Johannesen, J.T., Livbjerg, H., Jensen, A.D., and Christensen, T.S. (2004) *Combust. Flame*, **136**, 91.

10
Perspectives of the Automotive Industry on the Modeling of Exhaust Gas Aftertreatment Catalysts

Daniel Chatterjee, Volker Schmeißer, Marcus Frey, and Michel Weibel

10.1
Introduction

The environmental concerns and the increasing stringent emission limits worldwide have lead to the introduction of highly efficient exhaust gas aftertreatment systems by the automotive industry. Nearly all systems are based on catalytic systems to reduce the CO, HC (hydrocarbons), NO_x, and particulate matter emissions of vehicles. Especially for Diesel engines, current systems have become complex, consisting of combinations of different catalyst technologies and particulate filters (cf. Figure 10.1). Development cost and time of exhaust gas aftertreatment systems are now comparable to engines.

Modeling and simulation of heterogeneous catalytic processes related to exhaust gas aftertreatment systems have become an important part of the development process within the automotive industry [1–8]. Different aspects are analyzed by means of modeling and simulation. The ultimate objective is the prediction of the conversion efficiency of the total system. However, the prediction of temperatures, the identification of critical operating conditions, and the optimization of operation strategies are typical tasks. In addition, modeling and simulation is used to gain insights into the exhaust system not directly available by measurements.

One key challenge for modeling of heterogeneous catalysts in the automotive industry is the high variation in operation conditions due to the transient operation of the engine in the vehicle. In fact, modern exhaust aftertreatment systems are the most dynamic and complex chemical reactor systems in mass products. Further challenges arise due to the variation in length scales, for example, catalyst volumes starting from 0.5 l for passenger cars up to 45 l for heavy duty engines. Also, the amount of catalytic active material is subject to change during the development process. Moreover, even the state of the catalyst is changing during lifetime by aging. Clearly, these requirements for the simulation can be fulfilled only by models that are based on chemical and physical fundamentals.

The current models, applied by the automotive industry, are characterized by the limited knowledge of the detailed heterogeneous catalytic processes usually available

---

Modeling and Simulation of Heterogeneous Catalytic Reactions: From the Molecular Process to the Technical System,
First Edition. Edited by Olaf Deutschmann.
© 2012 Wiley-VCH Verlag GmbH & Co. KGaA. Published 2012 by Wiley-VCH Verlag GmbH & Co. KGaA.

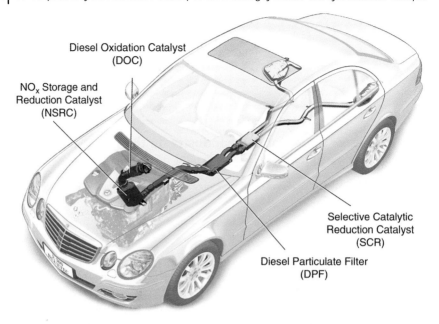

Figure 10.1 Example of a modern combined diesel exhaust gas aftertreatment system (Mercedes Benz E320 CDI) BlueTEC.

for the specific catalyst and the degree of complexity that can be handled in the models. For some catalysts, for example, three-way catalysts (TWC), the elementary steps are relatively well known and information on microkinetic reaction mechanisms including parameters can be found in the literature [9–11]. For other catalysts, for example, NO_x storage and reduction catalysts (NSRC), there is still a discussion in the literature on the elementary reaction steps. Therefore, a global reaction kinetic approach, "lumping" several microkinetic reaction steps together, is widely used in the automotive industry. It is important to note, however, that a detailed kinetic analysis of the specific catalysts is required to set up a chemically consistent global reaction kinetic scheme and respective rate expressions. Especially, all the relevant inhibition effects have to be included in the kinetic rate expressions.

10.2
Emission Legislation

Since the introduction of the first catalytic systems for automotive exhaust gas aftertreatment in the United States in 1974, emission standards have become more and more stringent. Today emission legislations are in place to regulate the CO, HC (Hydrocarbons), NO_x, and particulate matter emissions from passenger cars and heavy duty engines.

10.2 Emission Legislation

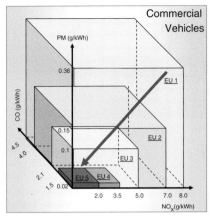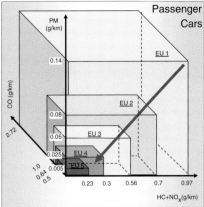

Figure 10.2 European emission legislation for commercial vehicles and Diesel passenger cars.

The European emission limits for passenger cars and commercial vehicles are shown in Figure 10.2. With the introduction of Euro5 in 2009, the NO_x emission of diesel passenger cars must meet 0.18 and 0.23 g/km for HC and NO_x, respectively. This requires a reduction of NO_x by 64% compared to Euro3 in 2000. Also, for commercial vehicles, NO_x was reduced by 60% with the introduction of Euro5 in 2008 compared to 2000. Particulate emissions were even more reduced since 2000. For commercial vehicles, these emissions have been reduced by 80% and for diesel passenger cars by 90%. In the United States, even stricter emission limits are in place; for example, Tier2 Bin5 for passenger cars (0.07 g/mi NO_x) requires an additional reduction of NO_x by 85% compared to Euro5. A further emission reduction, for example, with the introduction of Euro6 (2013 heavy duty engines, 2014 passenger cars) and EPA 2010 and EPA 2014 in the United States, will be required.

Like the emission limits, the certification procedure is different for passenger cars and commercial vehicles. Generally, test cycles are used, aiming to represent a combination of different local driving conditions as urban or freeway traffic. Also, different test cycles are applied, for example, in Europe, Japan, or United States.

In the case of a commercial vehicle, only the engine with the aftertreatment system is tested on an engine dynamometer. Steady-state test cycles and a transient test cycle are used. In Europe, the European stationary cycle (ESC) consists of 13 steady-state engine operating points in which the emissions are measured. Three additional points are chosen by the tester. The final emission result is obtained by averaging weighting factors over all measured points. Because of the relatively high engine load in the cycle, the exhaust temperatures are typically high. The same test is also applied under the name SET in the United States. In Europe, the stationary test is complemented by the European transient cycle (ETC) and European load response (ELR) tests. In the United States, the FTP-75 is used that operates only in a range of 20–25% of the maximum horsepower at given engine speed. Therefore, the exhaust temperatures are rather low. Global world harmonized test cycles (WHTC, WHSC) have been defined and will be used for future emission legislations [12].

306 | *10 Perspectives of the Automotive Industry on the Modeling of Exhaust Gas Aftertreatment Catalysts*

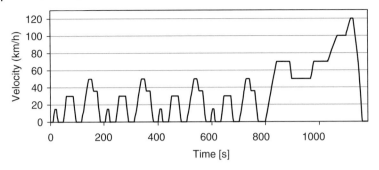

Figure 10.3 New European driving cycle [12].

The passenger car certification is done with the full vehicle tested on a chassis dynamometer. The boundary conditions are defined by the vehicle speed. In Europe, the new European driving cycle (NEDC) is used. It consists of four phases that are representative of urban driving, followed by one phase at higher speed for extra urban driving, as shown in Figure 10.3. In the United States, the FTP-75 is used, which is a sequence of the same test cycle, first with a cold engine and then by a repetition with a hot engine. Two additional test cycles are applied: the US06 to account for high speed driving and the SC03. Compared to the NEDC, the FTP-75 is more transient and covers a wider range of operating conditions [12].

10.3
Exhaust Gas Aftertreatment Technologies

To meet the low emission limits, the engine-out emissions have been continuously reduced. However, additional exhaust gas aftertreatment measures are required. Catalysts are the key components to achieve an efficient reduction of CO, HC, and NO_x in the exhaust line of a passenger car or a commercial vehicle. To reduce the particulate matter emission from the engine, particulate filters are used. In most cases, these filters are also coated with some catalytic active materials. Monolithic catalyst and filters based on ceramic or metallic materials are used nearly in all applications (cf. Figure 10.4). The channels of the monolithic structures are typically coated by a porous material, for example, based on $\gamma\text{-}Al_2O_3$, the so-called washcoat,

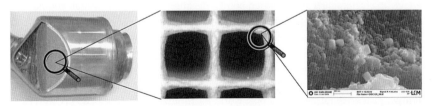

Figure 10.4 Structure of an automotive monolithic catalyst.

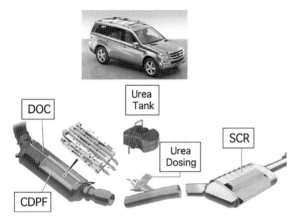

Figure 10.5 Urea-based exhaust system (BlueTEC II Mercedes Benz GL 320 CDI).

that contains the catalytic active material. In many cases, noble metals like platinum, rhodium, and palladium are used as catalytic active material.

The TWC is the state-of-the-art technology for the simultaneously removal of CO, HC, and NO_x from gasoline engines [13]. However, exhaust gas aftertreatment systems for Diesel engines have become more complex, consisting of combinations of different catalyst and filter technologies. Depending on the applications, that is, vehicles and markets, different technologies are used. In Figure 10.5, the BlueTEC II exhaust gas aftertreatment system of a Mercedes Benz GL320 CDI is shown as an example. It consists of a Diesel oxidation catalyst (DOC), a Diesel particulate filter (DPF), a urea solution injection module, and a selective catalytic reduction (SCR) catalyst. The single components focus on different aspects: CO and HC are oxidized to CO_2 in the DOC while NO is oxidized to NO_2, soot particles are trapped in the DPF, and NO_x is reduced in the SCR catalyst by NH_3, formed from the urea solution in the exhaust line [14]. The urea-SCR principle is in use since 2005 for commercial vehicles and since 2008 available for passenger cars.

A different system approach can be found, for example, in the Mercedes Benz E320 BlueTEC. As shown in Figure 10.6, the exhaust gas aftertreatment system of the E320 consists of DOC, NSRC, DPF, and SCR. The main advantage of this system is that no urea is required to operate it. Because of the NSRC (see Section 10.6), the engine has to operate periodically in lean and rich combustion modes. During lean engine operation, NO_x is stored in the NSRC, while during rich operation, the stored NO_x is reduced to N_2 and NH_3. The NH_3 is stored in the SCR catalyst and is used in the next lean phase to reduce NO_x [15, 16]. Figure 10.7 further illustrates the operating principle of this advanced aftertreatment system.

The design, application, and optimization of such advanced systems is a complex process involving the optimization of different parameters such as catalyst and filter coating, position, size, cell density of the monoliths, and, last but not least, the interaction between engine setup and aftertreatment system. The development and optimization of an exhaust gas aftertreatment system is further complicated due to

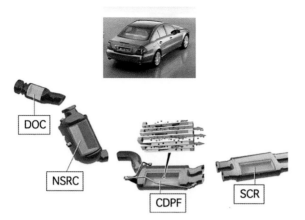

Figure 10.6 NSRC + SCR-based exhaust system (BlueTEC I Mercedes Benz E320 CDI).

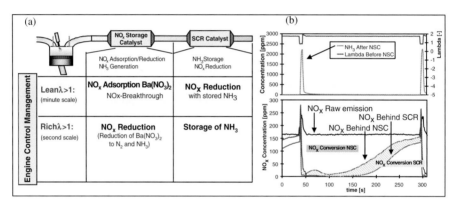

Figure 10.7 (a) NSRC and SCR functionalities in the BlueTEC I technology. (b) NO_x/NH_3 composition at different positions in the exhaust line.

the highly transient operating conditions. Figure 10.8 shows the engine-out exhaust mass flow and gas temperature before the first catalyst in the exhaust line of a Diesel passenger car within a NEDC test cycle. A critical part of the test cycle is the cold start in the first phase. Typically, temperatures higher than 150 °C are required to obtain significant conversion rates in automotive catalysts. Therefore, during cold start special engine operating strategies are applied to increase the exhaust temperature.

10.4
Modeling of Catalytic Monoliths

A chemically and physically based description of the processes in the catalytic monolith is a precondition for the reliability of the models over a wide range of

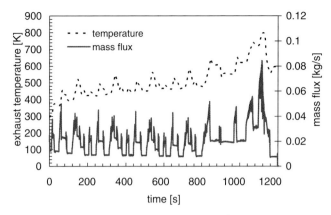

Figure 10.8 Diesel engine-out exhaust gas mass flows and temperatures in a NEDC test cycle.

operating conditions. However, the model should be as simple as possible with short computation times for an efficient application. Thus, over the years several ways of modeling of exhaust gas catalysts have become established, whose main aspects will be described in the following.

Owing to the monolithic structure of catalytic converters used for automotive exhaust gas aftertreatment, the catalyst can be represented by a single monolith channel (cf. Figure 10.9), assuming uniform gas flow distribution over the whole catalyst frontal area and adiabatic conditions for each channel [17–20].

Heat loss to the environment can be neglected in many cases due to the insulating mat between catalyst and its canning. As the catalytic active material itself is distributed as a thin layer (washcoat) at the channel wall, the channel can be mathematically described as a plug flow reactor with two phases: the gas phase with laminar flow under all relevant conditions and the surface (or solid) phase where heterogeneously catalyzed reactions take place [21, 22]. Although it has been reported that at very high temperatures (>600 °C) an increasing influence of homogeneous gas reactions can be observed [18], this chapter will outline only heterogeneous

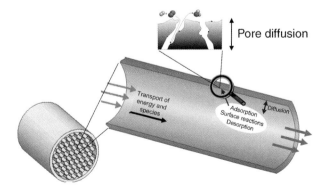

Figure 10.9 Schematic of a representative single monolith channel.

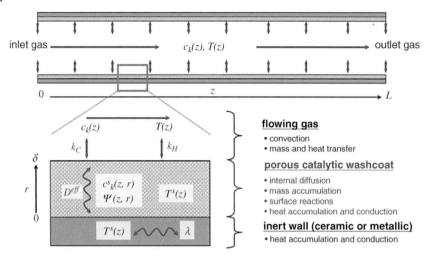

Figure 10.10 Schematic of the monolithic channel reaction system and its most important processes [23].

reaction systems. Mass and heat transport along the catalyst length takes place by convection (gas phase), mass and heat transfer between the gas and the washcoat by diffusion, perpendicularly to the axial direction. The latter can be described by a linear driving force approach with transfer coefficients derived from correlations using the dimensionless Nusselt and Sherwood numbers [17]. In the washcoat (solid phase), catalytic surface reactions and mass accumulation occur as well as heat accumulation and heat conduction in axial direction (Figure 10.10). The gas phase is treated as an ideal gas and isobaric along the catalyst length due to almost atmospheric conditions and low pressure drop (with the exception of Diesel particle filter). Mass or molar flow rate along the channel can be considered constant as changes in the total molar amount caused by reaction and accumulation are negligibly small. The velocity of the gas flow in the channel can be calculated from the exhaust mass flow, temperature, pressure, and open frontal area of the monolith:

$$v = \frac{\dot{V}^g}{A_{\text{ofa}}} = \frac{\dot{M}^g}{MW^g} \frac{RT^g}{p^g} \frac{1}{A_{\text{cat}}\varepsilon^g}. \tag{10.1}$$

Therefore, a precise determination of the monolith geometry including washcoat thickness is important, as it directly influences the component's residence time.

In simplified cases, the internal mass transport in the highly porous washcoat layer is neglected; that is, those effects are lumped into the reaction rate parameters [17, 18]. However, in several works, internal gas diffusion in the washcoat layer is reported to highly influence the catalyst behavior, not only at high temperatures [24–29]. Therefore, in general, the description of the molecular diffusion processes in the washcoat is required.

The reactions themselves can be described by a detailed formulation of the elementary steps, based on the molecular processes on the catalytic surface: adsorption and desorption of the molecules, their dissociation, surface diffusion, and recombination processes. Such a detailed model is applicable to explain observed phenomena or can be useful for specific examinations. The kinetic data for this approach can be estimated from laboratory experiments, surface science investigations, and quantum mechanical calculations [10, 25, 30–32]. However, those models are not yet fully developed and the high level of detail involves rather long computational times. Thus, such an approach is not yet suitable for the use in the development process of exhaust gas systems.

Instead, global reaction rate formulations of the Langmuir–Hinshelwood type are widely used for the description of heterogeneously catalyzed reactions [17, 18, 21, 33–37]. They cover the above-mentioned processes by one rate formulation for each reaction. Examples are shown in the following sections. Such a description requires less computational effort for its numerical solution, but kinetic parameters must be costly fitted to measurement data.

Thus, a 1D model considering the above described phenomena contains the following balances: mass balances in the flowing gas, mass balances in the washcoat pores and on the catalyst surface, enthalpy balance of the flowing gas, and enthalpy balance of the solid phase. The resulting system of differential algebraic equations basically looks as follows [35]:

$$\frac{\partial c_k(z,t)}{\partial t} = -\frac{\partial (v \cdot c_k)}{\partial z} + \frac{k_c a}{\varepsilon^g} c(y_k^s - y^k), \quad k = 1, \ldots, K, \tag{10.2}$$

$$\frac{\partial c_k^s(z,t)}{\partial t} = \frac{k_c a}{\varepsilon^s (1-\varepsilon^g) \phi^s} c(y_k - y_k^s) + \frac{1}{\varepsilon^s} \sum_{j=1}^{J} \nu_{k,j} R_j, \quad k = 1, \ldots, K, \tag{10.3}$$

$$\frac{\partial \psi_m(z,t)}{\partial t} = \frac{1}{\Psi_m^{\text{cap}}} \sum_{j=1}^{J} \nu_{m,j}^\psi R_j, \quad m = 1, \ldots, M, \tag{10.4}$$

$$\varrho c_p \frac{\partial T(z,t)}{\partial t} = -v \frac{\partial T}{\partial z} \varrho c_p + \frac{k_h a}{\varepsilon^g} (T^s - T), \tag{10.5}$$

$$\varrho^s c_p^s \frac{\partial T^s(z,t)}{\partial t} = \lambda^s \frac{\partial^2 T^s}{\partial z^2} + \frac{k_h a}{1-\varepsilon^g} (T - T^s) - \phi^s \sum_{j=1}^{J} \Delta H_{r,j} R_j. \tag{10.6}$$

The boundary conditions at the monolith inlet and outlet are

$$T(z=0) = T^{\text{in}}, \tag{10.7}$$

$$\frac{\partial T^s}{\partial z}(z=0) = \frac{\partial T^s}{\partial z}(z=L) = 0, \tag{10.8}$$

$$c_k(z=0) = c_k^{\text{in}}, \quad k = 1, \ldots, K. \tag{10.9}$$

If washoat diffusion has to be included in the model, a simple approach is the multiplication of the intrinsic reaction rates R_j with the so-called effectiveness factor η [4, 24–26]. The applicability of the effectiveness factor approach is quite limited in complex systems with competing reactions, surface deposition of reaction components, nonlinear rate laws, and under transient operation conditions. Typically, the effectiveness factor method can be used in the case of CO, H_2, and HC oxidation light-off in a DOC.

A more accurate description is possible by adding a 1D reaction diffusion equation [4, 27] to resolve the concentration gradients within the washcoat and extending the balance equations to

$$\frac{\partial c_k(z,t)}{\partial t} = -\frac{\partial (v c_k)}{\partial z} + \frac{k_c a}{\varepsilon^g} c(y_k^s|_{r=\delta} - y_k), \quad k = 1, \ldots, K, \tag{10.10}$$

$$\frac{\partial c_k^s(z,r,t)}{\partial t} = \frac{D_k^{\text{eff}}}{\varepsilon^s} \frac{\partial^2 c_k^s}{\partial r^2} + \frac{1}{\varepsilon^s} \sum_{j=1}^{J} v_{k,j} R_j, \quad K = 1, \ldots, k, \tag{10.11}$$

$$\frac{\partial \psi_m(z,r,t)}{\partial t} = \frac{1}{\Psi_m^{\text{cap}}} \sum_{j=1}^{J} v_{m,j}^\psi R_j, \quad m = 1, \ldots, M, \tag{10.12}$$

$$\varrho c_p \frac{\partial T(z,t)}{\partial t} = -v \frac{\partial T}{\partial z} \varrho c_p - \frac{k_h a}{\varepsilon^g} (T^s - T), \tag{10.13}$$

$$\varrho^s c_p^s \frac{\partial T^s(z,t)}{\partial t} = \lambda^s \frac{\partial^2 T^s}{\partial z^2} + \frac{k_h a}{(1-\varepsilon^g)} (T - T^s) - W(T^s - T^e)$$

$$- \frac{a}{(1-\varepsilon^g)} \sum_{j=1}^{J} \int_{r=0}^{\delta} \Delta H_j R_j dr. \tag{10.14}$$

The relation between the characteristic washcoat thickness δ and the volume of catalytic washcoat layer (represented in the 1D model by the volume fraction ϕ^s) is as follows:

$$\delta \cdot a = \phi^s (1 - \varepsilon^g). \tag{10.15}$$

The continuity between the external and the internal field at the gas/solid interface is granted by

$$D_k^{\text{eff}} \frac{\partial c_k^s}{\partial r}\bigg|_{r=\delta-} = k_c c(y_k - y_k^s|_{r=\delta+}), \quad k = 1, \ldots, K,$$

$$\frac{\partial c_k^s}{\partial r}\bigg|_{r=0} = 0, \quad k = 1, \ldots, K. \tag{10.16}$$

The effective diffusivities, $D_{\text{eff},j}$, can be evaluated from measurements [38, 39] or calculated from morphological data according to a modified Wakao–Smith random pore model [40]. Estimated values, for example, for NO and NH_3 effective intraporous diffusivities for V-based SCR catalyst, are on the order of $5 \times 10^{-6}\,m^2/s$.

The resulting system of parabolic partial differential equations can be solved by the finite difference method with adaptive timestep control using several modifications for an effective numerical solution [23]. These include the discretization of continuous coordinates (z, t), the application of difference approximations of the derivatives, the decomposition of the set of equations for T^s, T, c, and c^s, quasilinearization of the reaction terms R_j, and solving the resulting system of linear algebraic equations.

10.5
Modeling of Diesel Particulate Filters

Similar to the catalytic converters, Diesel particulate filters are also based on extrusion-based monolithic substrates. However, alternating plugging of neighboring flow-through channels at the upstream and downstream sides leads to a different flow distribution, as shown in Figure 10.11. The exhaust flow is enforced to pass through the porous and permeable wall (flow direction indicated by arrows), whereas entrained particles are captured on to top of the wall surface. The collection of particles on top of the wall causes the formation of a gradually increasing soot layer that has to be removed by frequent soot oxidation. To support the so-called passive soot regeneration (continuous soot oxidation by NO_2) [38], the walls are in many cases impregnated or coated with a catalytic washcoat containing noble metals such as platinum or palladium.

Based on the fundamental work of Bissett [41] in 1984, the so-called "single-channel" approach has been established to account for the complex flow distribution of DPF systems [8, 42–45] with characteristic heat and mass transfer. The control volume comprises a quarter of the inlet and the outlet channel with one-dimensional discretization in axial direction (blue grid indicated in Figure 10.11) to compute flow distribution, heat transfer, and soot and species reaction along the channel and inside to wall. In general, the model considers the governing equations for conservation of mass, momentum, and energy. The subscripts in the balance equation for the mass conservation (Eq. (10.17)) specifies the inlet (1) or outlet channel (2).

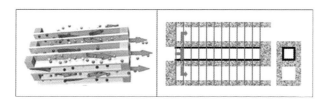

Figure 10.11 DPF single-channel approach.

$$\frac{\partial(\varrho_i v_i)}{\partial z} = (-1)^i \frac{4}{d_i} \varrho_i v_i \quad i = 1, 2. \tag{10.17}$$

The momentum balance of the exhaust gas in both channels (Eq. (10.18)) considers the viscous drag forces on the right-hand side.

$$\frac{\partial p_i}{\partial z} + \frac{\partial(\varrho_i v_i^2)}{\partial z} = -\frac{a_i \mu v_i}{d_i} \quad i = 1, 2. \tag{10.18}$$

The calculation of the local temperatures is based on the following energy balances:

Energy balance of the inlet channel gas: $c_p \varrho_1 v_1 |_z \frac{\partial T_1}{\partial z} = h_1 \frac{4}{d_1} (T^S - T_1),$

$$\tag{10.19}$$

Energy balance of the outlet channel gas: $c_p \varrho_2 v_2 |_z \frac{\partial T_2}{\partial z}$

$$= (h_2 + c_p \varrho_w v_w |) \frac{4}{d_2} (T^S - T_2), \tag{10.20}$$

Energy balance in the solid phase: $c_{p\varrho_s}^S \frac{\partial T^S}{\partial t} = \lambda_z^S \frac{\partial^2 T^S}{\partial x^2} + \lambda_z^S \frac{1}{r} \frac{\partial^2 T^S}{\partial r^2} + S.$

$$\tag{10.21}$$

The source term S in Eq. (10.21) accounts for the heat convection and the exothermal energy generated by the soot oxidation.

Energy balance source term: $S = H_{conv} + H_{wall} + H_{react}.$ (10.22)

Reaction exotherm: $H_{react} = s_F \int_{-w}^{w} f_x \sum_k R_k \Delta H_k dx.$ (10.23)

Convection of heat due to flow through wall: $H_{wall} = \varrho_s v_s s_F c_P (T_1 - T^S).$

$$\tag{10.24}$$

Convection of heat due to flow in channel:

$$H_{conv} = h_1 s_F (T_1 - T^S) + h_2 s_F (T_2 - T^S). \tag{10.25}$$

An important parameter for the application of a DPF is the generated pressure drop. This pressure drop is affected by

- the flow contraction losses at the DPF inlet face,
- the viscous drag losses of the exhaust gas in the inlet and the outlet channel,
- the pressure loss due to the gas flow through the DPF wall and soot layer, which can be described by Darcy's law and,
- the expansion of the gas flow at the DPF outlet face.

10.6
Selective Catalytic Reduction by NH_3 (Urea-SCR) Modeling

The selective catalytic reduction of NO_x by NH_3 is a very effective and established deNO$_x$ technology for stack gases from power plants and other stationary sources [46]. In the recent years, this technology as emerged as a promising tool for the reduction of NO_x emissions from Diesel engines [47, 48]. It was introduced in the European market with the Euro4 legislation in 2005 for commercial vehicles and in the United States for passenger car applications in 2008.

The application of the SCR technology to mobile applications is characterized by specific demands, for example, dynamic operation and a large functional temperature window. In particular, the low temperature in mobile applications represents a major development challenge.

As a source for the required NH_3, urea solution is injected in the exhaust line, where it is converted by thermolysis and hydrolysis to NH_3 [14, 49]. Different catalyst technologies are used for mobile applications. V-based catalysts are mostly used in the form of extruded monoliths, and Fe or Cu ion-exchanged zeolites are applied typically in the form of coated monoliths. It seems that a similar modeling approach can be used for the global modeling of the catalytic processes on these materials [34, 50, 51]. The global reactions listed in Table 10.1 describe the relevant chemical reactions occurring on a SCR catalyst.

The so-called fast SCR reaction (SCR5 in Table 10.1) is of special importance to increase the NO_x conversion efficiency at low exhaust temperatures for mobile applications. In the early 1980s, Kato *et al.* [52] found that the SCR reaction is much faster when NO_x is present as a mixture of NO and NO_2 in the field of low temperatures ($T < 300\,°C$). As can be seen from the stoichiometry of the reaction SCR5 in Table 10.1, an equimolar mixture of NO and NO_2 represents the optimal feed composition. In the practical application, an oxidation catalyst (DOC) or a coated Diesel particulate filter (CDPF) located upstream of the SCR catalyst is used to convert a fraction of the engine-out NO to NO_2 [34, 53, 54].

Table 10.1 Global reactions for SCR catalysts.

Reaction step		Reaction rate	No.
$NH_3 \rightarrow NH_3^*$	NH_3 adsorption	R_{ads}	SCR1
$NH_3^* \rightarrow NH_3$	NH_3 desorption	R_{des}	SCR2
$NH_3^* + 3/4\,O_2 \rightarrow 1/2\,N_2 + 3/2\,H_2O$	NH_3 oxidation	R_{ox}	SCR3
$NH_3^* + NO + 1/4\,O_2 \rightarrow N_2 + 3/2\,H_2O$	Standard SCR	R_{NO}	SCR4
$2\,NH_3 + NO + NO_2 \rightarrow 2\,N_2 + 3\,H_2O$	Fast SCR reaction	R_{Fast}	SCR5
$NH_3^* + 3/4\,NO_2 \rightarrow 7/8\,N_2 + 3/4\,H_2O$	NO_2 SCR	R_{NO_2}	SCR6
$2\,NH_3^* + 2\,NO_2 \rightarrow N_2O + 3\,H_2O + N_2$	N_2O formation	R_{N_2}	SCR7
$NO + 1/2\,O_2 \rightarrow NO_2$	NO oxidation	R_{NOox}	SCR8

10.6.1
Kinetic Analysis and Chemical Reaction Modeling

The SCR reactions were studied and modeled for years in the context of power plant applications; however, the effects revealed in the dynamic operation of SCR systems in automotive applications have lead to a more fundamental investigation of both standard and fast SCR reaction mechanisms [55–58]. Many of these investigations were triggered or at least complemented by kinetic modeling.

Some modeling approaches, used in the automotive industry, are very simplified and focused on control-oriented purposes [59] or are based on steady-state analysis [60, 61]. Reaction kinetic models based on dynamic studies of the SCR reaction, also in the presence of NO_2, are presented in Refs [27, 51, 54, 62]. The following reaction schemes and rate expressions are based on the work of Tronconi et al.

10.6.1.1 NH$_3$ Adsorption, Desorption, and Oxidation

The first step in the SCR reaction is the adsorption and storage of NH_3 onto the catalyst (SCR1 and SCR2 in Table 10.1). Based on indications in the literature [27, 63, 64], a nonactivated NH_3 adsorption process and a Temkin-type NH_3 desorption are usually assumed, that is,

$$R_{ads} = k_{ads} c^s_{NH_3} (1 - \theta_{NH_3}), \tag{10.26}$$

$$R_{des} = k_{0,des} \exp\left[-\frac{E_{des}}{RT}(1 - \alpha\theta_{NH_3})\right] \cdot \theta_{NH_3}. \tag{10.27}$$

The oxidation rate of NH_3 with O_2 to N_2 is described by a simple first-order rate expression based on the surface concentration of NH_3^*.

$$R_{ox} = k_{0,ox} \exp\left(-\frac{E_{ox}}{RT}\right)\left(\frac{y_{O_2}}{0.02}\right)^\beta \theta_{NH_3}. \tag{10.28}$$

10.6.1.2 NO-SCR Reaction

The so-called standard SCR reaction (SCR4 in Table 10.1) is the dominant SCR reaction if no oxidation catalyst is present upstream. Excess ammonia inhibits the SCR reaction, as already out pointed out by several authors [65–68]. This yields, for example, the temporally enhanced $deNO_x$ activity revealed by the data presented in Figure 10.12 at NH_3 shut off at 5000s.

The classical Eley–Rideal approaches [46] used in the steady-state simulation of SCR converters are unable to reproduce the dynamic effects at NH_3 shutoff. Therefore, a dual site modified redox (MR) rate law is proposed in the literature [62]. This rate law is based on the assumptions that two sites for NH_3 adsorption (acidic nonreducible sites) and for $NO + NH_3^*$ activation/reaction (redox sites) are present on the catalyst surface and that NH_3 can block the redox sites. Typically, a simplified form is used for automotive applications [27, 54].

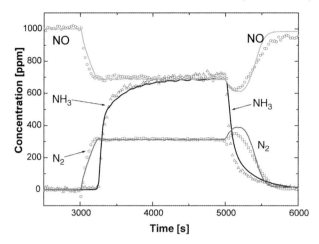

Figure 10.12 NH$_3$–NO/O$_2$ laboratory experiment at 200 °C, 1000 ppm NO, 1% O$_2$, 1% H$_2$O, 1000 ppm NH$_3$ pulse (from 3000 to 5000s), and GHSV = 90 000 h^{-1}. Reactor outlet concentration of ammonia (triangles), NO (circles), and N$_2$ (squares); solid lines—kinetic fit, V-based SCR catalyst. Adapted from SAE 2005-01-0965 [25].

$$R_{NO} = k_{0,NO} \cdot \exp\left(-\frac{E_{NO}}{RT}\right) \gamma_{NO} c_{NO}^S \theta_{NH_3} \left(\frac{y_{O_2}}{0.02}\right)^\beta,$$

$$\gamma_{NO} = \frac{1}{1 + K_{LH}(\theta_{NH_3}/(1-\theta_{NH_3}))}. \quad (10.29)$$

This rate expression also accounts for the moderate promoting effect of O$_2$ on the SCR activity [27, 62, 69, 70].

The oxidation of NO by O$_2$ to NO$_2$ is not very efficient on many SCR catalysts. However, the following rate expression can be used [34]:

$$R_{NOox} = k_{NOox}^0 \exp\left(-\frac{E_{NOox}}{RT}\right) \left(c_{NO}^S \sqrt{p_{O_2}} - \frac{c_{NO_2}^S}{K^{eq}}\right). \quad (10.30)$$

10.6.1.3 NH$_3$–NO–NO$_2$ Reactions

The presence of NO$_2$ in the feed is strictly connected to the use of the SCR process in mobile applications. The NO$_2$ formed by an upstream oxidation catalyst or coated Diesel particulate filter is one of the keys to overcome one of the biggest limitations of mobile SCR systems, namely, the low NO$_x$ conversion below 250 °C. However, the addition of NO$_2$ increases the complexity of the reaction network and may result in undesired by-products such as NH$_4$NO$_3$ and N$_2$O [71, 72]. Regarding the global reactions listed in Table 10.1, reactions SCR5–SCR7 are now also becoming a part of the reaction network.

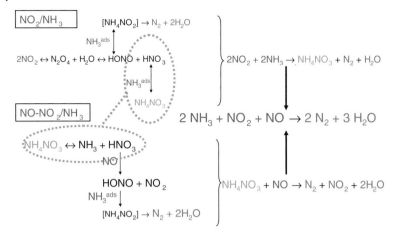

Figure 10.13 Chemistry of NH_3–NO_2 and of NH_3–NO/NO_2 SCR reacting systems over V-based and Fe–zeolite catalysts proposed by Tronconi et al. [4, 54, 55].

Recent findings have pointed that nitrates and nitrites on the surface play a critical role in the mechanism [56, 58, 71, 73, 74]. For V-based catalysts, a redox cycle is proposed, by Tronconi *et al.*, where the V-sites are reoxidized by O_2 (standard SCR) and nitrates (fast SCR) [55]. The overall reaction scheme proposed for V-based catalysts and Fe-based catalysts [56, 71, 73, 74] for the fast SCR is shown in Figure 10.13 and can be summarized as follows:

1) NO_2 is converted to nitrites and nitrates (represented by HNO_2 and HNO_3) by dimerization and disproportion.
2) NO reduces nitrates to nitrites.
3) NH_3 eventually converts the nitrites to N_2 via NH_4NO_2 as an intermediate.
4) In defect of NO, that is, NO_2/NO_x ratios higher than 50% in the inlet feed, the nitrates can react with NH_3 to form NH_4NO_3 that builds up onto the catalyst for low-temperature conditions ($T < 170\,°C$) or decomposes at higher temperatures and may lead to N_2O formation.

It is important to note that the addition of all reactions listed in Figure 10.13 leads to the correct stoichiometry of the global fast SCR reaction (Table 10.1, SCR5).

Based on the detailed analysis of the reaction mechanism, a respective global kinetic model is proposed [54]. This model is able to predict the conversion efficiency and product selectivity in a wide range of operating conditions (cf. Figure 10.14).

If NH_4NO_3 formation can be neglected, more simplified reaction schemes and models for the fast SCR and N_2O formation are available. In these models, rate expressions for the global reactions (SCR5) and (SCR7) are directly used to model the fast SCR and N_2O formation. For V-based and Fe–zeolite catalysts, the following rates are proposed [1, 34]:

$$r_{\text{Fast}} = k^0_{\text{Fast}} \exp\left(-\frac{E_{\text{Fast}}}{R}\left(\frac{1}{T} - \frac{1}{473}\right)\right) \cdot \frac{c^s_{NO_2}}{K_{NO_2} + c^s_{NO_2}} c^s_{NO} \theta_{NH_3}, \qquad (10.31)$$

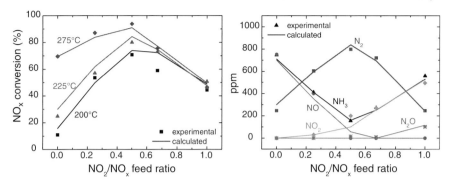

Figure 10.14 Measured and simulated steady-state data for varying NO_2/NO_x ratios. Feed: 1000 ppm NH_3, 1000 ppm NO_x, 1% H_2O, 2% O_2, balance He, SV = 210 000 h^{-1}. Right: experimental reactor outlet concentrations at $T = 225\,°C$ of NH_3 (triangles), NO (circles), NO_2 (diamonds), N_2 (squares), and N_2O (stars); solid lines—kinetic fit. Left: Steady-state SCR NO_x conversion. Symbols: measurements; lines: simulation. Adapted from SAE 2006-01-0468 [54].

$$R_{N_qO} = k^0_{N_2O} \exp\left(-\frac{E_{N_2O}}{RT}\right) c^S_{NO_2} \theta_{NH_3}. \qquad (10.32)$$

Comparable reaction rate models can be found in Refs [51, 60].

It should be noted that the fitting of all the presented rate expression constants requires a careful calibration strategy. The use of measurement data obtained with catalyst powder minimizes the incorporation of transport effects in the reaction kinetics. Furthermore, a sequential calibration strategy has to be used to minimize parameter correlations [4, 54].

10.6.2
Influence of Washcoat Diffusion

Under the typical operation conditions, the overall NO_x conversion can be limited not only by the reaction kinetics but also by diffusion effects. The difference between the intrinsic SCR reactivity and the overall reactivity of a coated Fe–zeolite is shown in Figure 10.15. In the temperature region below 225 °C, the overall rate is kinetically limited. Between 225 and 350 °C, the conversion efficiency of the catalyst powder is higher because the overall conversion is mass transfer limited.

To account for the diffusion within the washcoat, a 1D reaction diffusion equation yields a good description of the relevant effects [27].

10.7
Diesel Oxidation Catalyst, Three-Way Catalyst, and NO_x Storage and Reduction Catalyst Modeling

Among DOC, TWC, and NSRC, the latter is the most complex system, covering nearly all functionalities of the other two systems. The relevant functionalities having

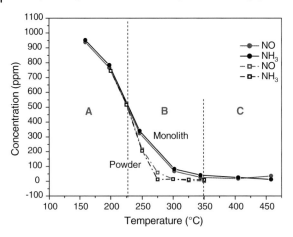

Figure 10.15 Steady-state reactor outlet concentrations for a Fe–zeolite catalyst in monolith and powder form. Conditions: 1000 ppm NH_3 and NO, 2% O_2, 1% H_2O, SV 45 000 h^{-1}. A, kinetic control; B, mass transfer control; C, total conversion [4].

considered by the respective model are described in the following section, starting with the simplest system.

10.7.1
Diesel Oxidation Catalyst

The main purpose of a DOC is the conversion (oxidation) of CO and hydrocarbons (HCs) with the excess O_2 in lean exhaust gas (e.g., Diesel exhaust gas) [75]. C_3H_6 is often considered representatively for the unburnt HCs in a model. In parallel, NO is oxidized to NO_2. An increased NO_2 content is useful for a downstream SCR (fast SCR reaction) or DPF (CRT effect) (Sections 10.6 and 10.8). Pt is used as catalytic active material, often in combination with Pd. The NO oxidation must be formulated as thermodynamic equilibrium reaction, considering the decrease in the NO_2 content toward higher temperatures [76, 77]. However, around the light-off temperature, a part of NO can also be reduced by HCs and CO to N_2 or N_2O (HC-deNO$_x$ effect) [78].

Furthermore, the DOC washcoat contains zeolites for a better cold start performance. As long as the catalyst has not reached the light-off temperature, HCs from the exhaust gas are trapped in the zeolite material. Later, toward light-off, the HCs will desorb due to the elevated temperature and immediately react with O_2 [75].

As Diesel exhaust gas exhibits rather low temperatures, another important effect must not be neglected: the partial reduction of NO_2 to NO. Around light-off temperature or even below, NO_2 entering the catalyst may be converted completely to NO when CO and HCs are present, even under lean conditions [79, 80]. This may affect the performance of a downstream DPF, SCR, or NSRC.

10.7.2
Three-Way Catalyst

The TWC has been successfully introduced in gasoline engine cars already in the 1970s. Owing to the long time of experience and development since then, the reaction mechanisms are well understood. Thus, sophisticated models of a TWC exist [10, 18, 25, 31, 81]. Unlike the DOC, the TWC is usually not used for lean exhaust gas, but for a stoichiometric exhaust gas composition. Under such conditions, CO and HCs are oxidized, while simultaneously NO is being reduced to N_2, CO_2, and H_2O. Therefore, the NO reduction reaction plays a more important role compared to the DOC, while NO oxidation can hardly be observed with a TWC under stoichometric conditions. To compensate variations around the stoichiometric gas composition, the TWC contains cerium (Ce) in the washcoat, enabling the catalyst to store O_2. Under net oxidizing conditions, O_2 will be stored, while under net reducing conditions, it will be released. Moreover, induced oscillations around the stoichiometric point have shown to increase the conversion efficiency of a TWC [82]. The ability of a TWC model to describe the O_2 storage and release behavior properly is therefore very important. Furthermore, owing to the higher exhaust temperatures of gasoline engines, H_2 production by water–gas shift and steam reforming reactions occur on the TWC and must be described by the model, also considered as equilibrium approach in the rate formulation.

10.7.3
NO_x Storage and Reduction Catalyst

Regarding its functionalities, the NSRC basically resembles a TWC with the additional feature of storing NO_x, enabled mostly by the addition of barium (Ba) to the washcoat. Along with Pt and Pd, it usually contains Rh for a better NO_x reduction performance [77] (cf. Figure 10.16). Hence, the set of surface reactions of the TWC model is used in the NSRC model, but must be expanded. Owing to the different operation mode of an NSRC—longer lean periods with short rich phases instead of a stoichiometric gas composition—the operating conditions of an NSRC cover a wide lambda range. While changing from lean to rich conditions for the

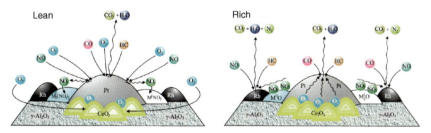

Figure 10.16 Catalytic processes on a NO_x storage–reduction catalyst under lean (left) and rich conditions (right) [http://www.vscht.cz/monolith].

purpose of regeneration, the released amount of O_2 from the O_2 storage material can decrease the effectivity of the reducing components for regeneration. Therefore, not only the correct amount of stored O_2 with its storage and release kinetics, but also the CO, HC, and H_2 oxidation have to be well described, with low content under lean conditions and high concentrations under rich conditions. As the storage of NO_2 is much more pronounced than NO storage [77], the content of NO_2/NO_x ratio in the gas phase along the catalyst strongly influences the storage and NO_x conversion performance. Thus, NO oxidation and the above-mentioned reduction of NO_2 to NO under lean conditions as long as CO and HCs are present are of great importance for the NSRC.

Beyond the surface reactions, the storage and regeneration reactions must be described. Although different forms or states of the storage material are possible, barium carbonate ($BaCO_3$) is the most probable one in presence of CO_2 and H_2O. However, often barium oxide (BaO) is used in model equations. In both cases the storage reactions show the same stoichiometry regarding NO_x. Furthermore, different mechanisms of NO_x storage are reported in literature. Among these, the two main storage paths must be included in the model: the storage of NO and of NO_2, both under participation of O_2. Another path can play a significant role, depending on the specific catalyst technology and operating conditions: the storage of NO_2 under release of NO (disproportionation mechanism) [77]. Under rich conditions, the catalyst is regenerated by H_2, CO, and HCs at elevated concentrations over the whole temperature range. Furthermore, if the model shall describe an NSRC optimized for the combination with an SCR catalyst (e.g., BlueTEC I system, see Section 10.3), the generation of NH_3 under rich conditions has to be modeled, as well as the regeneration and NO_x reduction by NH_3 [83, 84]. Independently, the decomposition of barium nitrate due to thermodynamic instability must be considered for elevated temperatures [77].

However, the processes on the NSRC are not yet fully understood. Several reaction mechanisms regarding storage and reduction are still being discussed in the literature [85–88]. Therefore, a greater variety of NSRC models and representations exist [20, 32, 35, 84, 89–91].

Table 10.2 displays all important surface reactions of an NSRC model, including the reactions of a DOC and TWC. As shown here, the O_2 release reaction can be formulated regarding the regeneration of the O_2 storage material explicitly by the reducing components CO, HC, and H_2. An alternative, more simple approach considers O_2 release independently from the concrete reducing agent, but depending on the atmospheric conditions (net oxidizing or net reducing), that is, on the λ value [18].

In contrast, as the regeneration of the NO_x storage material is reported to be most effective by H_2, followed by CO and HC, these regeneration reactions must be formulated individually for each reducing species to account for that. However, due to the fast and short regeneration process, the determination of the individual kinetic rate laws and parameters is difficult.

In addition to the set of surface reactions in Table 10.2, the HC storage in the DOC model is considered by the following:

10.7 Diesel Oxidation Catalyst, Three-Way Catalyst, and NO_x Storage and Reduction Catalyst Modeling

Table 10.2 Surface reactions and reaction rate formulation of an NSRC model. Adapted from Koči et al. [92].

Reaction	Rate	
$CO + \frac{1}{2}O_2 \rightarrow CO_2$	$R_{NSRC1} = k_{NSRC1} y_{CO} y_{O_2} \frac{1}{G_{1b}}$	(NSRC1)
$H_2 + \frac{1}{2}O_2 \rightarrow H_2O$	$R_{NSRC2} = k_{NSRC2} y_{H_2} y_{O_2} \frac{1}{G_1}$	(NSRC2)
$C_3H_6 + \frac{9}{2}O_2 \rightarrow 3\,CO_2 + 3\,H_2O$	$R_{NSRC3} = k_{NSRC3} y_{C_3H_6} y_{O_2} \frac{1}{G_1}$	(NSRC3)
$CH_4 + 2O_2 \rightarrow CO_2 + 2H_2O$	$R_{NSRC4} = k_{NSRC4} y_{CH_4} y_{O_2}$	(NSRC4)
$NO + \frac{1}{2}O_2 \leftrightarrows NO_2$	$R_{NSRC5} = k_{NSRC5} y_{NO} y_{O_2}^{0.5} \left(1 - \dfrac{y_{NO_2}}{y_{NO} y_{O_2}^{0.5} \cdot K_{y,NSRC5}^{eq}}\right) \dfrac{1}{G_{1c}}$	(NSRC5)
$CO + H_2O \leftrightarrows CO_2 + H_2$	$R_{NSRC6} = k_{NSRC6} y_{CO} y_{H_2O} \left(1 - \dfrac{y_{CO_2} y_{H_2}}{y_{CO} y_{H_2O} \cdot K_{y,NSRC6}^{eq}}\right)$	(NSRC6)
$C_3H_6 + 3H_2O \leftrightarrows 3\,CO + 6H_2$	$R_{NSRC7} = k_{NSRC7} y_{C_3H_6} y_{H_2O} \left(1 - \dfrac{y_{CO}^3 y_{H_2}^6}{y_{C_3H_6} y_{H_2O}^3 \cdot K_{y,NSRC7}^{eq}}\right)$	(NSRC7)
$CO + NO \rightarrow CO_2 + \frac{1}{2}N_2$	$R_{NSRC8} = k_{NSRC8} y_{CO} y_{NO}^{0.5} \frac{1}{G_1}$	(NSRC8)
$\frac{5}{2}H_2 + NO \rightarrow H_2O + NH_3$	$R_{NSRC9} = k_{NSRC9} y_{H_2} y_{NO}^{0.5} \frac{1}{G_1}$	(NSRC9)
$C_3H_6 + 9\,NO \rightarrow 3\,CO_2 + 3H_2O + \frac{9}{2}N_2$	$R_{NSRC10} = k_{NSRC10} y_{C_3H_6} y_{NO}^{0.5} \frac{1}{G_1}$	(NSRC10)
$NO_2 + CO \rightarrow NO + CO_2$	$R_{NSRC11} = k_{NSRC11} y_{NO_2} y_{CO} \frac{1}{G_6}$	(NSRC11)
$9\,NO_2 + C_3H_6 \rightarrow NO + 3CO_2 + 3H_2O$	$R_{NSRC12} = k_{NSRC12} y_{NO_2} y_{C_3H_6} \frac{1}{G_6}$	(NSRC12)
$Ce_2O_3 + \frac{1}{2}O_2 \rightarrow Ce_2O_4$	$R_{NSRC13} = k_{NSRC13} \Psi_{cap,Ce_2O_4} y_{O_2} \left(\psi_{O_2}^{eq} - \psi_{O_2}\right)$	(NSRC13)
$Ce_2O_4 + CO \rightarrow Ce_2O_3 + CO_2$	$R_{NSRC14} = k_{NSRC14} \Psi_{cap,Ce_2O_4} y_{CO} \psi_{O_2}$	(NSRC14)

(Continued)

Table 10.2 (Continued)

Reaction	Rate	
$Ce_2O_4 + H_2 \rightarrow Ce_2O_3 + H_2O$	$R_{NSRC15} = k_{NSRC15} \Psi_{cap,Ce_2O_4} y_{H_2} \psi_{O_2}$	(NSRC15)
$Ce_2O_4 + \frac{1}{9}C_3H_6 \rightarrow Ce_2O_3 + \frac{1}{3}CO_2 + \frac{1}{3}H_2O$	$R_{NSRC16} = k_{NSRC16} \Psi_{cap,Ce_2O_4} y_{C_3H_6} \psi_{O_2}$	(NSRC16)
$2NO_2 + BaO + \frac{1}{2}O_2 \rightarrow Ba(NO_3)_2$	$R_{NSRC17} = k_{NSRC17} \Psi_{cap,BaO} y_{NO_2} \left(\psi_{NO_x}^{eq} - \psi_{NO_x}\right)^2$	(NSRC17)
$2NO + BaO + \frac{3}{2}O_2 \rightleftarrows Ba(NO_3)_2$	$R_{NSRC18} = k_{NSRC18} \Psi_{cap,BaO} y_{NO} \left(\psi_{NO_x}^{eq} - \psi_{NO_x}\right)^2$	(NSRC18)
$3NO_2 + BaO \cdot bulk \rightarrow Ba(NO_3)_2 \cdot bulk + NO$	$R_{NSRC19} = k_{NSRC19} \Psi_{cap,Ba.bulk} y_{NO_2} \left(\psi_{NO_x,bulk}^{eq} - \psi_{NO_x,bulk}\right)^2$	(NSRC19)
$Ba(NO_3)_2 \cdot bulk \rightarrow Ba(NO_3)_2 + bulk$	$R_{NSRC20} = k_{NSRC20} \Psi_{cap,Ba.bulk} \left(\psi_{NO_x,bulk} - \psi_{NO_x}\right)$	(NSRC20)
$Ba(NO_3)_2 + 5CO \rightarrow N_2 + 5CO_2 + BaO$	$R_{NSRC21} = k_{NSRC21} \Psi_{cap,Ba} y_{CO} \psi_{NO_x}^2 \frac{1}{G_3}$	(NSRC21)
$Ba(NO_3)_2 + 8CO + 3H_2O \rightarrow 2NH_3 + 8CO_2 + BaO$	$R_{NSRC22} = k_{NSRC22} \Psi_{cap,Ba} y_{CO} \psi_{NO_x}^2 \frac{1}{G_3} \frac{1}{G_5}$	(NSRC22)
$Ba(NO_3)_2 + 8H_2 \rightarrow 2NH_3 + 5H_2O + BaO$	$R_{NSRC23} = k_{NSRC23} \Psi_{cap,Ba} y_{H_2} \psi_{NO_x}^2 \frac{1}{G_3} \frac{1}{G_5}$	(NSRC23)
$Ba(NO_3)_2 + \frac{5}{9}C_3H_6 \rightarrow N_2 + \frac{5}{3}CO_2 + \frac{5}{3}H_2O + BaO$	$R_{NSRC24} = k_{NSRC24} \Psi_{cap,Ba} y_{C_3H_6} \psi_{NO_x}^2 \frac{1}{G_3}$	(NSRC24)
$Ba(NO_3)_2 + 3CO \rightarrow 2NO + 3CO_2 + BaO$	$R_{NSRC25} = k_{NSRC25} \Psi_{cap,Ba} y_{CO} \psi_{NO_x}^2 \frac{1}{G_4}$	(NSRC25)
$Ba(NO_3)_2 + 3H_2 \rightarrow 2NO + 3H_2O + BaO$	$R_{NSRC26} = k_{NSRC26} \Psi_{cap,Ba} y_{H_2} \psi_{NO_x}^2 \frac{1}{G_4}$	(NSRC26)
$Ba(NO_3)_2 + \frac{1}{3}C_3H_6 \rightarrow 2NO + CO_2 + H_2O + BaO$	$R_{NSRC27} = k_{NSRC27} \Psi_{cap,Ba} y_{C_3H_6} \psi_{NO_x}^2 \frac{1}{G_4}$	(NSRC27)
$Ba(NO_3)_2 + \frac{10}{3}NH_3 \rightarrow \frac{8}{3}N_2 + 5H_2O + BaO$	$R_{NSRC28} = k_{NSRC28} \Psi_{cap,Ba} y_{NH_3} \psi_{NO_x}^2$	(NSRC28)
$2NH_3 + 3NO \rightarrow \frac{5}{2}N_2 + 3H_2O$	$R_{NSRC29} = k_{NSRC29} y_{NH_3} y_{NO}^{0.5}$	(NSRC29)
$NH_3 + \frac{3}{2}O_2 \rightarrow N_2 + 3H_2O$	$R_{NSRC30} = k_{NSRC30} y_{NH_3} y_{O_2}$	(NSRC30)
$2NH_3 + 3Ce_2O_4 \rightarrow N_2 + 3H_2O + 3Ce_2O_3$	$R_{NSRC31} = k_{NSRC31} \Psi_{cap,Ce_2O_4} y_{NH_3} \psi'_{Ce_2O_4}$	(NSRC31)
$H_2O \rightarrow H_2O^{ads}$	$R_{NSRC32} = k_{NSRC32} \Psi_{cap,H_2O} p_{H_2O} \left(\psi_{H_2O}^{eq} - \psi_{H_2O}\right)$	(NSRC32)
$H_2O^{ads} \rightarrow H_2O$	$R_{NSRC33} = k_{NSRC33} \Psi_{cap,H_2O} \left(\psi_{H_2O} - \psi_{H_2O}^{eq}\right) \left(p_{H_2O}^0 - p_{H_2O}\right)$	(NSRC33)

y_i denotes the molar fractions of the gas species within the washcoat pores.

$$C_3H_6 + Ze \rightarrow C_3H_6 \cdot Ze \quad R_{HCads} = k_{HCads} \Psi_{cap,Ze_{HC}} y_{C_3H_6} \left(\psi_{C_3H_6}^{eq} - \psi_{C_3H_6} \right) \quad \text{(HCads)}$$

$$C_3H_6 \cdot Ze \rightarrow C_3H_6 + Ze \quad R_{HCdes} = k_{HCdes} \Psi_{cap,Ze_{HC}} y_{C_3H_6}^2 \frac{1}{G_{HCdes}} \quad \text{(HCdes)}$$

The reaction rate formulations for the above system are of the Langmuir–Hinshelwood type. They can include inhibition terms in the denominator, accounting for surface coverage effects influencing the heterogeneously catalyzed reactions. These are often adopted from Ref. [33] and can also consider inhibition by adsorption on separated catalytic active sites. The rate formulations are also listed in Table 10.2. The corresponding formulation of the inhibition terms is depicted in Table 10.3.

The storage reaction rates are formulated depending on the overall storage capacity and the deviation of the surface coverage from equilibrium coverage (= effective storage capacity). The release rates are influenced by the overall capacity and the actual surface coverage.

The kinetic coefficients k_i of each reaction i are calculated via an Arrhenius expression to account for the temperature influence.

$$k_i = k_{0,i} \cdot \exp\left(\frac{-E_{a,i}}{RT^s} \right). \quad (10.33)$$

The inhibition coefficients $K_{\text{inh},l}$ are calculated according to

$$K_{\text{inh},l} = K_{\text{inh}0,l} \cdot \exp\left(\frac{E_l}{T^s} \right). \quad (10.34)$$

Reactions (NSRC5), (NSRC6), and (NSRC7) are formulated as thermodynamic equilibrium reactions. The solely temperature-dependant equilibrium constants $K_{y,i}^{eq}$ of those reactions can be accessed from the involved species entropy and enthalpy of formation.

Table 10.3 Inhibition terms. Adapted from Koči et al. [92].

$G_1 = \left(1 + K_{\text{inh},1} y_{CO} + K_{\text{inh},2} y_{C_3H_6}\right)^2 \cdot \left(1 + K_{\text{inh},3} y_{CO}^2 y_{C_3H_6}^2\right) \cdot \left(1 + K_{\text{inh},4} y_{NO_x}^{0.7}\right) \cdot T$	(Inh1)
$G_{1b} = \left(1 + K_{\text{inh},1} y_{CO} + K_{\text{inh},2} y_{C_3H_6}\right)^2 \cdot \left(1 + K_{\text{inh},3} K_{\text{inh},5} y_{CO}^2 y_{C_3H_6}^2\right) \cdot \left(1 + K_{\text{inh},4} y_{NO_x}^{0.7}\right) \cdot T$	(Inh2)
$G_{1c} = \left(1 + K_{\text{inh},6} y_{CO} + K_{\text{inh},7} y_{C_3H_6}\right)^2 \cdot \left(1 + K_{\text{inh},8} y_{CO}^2 y_{C_3H_6}^2\right) \cdot \left(1 + K_{\text{inh},9} y_{NO_x}^{0.7}\right) \cdot T$	(Inh3)
$G_3 = 1 + K_{\text{inh},10} y_{O_2}$	(Inh4)
$G_4 = \left(1 + 0.1 K_{\text{inh},10} y_{O_2}\right) \cdot \left(1 + K_{\text{inh},11} y_{NO_x}\right)$	(Inh5)
$G_5 = 1 + K_{\text{inh},12} y_{CO}$	(Inh6)
$G_6 = \left(1 + K_{\text{inh},13} y_{CO}\right) \cdot \left(1 + K_{\text{inh},14} y_{C_3H_6}\right)$	(Inh7)
$G_{HCdes} = 1 + K_{\text{inh},HCdes} y_{C_3H_6} T$	(Inh8)

10.7.3.1 Species Transport Effects Related to NSCR: Shrinking Core Model

The NO_x conversion performance of an NSRC shows in reality a typical, bell-shaped temperature-dependent behavior. Starting with little NO_x conversion at low temperatures, it passes through a maximum, usually around 300 °C, followed by a decreasing performance with higher temperatures. While the latter effect is caused by the decrease in the storage capacity due to the decline in the thermodynamic equilibrium of nitrate, this is not the reason for the little conversion at low temperatures. Instead, kinetic and even more important diffusion limitations in the storage material (Ba particle) are the cause, affecting both storage and regeneration reactions. However, the diffusion process in the particle, which precedes the storage reaction itself, is rarely modeled in detail. In this case, the low conversion in that temperature range can alternatively be accounted for by a reduced, temperature-dependant storage capacity ("effective storage capacity") [35].

The diffusion process into the storage particle can be described in addition via a linear driving force approach, resulting of course in a more complex model (shrinking core model) [90, 91, 93]. The diffusion of the gas species from the catalyst surface into the Ba particle is hindered by the emerging nitrate layer around the particle, whose thickness increases with increasing NO_x storage (cf. Figure 10.17).

The benefits of the enhanced complexity of the shrinking core approach are the following:

- Increasing diffusion barrier accounts for the characteristic storage behavior of an NSRC
- Poor performance at low temperatures is considered by the temperature-dependant diffusion coefficient, avoiding the introduction and tuning of the "effective storage capacity."
- Kinetic diversity in the storage and regeneration process can be explained by the breakup of the dense nitrate layer during regeneration, enabling a higher diffusion of the reducing species into the particle.
- Combination of the effects of diffusion and reaction in the storage and regeneration process results in a qualitatively good description of the catalyst behavior in case of incomplete regeneration during the cyclic lean/rich operation (cf. Figure 10.18).

The mechanism details, additional balances, and equations for modeling the shrinking core approach are not presented here due to space limitation, but can be found in Refs [90, 91].

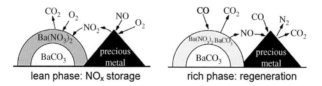

Figure 10.17 Shrinking core model: formation of a dense nitrate layer during storage (left) and its breakup during regeneration (right) [90].

10.7 Diesel Oxidation Catalyst, Three-Way Catalyst, and NO$_x$ Storage and Reduction Catalyst Modeling

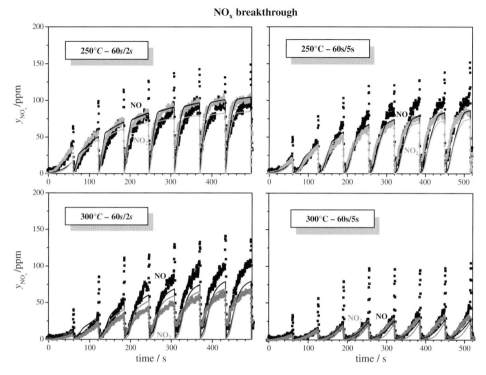

Figure 10.18 Simulated (shrinking core model) compared to measured NO$_x$ breakthrough curves [91].

10.7.3.2 NH$_3$ Formation During Rich Operation within a NSRC

When the NSRC is used in combination with a downstream SCR catalyst, as it is the case, for example, in the BlueTEC I system (see Section 10.2), the production of NH$_3$ in the NSRC is important for the overall NO$_x$ conversion efficiency of the exhaust gas system [94]. With the reactions listed in Table 10.2, the NH$_3$ generation under rich conditions and its subsequent reactions with gas-phase O$_2$ or stored O$_2$ as well as with NO can be described Ref. [84, 92]. The resulting complex behavior is shown in Figure 10.19 in form of the development of the axial concentration profiles of

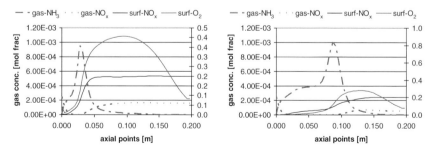

Figure 10.19 NH$_3$ formation within an NSRC monolith channel after 2 s rich phase (left) and 4 s rich phase (right). Adapted from [84].

gas-phase NH_3 and NO_x, as well as surface (stored) NO_x and O_2. The two graphs show the profiles at 2 and 4 s of the rich phase. It can be seen that produced amount of NH_3 is highest at the location of the NO_x and O_2 regeneration front. Behind the regeneration front, NH_3 is consumed by O_2 and NO, released from the surface. Thus, the resulting NH_3 peak in the channel moves toward the end of the monolith with the proceeding of the regeneration process, where it will finally break through and can be stored and used in a downstream SCR.

10.8
Modeling Catalytic Effects in Diesel Particulate Filters

To increase the reaction rates for soot oxidation, a catalytic coating, containing platinum and palladium, can be impregnated into the walls of the DPF channels. Basically all the chemical phenomena such as species oxidation, reduction, adsorption, and desorption can be adapted from flow-through catalyst modeling (Table 10.3, reactions NSRC1–NSRC7) and implemented in the single-channel approach to model coated DPF systems. In addition, the global reactions listed in Table 10.4 have to be considered to describe the relevant soot oxidation reactions and the passive regeneration [95, 96].

The coupling between chemical kinetics and species transport in the flow-through situation can lead to specific situations. An example is the formation and reduction of NO_2 (see Section 10.7, NSRC5).

$$2\,NO + O_2 \leftrightarrow 2\,NO_2$$

This functionality is required to describe the continuous soot regeneration in the presence of NO_2 and to account for the thermodynamic equilibrium.

An additional discretization of the one-dimensional model approach in radial direction enables the calculation of species concentration profiles through the catalytically coated wall and, if existing, the soot layer. This approach has been introduced by Haralampous and Koltsakis [97] to model the so-called "back-diffusion" of NO_2. NO is oxidized to NO_2 while passing catalytic sites in the filter

Table 10.4 Soot oxidation reactions.

Reaction	
$C + O_2 \rightarrow CO_2$	(DPF1)
$C + \frac{1}{2}O_2 \rightarrow CO$	(DPF2)
$C + 2\,NO_2 \rightarrow CO_2 + 2\,NO$	(DPF3)
$C + NO_2 \rightarrow CO + NO$	(DPF4)
$C + \frac{1}{2}O_2 + NO_2 + (NO_2) \rightarrow CO_2 + NO + (NO_2)$	(DPF5)
$C + \frac{1}{2}O_2 + (NO_2) \rightarrow CO + (NO_2)$	(DPF6)

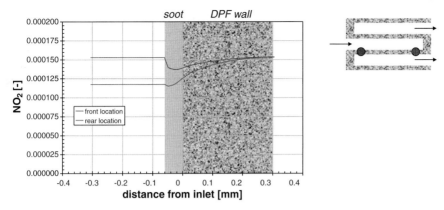

Figure 10.20 Radial NO$_2$ concentration profile near DPF inlet (blue) and outlet (red) under steady-state conditions (5 g$_{soot}$/l, c_{NO2} = 153 ppm).

wall. Depending on the NO$_2$ concentration in the inlet and the outlet channel, diffusion of NO$_2$ in counterflow direction can be observed, which improves the oxidation of soot by NO$_2$. Furthermore, such a model approach accounts for the "NO$_2$ recycling" effect, which is another enhancement of the passive regeneration: While passing through the wall, NO is repeatedly oxidized to NO$_2$, which is immediately reduced by soot deposits in the wall to again NO. Both phenomena affect the effective radial NO$_2$ concentration profile as shown in Figure 10.20. The graph on the left-hand side indicates the NO$_2$ profile in the inlet channel (<−0.06 mm), the soot layer (−0.06–0 mm), the DPF wall (0–0.3 mm), and the outlet channel (>0.3 mm). Near the DPF inlet (blue location), the NO$_2$ concentration is reduced to NO when entering the soot layer because of soot oxidation with subsequent NO$_2$ formation in the catalytically coated DPF wall. The NO$_2$ formation inside the DPF wall in conjunction with its diffusion transport in counterflow direction (= back-diffusion) typically affects the NO$_2$ concentration profile.

To optimize the species conversion of coated DPF systems, the local distribution of washcoat and catalysts can be nonuniform; for example, to improve the light-off performance of catalysts, higher catalyst concentrations might be placed in the upstream part of the DPF. The one-dimensional model can describe such a "zone coated" system as described by Koltsakis et al. [97] and applied for a model-based optimization of the catalyst distribution.

10.9
Determination of Global Kinetic Parameters

The high number of reactions, each combined with its number of kinetic parameters including inhibition coefficients, leads to a considerable amount of parameters that have to be determined. Therefore, appropriate experimental measurements with the respective catalyst have to be carried out. The simulation results must then be fitted to the measured data, determining the correct set of kinetic parameters. This task can be supported by the use of a parameter estimation tool. However, as the processes on the catalyst are very complex, especially when they cannot be observed isolated, for example, in the case of inhibition effects, the design of appropriate experimental conditions is not trivial. Thus, the determination of the model parameters is an iterative and costly process whose demand of time is often underestimated. This is especially the case if the model shall be enabled to cover a preferably wide temperature and concentration/lambda range.

One drawback of a model with global reaction kinetics is that changes in the catalyst properties always make changes in the kinetic parameters necessary, which means the determination of a new set of kinetic parameters. The catalyst properties may be subject to change, for example, due to a different loading of precious metal or storage material. Another very important effect is the aging of the catalyst due to its exposure to high temperatures or due to poisoning (e.g., by SO_2), leading to a decrease in activity. The description of the aging process itself is not possible with such a model. Instead, it would be helpful to know at least which reactions are affected and in which way. Therefore, correlations between precious metal loading [1] or aging and kinetic parameters must be developed, considering also changes in the material properties, which can be received from catalyst characterization.

10.10
Challenges for Global Kinetic Models

Limitations of current global kinetic models of exhaust gas catalysts, especially regarding the determination of kinetic parameters, are the change in the catalytic activity depending on the operating conditions of the catalyst. Higher reaction rates can be observed under transient conditions than under stationary conditions [93, 99, 100]. This effect can be caused by changes in the catalyst surface structure, going beyond the above-mentioned surface coverage effects. This complicates the generation of appropriate kinetic measurement data, as they are usually received under stationary conditions.

Another problematic issue are hysteresis effects during catalyst light-off and extinction due to heat up and cooling down, respectively. Depending on the direction of temperature change, the surface coverage, and thus the catalyst activity, shows differences that cannot be represented by a global kinetic model. While in the past the catalyst used to pass light-off after cold start and used to be operated above the light-off temperature for the rest of the driving cycle, modern Diesel engines exhibit very low exhaust gas temperatures. Furthermore, start–stop systems in the vehicle or its hybridization lead to occasional stops of the engine and the exhaust gas flow, resulting in the cooling down and eventually extinction of the aftertreatment catalysts. More detailed models for the reaction kinetics, directly correlating material properties with kinetic parameters (e.g., elementary step-based models), are required in the future to overcome the limitations of the global kinetic models.

10.11
System Modeling of Combined Exhaust Aftertreatment Systems

The modeling and simulation of combined exhaust gas aftertreatment systems requires the combination of the individual component models together with the connecting piping in a common software environment. Besides commercial tools such as AVL Boost [7, 101] or GT Power [102, 103], in-house developments such as the *ExACT* tool of Daimler [4, 34, 104, 105] are widely used in the automotive industry. By means of a system simulation environment, it is possible to consider the complete exhaust line in the simulation (cf. Figure 10.21). The connecting piping has also to be included in the simulation model to account for heat losses and the additional thermal mass. Measured engine-out raw emissions (exhaust mass flow, temperatures, and species concentrations) are typically used as a model input. As a result of the simulation, temperatures and species concentrations between the catalysts and the end of pipe can be analyzed. Detailed postprocessing also allows the analysis of the conditions inside the catalyst or Diesel particulate filter, for example, temperature and species concentration profiles.

The modeling principles discussed in the previous sections yields a good correlation between measured and calculated data if the basic assumptions, for example, spatial homogeneous inlet conditions at catalyst inlet, are fulfilled. As an example, Figure 10.22 presents a comparison between measured and calculated temperatures and CO, HC, and NO_x emission for an exhaust gas aftertreatment system consisting of a DOC, CDPF, and NSRC in series.

By means of the exhaust aftertreatment system simulation, different analyses can be conducted. Typical examples are the optimization of catalyst volumes, monolith cell densities, operating strategy (e.g., engine thermomanagement), optimization of component positions, and so on. Some illustrative examples are presented in the following sections.

Example 10.1: Optimization of Component Order

The analysis of Figure 10.22 reveals only a low NO_x conversion efficiency for the DOC + CDPF + NSRC system. This is because of the low exhaust temperatures of

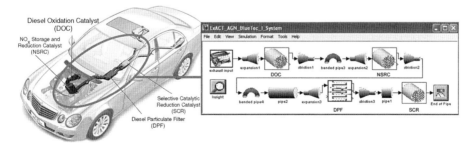

Figure 10.21 Representation of a complete exhaust system consisting of DOC + NSRC + DPF + SCR within a system simulation environment (*ExACT*).

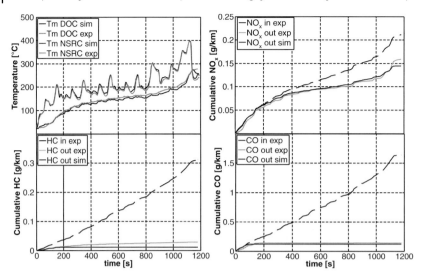

Figure 10.22 Simulated and measured DOC and NSRC temperatures and cumulated end of pipe CO, HC, and NO_x emissions within a NEDC test cycle. Exhaust system: DOC + CDPF + NSRC; engine with low NO_x and low exhaust temperature calibration.

that setup. However, if the NSRC is placed upstream of the CDPF, a higher NO_x conversion efficiency is revealed by the simulation (cf. Figure 10.23). The upstream position leads to significant higher NSRC temperatures in the NEDC. Owing to the higher NO_x storage capacity of the NSRC at elevated temperatures, a higher NO_x

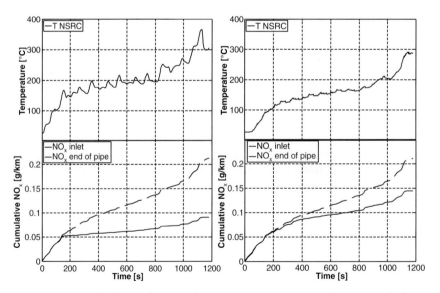

Figure 10.23 Simulated spatially averaged NSRC temperatures and cumulated NO_x end of pipe emissions for a DOC + NSRC + CDPF (left) and a DOC + CDPF + NSRC (right) system.

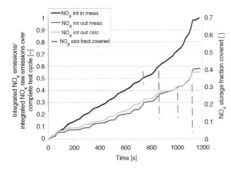 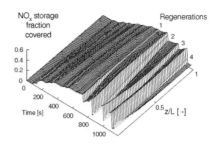

Figure 10.24 NSRC performance within a NEDC test cycle. *Left*: Measured (meas) and calculated (calc) cumulative NO_x emissions and spatially averaged covered fraction of the NO_x storage capacity. *Right*: Computed spatiotemporal concentration profile of the stored NO_x. The z/L stands for the dimensionless spatial coordinate along the monolith: 0 is at the inlet, 1 at the outlet. Adapted from SAE 2007-01-1117 [34].

conversion efficiency is possible. This is confirmed by the lower end of pipe NO_x emissions shown in Figure 10.23. Therefore, a clear advantage is identified for the DOC + NSRC + DPF configuration. For engines with higher exhaust temperatures, a different result is possible.

Example 10.2: Analysis of NSRC Reduction Efficiency

If the NO_x storage capacity of the NSRC is exhausted, regeneration is initiated by operating the engine in fuel-rich mode for a few seconds. Under fuel-rich exhaust conditions, the stored NO_x is reduced, giving free storage sites for NO_x adsorption in the following phase (see Section 10.7). The situation during a NEDC cycle is depicted in Figure 10.24. Within that cycle regeneration, four events were conducted, indicated by the rapid drop in the spatially averaged NO_x storage fraction at 850, 1000, and 1150s. The corresponding calculated spatiotemporal concentration profiles of the stored NO_x reveal that during the first regeneration phase only the front of the catalyst is regenerated. Only during the last regeneration phase, the catalyst is fully regenerated. This is due to a longer rich phase and higher catalyst temperatures. Based on these simulation results, a further optimization is possible, leading to higher catalyst regeneration efficiency.

Example 10.3: Analysis of SCR Performance

The NO_x conversion efficiency of an SCR system can adequately be predicted by the models discussed in Section 10.6. The result of a fully transient simulation of an ETC and ESC test cycle using a 18 l heavy duty SCR catalyst is presented in Figure 10.25. Urea is dosed from the beginning of the test cycles and is assumed to be converted

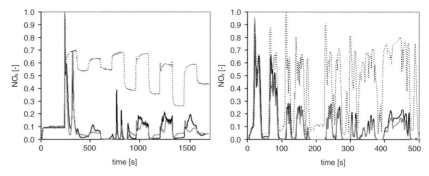

Figure 10.25 Normalized NO_x concentration at SCR catalyst inlet and outlet of an ESC (left) and the first part of a ETC (right) test cycle. Dotted black lines, inlet values; solid black lines, outlet measurement; gray lines, outlet simulation. Adapted from SAE 2005-01-1965 [27].

completely to NH_3. The increase in the NO_x conversion with time can be explained by the fact that the SCR catalyst temperature is below 200 °C at engine start. During the test cycle, the SCR heats up to 350 °C, which increases the SCR reaction rates. The corresponding axial profiles at different phases in the ETC test cycle are shown in Figure 10.26, revealing that mainly the first part of the catalyst is active in NO_x reduction. However, the large catalyst volume is necessary as a NH_3 buffer during temperature transients to avoid NH_3 slip. Furthermore, Figure 10.26 reveals the strong concentration gradients present in the catalytic walls of an extrude SCR catalyst. Based on our experience, the overall test cycle SCR NO_x conversion efficiency can be predicted within the error range of 3–4% [27, 71].

Example 10.4: SCR versus DOC + SCR Performance

The addition of a DOC and CDPF upstream of the SCR may significantly increase the NO_x conversion efficiency. This is due to the formation of NO_2 in the DOC or CDPF

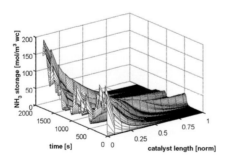
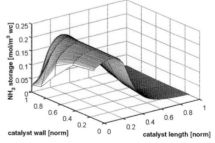

Figure 10.26 Simulated profiles of stored NH_3^* within a V-based extruded SCR catalyst. *Left*: Axial profiles versus time (left, first part of an ETC). *Right*: radial profiles (right, at 365 s within an ESC). All spatial coordinates are normalized. Gas flow is from axial position 0 to 1. The radial coordinate points away from

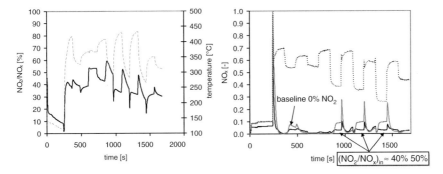

Figure 10.27 Left: Simulated transient NO$_2$/NO$_x$ ratio (solid black line) and temperature (dashed gray line) at DOC outlet during an ESC test cycle. Right: Normalized NO$_x$ concentrations before and after SCR during an ESC test cycle simulation. Inlet concentration, dotted black line; simulated outlet concentration without DOC, dashed gray line; simulated outlet concentration with DOC, solid black line. Adapted from SAE 2006-01-0468 [54].

that leads to the onset of the so-called fast SCR reaction (see Section 10.5). On the other hand, the addition of a component upstream of the SCR catalyst delays the temperature increase in the SCR catalyst because of the additional thermal mass.

The simulation results depicted in Figure 10.27 [54] highlights the impact of a platinum-based DOC upstream of an 18 l V-based SCR catalyst on the NO$_x$ conversion within an ESC test cycle. The simulation reveals that the DOC improves the NO$_x$ conversion significantly due to the formation of NO$_2$ and hence the reduction of the NO$_x$ breakthroughs in the later part of the test cycle. In those critical phases, a NO$_2$/NO$_x$ ratio in the range of 40–50% is generated by the DOC that is close to the optimum value of 50% (see Section 10.5). This confirms that the chosen DOC is quite well adapted to the specific application. However, not only the DOC volume has to be optimized, but also the platinum loading needs to be considered [1].

10.12
Conclusion

The stringent emission limits worldwide require the reduction of engine-out emissions and the application of highly efficient exhaust gas aftertreatment systems. Especially, the demand for a simultaneous reduction of different species, for example, particulate matter and NO$_x$, has led to the development of combined aftertreatment systems, consisting of different catalyst technologies and particulate filters. Simulation of such systems has become essential for the automotive industry with respect to development time and cost.

The main challenge in the modeling of complete exhaust systems is the highly transient operation and the number of different aftertreatment components. The transient operation leads to a large variation in operating conditions. A chemical- and physical-based modeling approach is the most promising one. Models for the heterogeneous catalytic processes are key elements in setting up system level models that are able to predict the conversion of CO, HC, and NO_x within the exhaust line. In addition, transport effects such as washcoat diffusion are also to be considered in the models.

In the past 20 years, a significant progress was made in modeling the catalytic processes within automotive catalysts. However, for most automotive catalyst technologies, for example, NSRC, reliable models of the detailed elementary reaction steps are still lacking or the reaction mechanism itself is still under discussion. Hence, the most commonly used models in the automotive industry are based on global chemistry approaches, aiming to include the relevant inhibition and dynamic effects. The main drawback of this approach is that a change in the catalyst composition or structure requires at least a complete recalibration of the reaction kinetics.

Because of this gap in connecting the catalytic material itself with the conversion performance, hysteresis effects due to changes in the catalyst state, for example, palladium or platinum oxidation, or catalyst aging cannot be considered. Also, with the present models, only limited conclusions can be made from system performance to the required optimization of the catalyst material. Therefore, still a large number of experiments are required to define the best catalyst composition for a dedicated application.

Surface science and more detailed models based on elementary kinetics offer the potential to overcome the above-mentioned limitations in the future. At the present state, they are an important source of information for setting up chemically consistent global kinetic models in the automotive industry. The ongoing development and application of new catalyst technologies for exhaust gas aftertreatment systems, such as NH_3 slip or urea hydrolysis catalysts, lead to new challenges to the modeling of heterogeneous catalytic processes.

Acknowledgments

Bernd Krutzsch and Günter Wenninger are acknowledged for supporting our work. Enrico Tronconi, Isabella Nova, Milos Marek, Petr Kočí, Grigorios Koltsakis, and Gerhart Eigenberger are acknowledged for their fruitful and important scientific collaboration over the past 10 years. Last but not least, Thomas Burkhardt, Oscar Montoya, and Anke Traebert are acknowledged for their support.

List of Symbols

a	density of external surface area in monolith, m^2/m^3
c	concentration (bulk gas), mol/m^3

10.12 Conclusion

c^s	concentration in washcoat pores, mol/m^3
c_p	specific heat capacity of gas, J/(kg K)
c_p^s	effective specific heat capacity of solid phase, J/(kg K)
d	monolith channel diameter, m
D	diffusivity, m^2/s
E	activation energy of reaction, J/mol
f_x	geometrical parameter (local soot thickness/hydraulic diameter)
G	kinetic inhibition term
ΔH_r	standard reaction enthalpy, J/mol
J	number of reactions
k	kinetic constant of reaction, dimension depends on the reaction
k_0	pre-exponential factor of reaction, dimension depends on the reaction
k_c	mass transfer coefficient, m/s
k_h	heat transfer coefficient, J/(m^2 K s)
K	number of gas components
K_a, K_{LH}	kinetic inhibition constants
K^{eq}	thermodynamic equilibrium constant
L	length of monolith, m
M	number of surface-deposited components
p	pressure, Pa
r	transverse spatial coordinate in catalytic washcoat layer, m
R_j	reaction rate (related to washcoat volume), mol/(m^3 s)
R	universal gas constant, 8.31434 J/(mol K)
S_w	washcoat thickness, m
S_F	Filter specific area (total filtration surface/total volume of filter monolith), m^{-1}
SV	gas hourly space velocity (defined at standard $T = 273.15$ K and $p = 101325$ Pa), 1/s
t	time, s
T	temperature of gas, K
T^s	temperature of solid phase, K
y_k	species mole fraction
v	linear gas velocity, m/s
V	volume, m^3
$\dot{V}$	volumetric flow rate, m^3/s
w	weight of component in sum of squares
w	soot layer thickness, m
y	molar fraction
z	spatial coordinate along monolith (axial), m

Greek Letters

α	NH$_3$/NO$_x$ dosing ratio in NH$_3$-SCR
α	parameter for surface coverage dependence in NH$_3$-SCR kinetics

α_i	constant in pressure drop calculation
β	O_2 reaction order in NH_3-SCR kinetics
γ	inhibition term in NH_3-SCR kinetics
ε^g	open frontal area fraction (monolith)
ε^s	porosity of catalytic washcoat layer
η	effectiveness factor
θ	surface coverage of adsorbed component
μ	dynamic viscosity, Pa s
$\nu_{k,j}$	stoichiometric coefficient, mol/mol
λ	heat conductivity, J/(m K s)
ϱ	gas density, kg/m^3
ϱ^s	apparent density of solid phase (including pores), kg/m^3
ϕ^s	volume fraction of catalytic washcoat in entire solid phase
ψ	relative surface concentration of stored component
Ψ_{cap}	storage capacity (related to catalytic washcoat volume), mol/m^3

Subscripts and Superscripts

ads	denotes NH_3 adsorption in NH_3-SCR
calc	calculation
des	denotes NH_3 desorption in NH_3-SCR
eq	equilibrium
exp	experiment
FAST	denotes the $NO_2 + NO + NH_3$ reaction in NH_3-SCR
g	gas
in	inlet
j	index of reaction
k	index of gas component
LH	denotes inhibition constant in NH_3-SCR kinetics
m	index of surface-deposited component
meas	measurement
NO	denotes NO-deNO$_x$ reaction in NH_3-SCR
NO$_2$	denotes direct NO_2-deNO$_x$ reaction in NH_3-SCR
N$_2$O	denotes N_2O formation in NH_3-SCR
μ	micro
out	outlet
ox	denotes NH_3 oxidation in NH_3-SCR
ref	reference
s	solid phase (washcoat and monolith substrate)
sim	simulation
vol	volume
w	washcoat
wt	weight

Abbreviations

1D	spatially one dimensional
cpsi	channels per square inch (cross-sectional monolith channel density)
CDPF	coated Diesel particulate filter
DOC	Diesel oxidation catalyst
DPF	Diesel particulate filter
deNO$_x$	abatement of nitrogen oxides
EPA	Environmental Protection Agency (United States)
ESC	European stationary driving cycle (for heavy duty vehicles)
ETC	European transient driving cycle (for heavy duty vehicles)
ExACT	Exhaust Aftertreatment Components Toolbox
FTP	Federal test procedure, US driving cycle
g/bhp-h	grams per brake horsepower-hour
HC	hydrocarbon(s)
LEV	low-emission vehicle
LNT	lean NO$_x$ trap (equivalent to NSRC)
MR	modified redox mechanism in SCR
NEDC	new European driving cycle
NM	noble metal(s)
NMHC	nonmethane hydrocarbons
NO$_x$	nitrogen oxides, NO and NO$_2$ only
NSRC	NO$_x$ storage and reduction catalyst
PGM	precious group metals (equivalent to NM)
PM	particulate matter
SC03	US driving cycle with air conditioning device switched on
SCR	selective catalytic reduction of NO$_x$
SOF	soluble organic fraction in TPM
TPD	temperature programmed desorption (temperature ramp)
TPM	total particulate matter
TPR	temperature programmed reaction (temperature ramp)
TRM	transient response method
TWC	three-way catalyst
US06	US highway driving cycle

References

1 Chatterjee, D., Burkhardt, T., Rappe, T., Güthenke, A., and Weibel, M. (2008) SAE Tech. Paper 2008-01-0867.

2 Braun, J., Hauber, T., Toebben, H., Windmann, J., Chatterjee, D., Correa, C., Deutschmann, O., Maier, L., Tischer, S., and Warnatz, J. (2002) SAE Tech. Paper 2002-01-0065.

3 Pischinger, S., Körfer, T., Wiartalla, A., Schnitzler, J., Tomazic, D., and Tatur, M. (2007) SAE Tech. Paper 2007-01-1128.

4 Güthenke, A., Chatterjee, D., Weibel, M., Krutzsch, B., Kočí, P., Marek, M.,

Nova, I., and Tronconi, E. (2008) *Adv. Chem. Eng*, **33**, 104–211.
5 Koltsakis, G.C., Konstantinidis, P.A., and Stamatelos, A.M. (1997) *Appl. Catal. B*, **12**, 161–191.
6 Büchner, S., Santos Lardies, S., Degen, A., Donnerstag, A., and Held, W. (2001) SAE Tech. Paper 2001-01-0940.
7 Wurzenberger, J.C., Wanker, R., and Schüßler, M. (2008) SAE Tech. Paper 2008-01-0865.
8 Konstandopoulos, A.G., Kostoglou, M., Vlachos, N., and Kladopoulou, E. (2005) SAE Tech. Paper 2005-01-0946.
9 Braun, J., Hauber, T., Toebben, H., Zacke, P., Chatterjee, D., Deutschmann, O., and Warnatz, J. (2000) SAE Tech. Paper 2000-01-0211.
10 Chatterjee, D., Deutschmann, O., and Warnatz, J. (2001) *Faraday Discuss.*, **119**, 371.
11 Hoebink, J.H.B.J., van Gemert, R.A., van den Tillaart, J.A.A., and Marin, G.B. (2000) *Chem. Eng. Sci.*, **55**, 1573–1581.
12 Dieselnet (2009) http://www.dieselnet.com/standards/cycles/ece_eudc.html
13 Koltsakis, G.C. and Stamatelos, A.M. (1997) *Prog. Energy Combust. Sci.*, **23**, 1–39.
14 Birkhold, F., Meingast, U., Wassermann, P., and Deutschmann, O. (2006) SAE Tech. Paper 2006-01-0643.
15 Wunsch, R., Binder, K., Günther, J., Hertzberg, A., Konrad, B., Krutzsch, B., Ölschlegel, H.-J., Renfftlen, S., Voigtländer, S., Weibel, M., Weirich, M., and Wenninger, G.,DaimlerChrysler AG patent DE 10131588 A1, Priority 03.07.2001.
16 Konrad, B., Krutzsch, B., Voigtländer, V., and Weibel, M.,DaimlerChryslerAG patent EP 957 242 B1, Priority 09.05.1998.
17 Kirchner, T. (1997) *Experimentelle Untersuchungen und dynamische Simulation der Autoabgaskatalysezur Verbesserung des Kaltstartverhaltens*, Dissertation, Institut für Chemische Verfahrenstechnik, University of Stuttgart, VDI-Fortschrittberichte, Reihe 12, Nr. 331.
18 Brinkmeier, C. (2006) Automotive Three Way Exhaust Aftertreatment under Transient Conditions - Measurments, Modeling and Simulation, PhD thesis, Institut für Chemische Verfahrenstechnik, University of Stuttgart, Germany.
19 Hayes, RE., Liu, B., Moxom, R., and Votsmeier, M. (2004) *Chem. Eng. Sci.*, **59**, 3169.
20 Scholz, C.M.L., Gangwal, V.R., de Croon, M.H.J.M., and Schouten, J.C. (2007) *J. Catal.*, **245**, 215–227.
21 Jirát, J., Kubíček, M., and Marek, M. (1999) *Catal. Today*, **53**, 583–596.
22 Tischer, S. and Deutschmann, O. (2005) *Catal. Today*, **105**, 407–413.
23 Kočí, P. (2005) Catalytic Monolith Reactors with Surface Deposition of Gas Components - Applications in Automobile Exhaust Gas Conversion, PhD thesis, Institute of Chemical Technology, Prague.
24 Hayes, RE., Liu, B., and Votsmeier, M. (2005) *Chem. Eng. Sci.*, **60**, 2037.
25 Hayes, R.E., Mukadi, L.S., Votsmeier, M., and Gieshoff, J. (2004) *Top. Catal.*, **30/31**.
26 Hayes, R.E. and Kolaczkowski, S.T. (1997) *Introduction to Catalytic Combustion*, Gordon and Breach Science Publishers.
27 Chatterjee, D., Burkhardt, T., Bandl-Konrad, B., Braun, T., Tronconi, E., Nova, I., and Ciardelli, C. (2005) SAE Tech. Paper 2005-01-1965.
28 Kočí, P., Kubíček, M., and Marek, M. (2004) *Ind. Eng. Chem. Res.*, **43**, 4503.
29 Kočí, P., Kubíček, M., and Marek, M. (2004) *Chem. Eng. Res. Des.*, **82** (A2), 284–292.
30 Tischer, S., Correa, C., and Deutschmann, O. (2001) *Catal. Today*, **69**, 57–62.
31 Kočí, P., Kubíček, M., and Marek, M. (2004) *Catal. Today*, **98**.
32 Olsson, L., Fridell, E., Skoglundh, M., and Andersson, B. (2002) *Catal. Today*, **73**, 263–270.
33 Voltz, S.E., Morgan, C.R., Liederman, D., and Jacob, S.M. (1973) *Ind. Eng. Chem. Prod. Res. Dev.*, **12** (4), 294–301.

34 Chatterjee, D., Burkhardt, T., Weibel, M., Nova, I., Grossale, A., and Tronconi, E. (2007) SAE Tech. Paper 2007-01-1136.

35 Kočí, P., Schejbal, M., Trdlička, J., Gregor, T., Kubíček, M., and Marek, M. (2007) *Catal. Today*, **119**, 64–72.

36 Olsson, L., Blint, R.J., and Fridell, E. (2005) *Ind. Eng. Chem. Res.*, **44** (9), 3021–3032.

37 Koltsakis, G.C. and Stamatelos, A.M. (2001) Storage of chemical species in emission control systems: the role of mathematical modeling. GPC 2001 Global Powertrain Congress, June 5–7, Detroit, MI.

38 Hayes, RE., Kolaczkowski, S.T., Li, P.K.C., and Awdry, S. (2000) *Appl. Catal. B*, **25**, 93.

39 Zhang, F., Hayes, RE., and Kolaczkowski, S.T. (2004) *Chem. Eng. Res. Des.*, **82**, 481.

40 Wakao, N. and Smith, J.M. (1962) *Chem. Eng. Sci.*, **17**, 825.

41 Bissett, E.J. (1984) *Chem. Eng. Sci.*, **39** (7/8), 1233–1244.

42 Koltsakis, G.C. and Stamatelos, A.M. (1997) *Ind. Eng. Chem. Res.*, **36**, 4255–4265.

43 Haralampous, O.A., Koltsakis, G.C., and Samaras, Z.C. (2003) Partial regenerations in diesel particulate filters, SAE Tech. Paper 2003-01-1881.

44 Stratakis, G. and Stamatelos, A.M. (2004) *Proc. I MechE D J. Auto. Eng.*, **218** (2), 203.

45 Konstandopoulos, A., Kostoglou, M., Vlachos, N., and Kladopoulou, E. (2005) Progress in diesel particulate filter simulation, SAE Tech. Paper 2005-01-0946.

46 Forzatti, P., Lietti, L., and Tronconi, E. (2002) Nitrogen oxides removal, in *Industrial Encyclopedia of Catalysis* (ed. I.T Horvath), 1st edn, John Wiley & Sons, Inc., New York.

47 ACEA final report on selective catalytic reduction, June 2003, http://europa.eu.int/comm/enterprise/automotive/mveg_meetings/meeting94/scr_paper_final.pdf.

48 Heck, RH., Farrauto, RJ., and Gulati, S.T. (2002) *Catalytic Air Pollution Control*, 2nd edn, John Wiley & Sons, Inc., New York.

49 Koebel, M. and Strutz, E.O. (2003) *Ind. Eng. Chem. Res.*, **42**, 2093–2100.

50 Malmberg, S., Votsmeier, M., Gieshoff, J., Söger, N., Mußmann, L., Schuler, A., and Drochner, A. (2007) *Top. Catal.*, **33**, 42.

51 Olsson, L., Sjövall, H., and Blint, R. (2008) *Appl. Catal. B*, **81**, 203.

52 Kato, S., Matsuda, S., Kamo, T., Nakajima, F., Kuroda, H., and Narita, T. (1981) *J. Phys. Chem.*, **85**, 4099.

53 Koebel, M., Madia, G., and Elsener, M. (2002) *Catal. Today*, **73**, 239.

54 Chatterjee, D., Burkhardt, T., Weibel, M., Tronconi, E., Nova, I., and Ciardelli, C. (2006) SAE Tech. Paper 2006-01-0468.

55 Tronconi, E., Nova, I., Ciardelli, C., Chatterjee, D., and Weibel, M. (2007) *J. Catal.*, **245**, 1.

56 Grossale, A., Nova, I., Tronconi, E., Chatterjee, D., and Weibel, M. (2008) *J. Catal.*, **256**, 312–322.

57 Koebel, M., Elsener, M., and Madia, G. (2001) *Ind. Eng. Chem. Res.*, **40**, 52–59.

58 Yeom, Y.H., Henao, J., Li, M.J., Sachtler, W.M.H., and Weitz, E. (2005) *J. Catal.*, **231**, 181–193.

59 Schär, J.P., Onder, C.H., Geering, H.P., and Elsner, M. (2004) SAE Tech. Paper 2004-01-0153.

60 Winkler, C., Flörchinger, P., Patil, M.D., Gieshoff, J., Spurk, P., and Pfeifer, M. (2003) SAE Tech. Paper 2003-01-0845.

61 York, A.P.E., Watling, T.C., Cox, J.P. et al. (2004) SAE Tech. Paper 2004-01-0155.

62 Nova, I., Tronconi, E., Ciardelli, C., Chatterjee, D., and Bandl-Konrad, B. (2006) *AIChE J.*, **52**, 3222.

63 Lietti, L., Nova, I., Camurri, S., Tronconi, E., and Forzatti, P. (1997) *AIChE J.*, **43**, 2559.

64 Lietti, L., Nova, I., Tronconi, E., and Forzatti, P. (2000) Unsteady-state kinetics of $DeNO_x$-SCR catalysis, in *Reaction Engineering for Pollution Prevention* (eds M.A. Abraham and R.P. Hesketh), Elsevier Science, pp. 85–112.

65 Kapteijn, F., Singoredjo, L., Dekker, N.J.J., and Moulijn, J.A. (1993) *Ind. Eng. Chem. Res.*, **32**, 445.

66 Koebel, M. and Elsener, M. (1998) *Chem. Eng. Sci.*, **53**, 657.

67 Nova, I., Lietti, L., Tronconi, E., and Forzatti, P. (2000) *Catal. Today,* **60**, 73.
68 Willey, RJ., Lai, H., and Peri, J.B. (1991) *J. Catal.,* **130**, 319.
69 Ciardelli, C., Nova, I., Tronconi, E., Konrad, B., Chatterjee, D., Ecke, K., and Weibel, M. (2004) *Chem. Eng. Sci.,* **59**, 5301.
70 Tronconi, E., Nova, I., Ciardelli, C., Chatterjee, D., Burkhardt, T., and Bandl-Konrad, B. (2005) *Cat. Today,* **105**, 529.
71 Ciardelli, C., Nova, I., Tronconi, E., Bandl-Konrad, B., Chatterjee, D., Weibel, M., and Krutzsch, B. (2007) *Appl. Catal. B,* **70**, 80.
72 Madia, G., Koebel, M., Elsener, M., and Wokaun, A. (2002) *Ind. Eng. Chem. Res.,* **41**, 351.
73 Nova, I., Tronconi, E., Ciardelli, C., Chatterjee, D., and Bandl-Konrad, B. (2006) *Catal. Today,* **114**, 3.
74 Devdas, M., Kröcher, O., Elsner, M., Wokaun, A., Söger, N., Pfeifer, M., Demel, Y., and Mussmann, L. (2006) *Appl. Catal. B,* **67**, 187–196.
75 Eastwood, P. (2001). *Critical Topics in Exhaust Gas Aftertreatment,* Research Studies Press Ltd., England.
76 Després, J., Elsener, M., Koebel, M., Kröcher, O., Schnyder, B., and Wokaun, A. (2004) *Appl. Catal. B,* **50**, 73–82.
77 Epling, W.S., Campbell, L.E., Yezerets, A., Currier, N.W., and Parks, J.E. II (2004) *Catal. Rev.,* **46** (2), 163–245.
78 Burch, R., Sullivan, J.A., and Watling, T.C. (1998) *Catal. Today,* **42**, 13–23.
79 Erkfeldt, S., Jobson, E., and Larsson, M. (2001) *Top. Catal.,* **16/17**, 127–131.
80 Schmeißer, V., de Riva Pérez, J., Tuttlies, U., and Eigenberger, G. (2007) *Top. Catal.,* **42–43**.
81 Harmsen, J.M.A. (2001) Kinetic modelling of the dynamic behavior of an automotive three-way catalyst under cold-start conditions, PhD thesis Technical University Eindhoven, ISBN 90-386-2932-X.
82 Matsunaga, S., Yokota, K., Muraki, H., and Fujitani, Y. (1987) Improvement of engine emissions over three-way catalyst by the periodic operations, SAE Tech. Paper 872098.
83 Epling, W.S., Yezerets, A., and Currier, N.W. (2007) *Appl. Catal. B,* **74**, 117–129.
84 Kočí, P., Plát, F., Stepánek, J., Kubíček, M., and Marek, M. (2008) *Catal. Today,* **137**, 253–260.
85 Fridell, E., Persson, H., Olsson, L., Westerberg, B., Amberntsson, A., and Skoghlund, M. (2001) *Top. Catal.,* **16/17** (1–4), 133–137.
86 Cant, N.W. and Patterson, M.J. (2002) *Catal. Today,* **73** (3–4), 271–278.
87 Després, J., Koebel, M., Kröcher, O., Elsener, M., and Wokaun, A. (2003) *Appl. Catal. B,* **43**, 389–395.
88 Nova, I., Castoldi, L., Lietti, L., Tronconi, E., Forzatti, P., Prinetto, F., and Ghiotti, G. (2004) *J. Catal.,* **222** (2), 377–388.
89 Koltsakis, G.C., Margaritis, N., Haralampous, O., and Samaras, Z.C. (2006) SAE Tech. Paper 2006-01-0471.
90 Tuttlies, U. (2006) Experimentelle Untersuchung und Modellbildung von NOx-Speicherkatalysatoren, PhD thesis, Institut für Chemische Verfahrenstechnik, University of Stuttgart, Germany.
91 Schmeißer, V. (2009) NOx-Speicherkatalysatoren: Einfluss der katalytischen Bestandteile und mathematische Modellierung, PhD thesis, Institut für Chemische Verfahrenstechnik, University of Stuttgart, Germany.
92 Kočí, P., Plát, F., Štěpánek, J., Bártová, Š., Marek, M., Kubíček, M., Schmeißer, V., Chatterjee, D., and Weibel, M. (2009) Global kinetic model for the regeneration of NOx storage catalyst with CO, H_2 and C_3H_6 in the presence of CO_2 and H_2O. *Catal. Today,* **147S**, S257–S264.
93 Schmeißer, V., Tuttlies, U., and Eigenberger, G. (2007) *Top. Catal.,* **42–43**, 15–19.
94 Chatterjee, D., Kočí, P., Schmeißer, V., Marek, M., Weibel, M., and Krutzsch, B. (2010) Modelling of a combined NOx storage and NH_3-SCR catalytic system for Diesel exhaust gas aftertreatment. *Catal. Today,* **151**, 395–409.

95 Jacquot, F., Logie, V., Brilhac, J.F., and Gilot, P. (2002) *Carbon*, **40**, 335–343.
96 Frey, M., Wenninger, G., Krutzsch, B., Koltsakis, G.C., Haralampous, O., and Samaras, Z.C. (2007) *Top. Catal.*, **42–43** (1–4), 237–245.
97 Haralampous, O.A. and Koltsakis, G.C. (2004) *Ind. Eng. Chem. Res.*, **43** (4), 875–883.
98 Koltsakis, G.C., Dardiotis, C.K., Samaras, Z.C., Frey, M., Wenninger, G., Krutzsch, B., and Haralampous, O. (2008) Model-based optimisation of catalyst zoning in diesel particulate filters, SAE Tech. Paper 2008-01-0445.
99 Skoglundh, M., Thormalen, P., Fridell, E., Hajbolouri, F., and Jobson, E. (1999) *Chem. Eng. Sci.*, **54** (20), 4559–4566.
100 Carlsson, P.A., Mollner, S., Arnby, K., and Skoglundh, M. (2004) *Chem. Eng. Sci.*, **59** (20), 4313–4323.
101 AVL, BOOST (2009) http://www.avl.com.
102 GT-POWER (2009) http://www.gtisoft.com/img/broch/broch_gtpower.pdf.
103 Tang, W., Wahiduzzaman, S., Wenzel, S., Leonard, A., and Morel, T. (2008) SAE Tech. Paper 2008-01-0866.
104 Güthenke, A., Chatterjee, D., Weibel, M., Waldbüßer, N., Thinschmidt, B., Kočí, P., Marek, M., and Kubíček, M. (2007) SAE Tech. Paper 2007-01-1117.
105 Güthenke, A., Chatterjee, D., Weibel, M., Waldbüßer, N., Kočí, P., Marek, M., and Kubíček, M. (2007) *Chem. Eng. Sci.*, **62**, 5357–5363.

Index

a

ab initio interaction potential 45
ab initio molecular dynamics (AIMD) simulations 41, 56
activation energy 31, 116, 121, 202, 226, 290
adiabatic approximation approximation.
  See Born–Oppenheimer approximation
adsorbate–adsorbate interactions 21, 165
adsorption dynamics, on precovered surfaces 55–59
– AIMD simulations of complex systems 56, 58
– dynamical effects on adsorption 57
– isotope exchange experiments 59
– sticking probabilities 56
– trajectory calculated in ab initio molecular dynamics simulations 58
adsorption energies 62, 291
ammonia synthesis 291–293
– catalyst for 293
– DFT calculations 293
– microkinetic analysis 292
– radial flow converters 292
– volcano plot 293
anode-supported tubular fuel cell 187
Arrhenius-like rate expression 115
ATR reactor 297
automotive monolithic catalyst 306
AVL Boost 331

b

barium carbonate (BaCO$_3$) 322
barium oxide (BaO) 322
benzene 160
benzene alkylation 171
BET approach 157
Bethe lattices 154
BET surface area 152

bidisperse catalysts 168
bimodal catalyst 156
bimodal pore structure 152
binary exchange coefficients 163
BKL algorithm 78
– important aspect of 79
BKL method 120
Born–Oppenheimer (BO) approximation 3, 4, 13, 40, 41, 43, 62
Bosanquet approximation 155
Brønsted–Evans–Polanyi-type relationships 90, 165
Butler–volmer formulation 204–206
button cells 207, 208
– anode gas-phase profiles 212, 213
– anode surface species profiles 213
– applicability 214
– charged species fluxes 208, 210–212
– electric potentials 208, 210–212
– extensibility 214
– physical dimensions, and model parameters 209
– polarization characteristics 208, 209

c

carbon fluxes 227
carbon nanotubes (CNTs) 158
Car–Parrinello ab initio molecular dynamics (CPMD) 17
catalyst
– automotive monolithic 306
– bench-scale testing of 285
– bimodal catalyst supports 156
– boundary layer 221
– catalyst deactivation models 155
– catalyst–support interactions 143
– characterization 330
– chemical 286

Modeling and Simulation of Heterogeneous Catalytic Reactions: From the Molecular Process to the Technical System,
First Edition. Edited by Olaf Deutschmann.
© 2012 Wiley-VCH Verlag GmbH & Co. KGaA. Published 2012 by Wiley-VCH Verlag GmbH & Co. KGaA.

- coatings 269
- diesel oxidation 307
- dispersion of material, in porous layers 260
- effects of hydrothermal aging 259
- gauze catalyst
-- to derive intrinsic kinetics 273
- heterogeneous 128
- iron-based 293
- mesoporous supports 158, 174
- metal 290
- microporous supports 163
- modeling, structure of 259, 260
- models, for supports 154
- monolithic structure 271
- nanoparticle 142
- Ni-based 134
- NO_x storage and reduction catalyst 321
- optimization of pore structures 174
- oxide 149
- particles 120, 125, 139, 141, 142, 149, 152, 153, 158, 166, 173, 272, 274, 292
- pellets 157, 166, 167, 272, 287, 288
- performance 173
- poison 46, 299
- porous
-- layer 245
-- particles 149, 150, 158
- promotion 294, 295
- properties 330
- real 143, 149
- Rh/ZrO_2 technical catalysts 231, 239, 240
- Ru catalysts 88, 292
- SCR 307, 313, 315–317
- single-crystal model catalysts 108
- structure
-- simple approach for modeling 259, 260
-- spatial resolution of 258, 259
- support manufactured by compression of microparticles 152
- surface
-- under steady-state operation 90
-- that accelerates reaction 40
- three-way 319, 321
- total deactivation 228
- turns to a sink of OH radicals 231
- V-based and Fe–zeolite 318
- zeolite 150, 158, 160, 169, 170
catalytically stabilized thermal combustion (CST) 222
catalytic channel-flow reactor 224
catalytic combustion monolith 250, 251
catalytic monoliths, modeling of 308–313
catalytic partial oxidation
- of gasoline surrogate *iso*-octane 267, 268
- of hydrocarbons 197
- of light alkanes 273
- of methane 130, 238–242
- reactor 197
catalytic reactors 115, 251
- with multiphase fluids 273–275
- simulations of 115
catalytic systems 41
- as black box 129
CFD codes 116, 129, 246, 256, 257, 263, 264, 273, 278
CFD simulation 125, 131, 141–143, 252, 253, 257, 258, 263, 264, 273, 274, 276, 277, 278, 298
CFD software 264
charge conservation 194, 195
- boundary conditions 195
- cell potential 196
- current density 196
- effective properties 195
- equations 194
charge transfer kinetics 198–204
chemical accuracy 84
chemical kinetics 196, 197
chemical potential gradient 162
chemical vapor deposition (CVD) 139, 275, 276
chemical vapor infiltration (CVI) 275, 276
chemisorption 151, 165, 166
coated diesel particulate filter (CDPF) 315
collective diffusion 158, 159
combined diffusion/reaction computation, approaches for 172
combined exhaust aftertreatment systems 331
- component order, optimization of 331–333
- NSRC reduction efficiency, analysis of 333
- SCR performance, analysis of 333, 334
- SCR *vs.* DOC + SCR performance 334, 335
complex kinetic schemes, simplifications 141, 142
computational algorithm 207
computational fluid dynamics (CFD) modeling 106, 225, 252
- of macroscale flow structures 252
- three-dimensional 306
computer reconstruction approaches 157
CO oxidation, at RuO_2 88
correlated random walk theory (CRWT) 167
corrugation reducing procedures 45
Coulomb repulsion 8
CPO-based gas turbine honeycomb reactors 240
cubic random pore network models 156

cylindrical pore interpolation model (CPIM) 160
cylindrical pore model 153

d
Darcy's law 257
Darken equation 159
n-decane hydroconversion 170
– free energy model applied to 171
– free energy of formation of intermediates
– – as a function of zeolite structure 170, 171
degree of rate control (DRC) approaches 102
density-based formulations 10
density-based methods 6, 7
– exchange–correlation functionals 10–12
– Hohenberg–Kohn theorems 7, 8
– Kohn–Sham equations 9, 10
– Thomas–Fermi model 7
DFT calculations 134
– adiabatic 67
– electronic effect, quantified by 290
– for heterogeneous catalysis 166
– to obtain kinetic parameters 72
– on periodic systems 165
– poisoning of sulfur due to 46
– static 52
1,2-dichloropropane 167
diesel oxidation catalyst (DOC) 307, 312, 319, 320, 331, 334, 335
diesel particulate filters, modeling of 313, 314
differential algebraic equation (DAE) system 264
diffusion 157
diffusional resistances 163
diffusion coefficients 76, 158, 161, 162, 174, 254, 326
2,4-dimethyl octane 170
direct numerical simulation (DNS) 243, 256, 271, 272
direct simulation Monte Carlo (DSMC) methods 222, 246
dissipative particle dynamics (DPDs) 166
dissociated H_2 molecules, relaxation dynamics of 59–62
– AIMD trajectories, determined for fixed substrate 61
– cannonball mechanism 62
– hot atom dynamics 61
– hot atom motion of hydrogen 62
– trajectory of hydrogen atoms 60
DRC sensitivity analysis 103
Dubbeldam's approach 160
dusty gas model (DGM) 159, 167, 191, 261, 262
dynamically corrected transition theory 160

e
effective core potential (ECP) 23
effective diffusivities 167
effective medium theory (EMT) 167
effectiveness factor 168
– as a function of particle radius 172
electric current 193
electric potential differences 190
electrocatalytic oxygen reduction reaction, on'Pt(111) 22–24
– binding energies, and dissociation barriers 24
– including electrode potential 35, 36
– including thermodynamic quantities 32–35
– molecules adsorbed on 23
– reduction reaction potentials 22
– simulations including water solvation 28–30
– – Eley–Rideal reactions 31, 32
– – Langmuir–Hinshelwood mechanisms 30, 31, 33
– water formation, from gaseous O_2 and H_2 24
– – HOOH formation 28
– – O_2 dissociation 25–27
– – OOH formation 27, 28
– – reaction pathways 25
electrochemical potential 193
electrochemical processes 187
electrode materials 190
electrode potential 22, 35, 36
electrolyte electric potential 190
electrolyte material (YSZ) 190
electron–electron interaction 8
electronically nonadiabatic reaction dynamics 62–66
– MD simulations 65
– spin effects 63
– sticking probability 64, 66
– triplet–singlet transitions 65
electronic degeneracies 16
electronic density 1
electron transport 192–194
electrostatic interaction 43
Eley–Rideal-type reactions 29, 31
– mechanisms 29, 31
– in water solvent 32
– – under ambient conditions 34
emission legislation 304–306

– European, for commercial vehicles and   305
energy barrier   83
enthalpy   26, 133, 254, 255, 257, 258, 311, 325
entropy   16, 83, 94, 131, 160, 325
– configuration, and size entropy effects in zeolites   164
– of rotation   17
– of translation   16
equilibrium thermodynamics   39
errors. *See also* root mean square error (RMSE)
– analysis and CFD simulation   263
– in corresponding first-principles rate constants   103, 107
– in DFT-LDA or DFT-GGA barriers   85
– estimation   278
– introduced by approximation   20
– range   334
– in rate constants propagates to   102
– self-interaction   12
– statistical error ΔS 52
Euclidian pore space   157
Euler–Lagrange equation   8, 9
European emission limits for passenger cars   305
European load response (ELR) test   305
European stationary cycle (ESC) test   305
European transient cycle (ETC) test   305
exact exchange energy (EXX)   12
ExACT.........tool of Daimler   331
exchange–correlation (xc) energy   9
exhaust gas aftertreatment technologies   306–308
– automotive monolithic catalyst   306
– BlueTEC I technology   308
– NSRC + SCR-based exhaust system   308
– urea-based exhaust system   307
exhaust gas recycle (EGR)   238

f

FAU, as large-pore zeolite   170
Favre-averaged transport equation   243
FeCr alloy honeycomb reactor   240
FEMLAB software package   264
Fick's first law   158
first-order approximations   32
first passage time (FPT)   155
first-principles kMC simulations   72, 81, 83, 85, 88, 92, 94, 97–100, 102–104, 107, 108
first-principles rate constants   84–87
– computational cost to determine accurate   107
– costs of setting up catalog of   105
– electronic structure theories to generate   84
– propagation of error in   107

– temperature-dependent   99
– TPD/TPR quantity, and slight inaccuracies   99
first reaction method (FRM)   119, 120
Fischer–Tropsch synthesis slurry bubble column reactors   274
flow laminarization domain, in catalytic combustion   245
fluid flow simulations   115
fluid–solid interphase   259
fluid transport   222
force fields   44
free energy   18, 32, 171
Freundlich isotherm   151, 165
Freundlich model   165
fuel cell technology   1
fuel electrode   205
fuel-lean combustion
– of hydrogen/air on platinum   234–238
– of methane/air on platinum   225
– – gas-phase kinetics   228–231
– – heterogeneous kinetics   225–228
– of propane/air on platinum   231–234
fuel-rich combustion
– homogeneous ignition   239
– kinetic corrections   239
– of methane/air on rhodium   238–240

g

gas diffusivities   160
gas hourly spatial velocities (GHSV)   223
gas-phase conditions   103
gas-phase conservation equations   190, 191
– boundary conditions   192
– chemical reaction rates   191, 192
– gas-phase transport   191
gas-phase reaction mechanisms   116
gas reactions catalyzed by solid materials   152
gas–solid flows, in chemical reactors   251
gas–surface interface   258
generalized gradient approximation (GGA)   11, 12, 84
generalized Maxwell–Stefan equations   167
generalized multicomponent adsorption isotherm   164, 165
geometry optimizations   13, 14
– common procedures for   13
– PES expressions   13
Gibbs adsorption equation   164
Gillespie algorithm   78
global kinetic models, challenges for   330
global kinetic parameters, determination of   329, 330
global world harmonized test cycles   305

Graham's law   160
GTPower   331

h

Hamilton operator   2
harmonic TST (hTST)   82, 83, 85
heat transfer in three-phase flow   275
heterogeneous catalytic reactions   40, 43, 149, 166, 221, 287
– approaches for modeling rates of   114
– chemical reaction rate   119
– DFT calculations for   166
– as kinetic phase transition phenomenon   95
– in porous catalysts   166
– in surface science   24
heterogeneous chemical reactions   115
– CH_4 reforming on a Ni-based catalyst, mechanism for   198
– coupling of flow field with   258–262
heterogeneously catalyzed gas–liquid reactions   274
heterogeneously catalyzed gas-phase reactions   113
hetero/homogeneous chemical reaction schemes. *See* fuel-lean combustion; fuel-rich combustion
hierarchy of multiscale modeling   2
highest occupied molecular orbital (HOMO)   44
high-pressure scanning tunnel microscopy   222
Hohenberg–Kohn theorem   7, 8
hole model   46
homogeneous chemical reactions   115
HOOH formation   34
hybrid DFT, with exact exchange   12
hydrogen evolution reaction (HER)   35

i

ideal adsorbed solution theory (IAST)   171
ideal gas approximation   32
industrial challenge, on modeling catalysis   283–285
– environmental challenges   285
– industrial R&D   284
– integral tests to verify rate expression   286
– for large-scale operation   284, 285
– modeling as useful tool   285
– plants, more integrated   283
– production costs   283
– technology on S-curve   284
industrial R&D in catalysis   299
in situ gas-phase diagnostics   223–225

internal energy, associated with molecular rotations   16
intraphase transport, modeling of   245, 246
ion transport   192–194
iron-based catalyst   293

j

Jacob's ladder   10
Jaguar code   23

k

Kerkhof.........'s arguments   159
kinetic energy   3, 7, 8, 9, 41, 43, 50, 52, 53, 55, 57, 60, 64, 66
kinetic models   240, 318, 330
kinetic Monte Carlo simulations   40, 54, 72, 80, 116, 121, 155
– configurational-bias   172
– first principles (*See* first-principles kMC simulations)
– master equation-based algorithm   72
– of surface chemistry   40, 54, 72, 82, 121, 139
kinetic phase transition   95
kinetic schemes
– applications, in models for technical systems   240
–– CPO-based gas turbine honeycomb reactors   240
–– FeCr alloy honeycomb reactor   240
–– measured and predicted temporal profiles   241, 242
– simplifying complex   141, 142
kMC algorithm   75–80, 92, 101
– master equation-based   102
kMC trajectories   75–78, 80, 92
Knudsen diffusivity   158
Koch curve   157
– molecules of different size on   157
Kohn–Sham equations   10, 62
k-points   20

l

laminar burning rates   237
laminar interphase transport   242
Langmuir adsorption constants   164
Langmuir–Hinshelwood–Hougen–Watson (LHHW) model   113, 114
Langmuir–Hinshelwood (LH) kinetics   151, 164, 166, 288
Langmuir–Hinshelwood rate expressions   163, 172
Langmuir–Hinshelwood reaction   29
– gas-phase   27
– mechanisms   29–31

– – HOOH formation 31
– – O_2 dissociation 30
– – OOH formation: 31
– in water solvent 30
Langmuir isotherm 164
laser-induced fluorescence (LIF) 221, 229, 231, 233, 234, 236
lattice approximation 84–87
lattice-based models 157
lattice–Boltzmann methods (LBMs) 263
lattice Boltzmann models 245
lattice mapping
– in conjunction with 85
– together with rate constant catalog 86
local (spin) density approximation (L(S)DA) 10, 11
lowest unoccupied molecular orbital (LUMO) 44
low Reynolds (LR) number models 243
LR turbulence models 243
lumping 304

m

macrokinetics approach 129
MARI approximation 93
Markovian state-to-state dynamics 75
material gap 149
Maxwell–Stefan (MS) diffusion coefficient 158
Maxwell–Stefan equations 171
MC trajectories 82
MD techniques 105
MD trajectories 60, 74, 76
mean field approximation 62, 93, 98, 113, 114, 116, 128–131, 138, 151, 252
– in catalytic partial oxidation of methane over Rh 130
– heterogeneous reaction mechanisms using 140
– potentials, and limitations 139, 141
mean field assumption 289
mean pore transport model (MPTM) 159
Menger sponge 157
mercury porosimetry 157
meta-generalized gradient approximation (MGGA) 12
metal–organic frameworks (MOFs) 150
microkinetic models 165
microkinetic reaction mechanisms 304
microkinetics analysis 289
microkinetics approach 113
micro/macroporc model 153
microreactor technology 223

mixed ionic electronic conductors (MIECs) 194
modeling catalytic effects, in diesel particulate filters 328, 329
modeling catalytic reactions, technical aspects 13
– geometry optimizations 13
– solvation methods 17–19
– thermodynamic treatments of molecules 16, 17
– transition-state optimizations 14
– vibrational frequencies 14–16
modeling phenomena, within porous catalyst particles 149–151
modeling, role of 287
– reactor models 287, 288
– surface science
– – and breakdown of the simplified approach 288–290
– theoretical methods 290, 291
model representation, for surface-specific problems 19
– cluster approach 21, 22
– slab/supercell approach 19–21
models
– of Foster and Butt 153
– of Wakao and Smith 153
modern combined diesel exhaust gas aftertreatment system 304
modified Taylor expansion 44
molecular degrees of freedom 44, 46
molecular diffusion 158
molecular dynamics (MD) simulations 43, 74, 161
– limitations 166
molecular vibrations 17
molecule–molecule collisions 158
molecule–wall collisions 158
Møller–Plesset perturbation theory 6
monolithic reactors 251
most abundant reaction intermediates (MARIs) 92, 164
multistep surface reaction mechanisms 134–139
– catalytic ignition of CH_4 oxidation, on platinum foil
– – sensitivity of uncovered surface area on rate coefficients 137
– heterogeneous ignition of $CH_4/O_2/N_2$ mixtures, on platinum foil 137
– – calculated time-dependent surface coverage and 137
– mechanisms established, for more complex reaction system 139

– methodology 136
– sensitivity coefficients 136
– stringent test, for model evaluation 138

n

nanoparticles
– extension of MC simulations to 120–124
– reshaping of supported 128
– in technical catalyst 124
– two supported nanoparticles surface model 123
nanoporous materials 158, 160
Navier–Stokes equations 106, 267
new European driving cycle (NEDC) 306
NH_3 (urea-SCR) modeling 315
– chemical reaction modeling 316
– global reactions 315
– kinetic analysis 316
– NH_3 adsorption, desorption, and oxidation 316
– NH_3–NO–NO_2 reactions 317–319
– NO-SCR reaction 316, 317
– selective catalytic reduction by 315
– washcoat diffusion, influence of 319
nickel and yttria-stabilized zirconia (Ni–YSZ) 187
nitrogen adsorption 157
Ni-YSZ anodes 197
nondissociative molecular adsorption dynamics 49–55
– *ab initio* molecular dynamics simulations 52
– analysis of trajectories 54
– chemisorbed O_2 species on Pt(111) 50
– conversion of the molecular kinetic energy 55
– elbow plots of O_2/Pt(111) 52, 54
– hard cube model (HCM) 50
– kinetic Monte Carlo simulations 54
– simulation, adsorption dynamics of O_2/Pt(111) 51
– sticking probability 49, 50, 52
– – at higher kinetic energies, leveling off 54
nonlinearity 116
non-Markovian behavior 106
NO_x storage and reduction catalysts (NSRC) 304, 321–325
– inhibition terms 325
– NH_3 formation, during rich operation within 327, 328
– species transport effects 326
– surface reactions, and reaction rate formulation 323, 324
nuclei–nuclei interaction 13

Nudge elastic band (NEB) methods, to find transition state 15
numerical methods
– and computational tools 262, 263
– for solution of governing equations 263, 264
numerical models 221

o

O_2 dissociation 33
OH-LIF measurements 232
Ohmic heating 196
Onsager transport coefficient 158
O–O bond 24
O–OH dissociation 34
open-circuit voltage (OCV) 206
optical arrangement, of the Raman/LIF setup 224
optical sum-frequency generation 222
optimization
– catalyst distribution 329
– catalyst material 336
– catalyst volumes 331
– component order 331
– different parameters 307
– geometry 13, 15
– pore structure 152, 173, 174
– reactors, and processes 278
– transition-state 14
ordinary differential equation (ODE) system 264
oxidative dehydrogenation 115
oxygen reduction reaction (ORR) 2, 29, 33

p

parallel pore model 153
parameterization 43–45
partial differential and algebraic equations (PDE) 262
partial oxidation. *See* catalytic partial oxidation
particle–particle correlations 160
percolation theory 154, 155, 195
PES. *See* potential energy surface (PES)
phase transitions 164
photoelectron emission microscopy 222
physisorption 151
Poisson–Boltzmann equation 18
Poisson distribution 77, 78, 80
polyaromatic hydrocarbons 197
pore connectivity 157
pore sizes 158
– for optimal catalytic activity 169
pore structure optimization 173, 174
porous solids 152

potential energy surfaces
- at arbitrary points of the configuration space 41
- to assist visualizing effect of electric potential difference on 202
- from atom–surface interaction 45
- of complex catalytic reactions 39
- interpolation of 43–45
- two-dimensional cuts 40
-- $H_2/S(2\times2)/Pd(100)$ derived from 44
-- $O_2/Pt(111)$ determined by 51
ppm reactions 288
propane dehydrogenation 272
pyrolysis 116

q

quantum chemical calculations 165
quantum dynamical (QD)
- determined by solving atomic Schrödinger equation 42
- of reactions at surfaces 45–49
- simulations of 43
quantum mechanics (QM) 1, 2, 6
- simulations 29

r

random site method (RSM) 120
random three-dimensional networks
- advantages 156, 157
rare event dynamics 72, 73
rational catalyst design 149
reaction diffusion equations 260, 261
reactive flows, modeling of 253
- momentum, and energy equations
-- for porous media 257, 258
- multicomponent flows, governing equations of 253–255
- three-phase flow 256, 257
- turbulent flows 256
reactor models 287
- two-dimensional 288
reactor simulations 264
- catalytic reactors, with multiphase fluids 273–275
- electrocatalytic devices 277, 278
- fixed bed reactors 271–273
- flow through channels 265–268
- material synthesis 275–277
- monolithic reactors 268–271
- wire gauzes 273
realistic electrochemical systems 35
rejection-free BKL algorithm
  basic steps 79
residence time algorithm 78

reverse MC analysis 126
Rh/ZrO$_2$ catalysts 240
- catalytic partial oxidation of methane 241
root mean square error (RMSE) 43, 45
Ru catalysts 88
RuO$_2$ surface 88
- calculated DFT-PES
-- and CO oxidation reaction mechanisms 89
- computed PES mappings 89
- cus and br site 88
- DFT computed lateral interactions 89
- first-principles kMC determined surface composition 94
-- equivalent plot 94
- MARI approximation 93
- modeled as associative desorptions 89
- O-rich feeds 88
- oxygen–metal bond breaking step 90
- parameter-free turnover frequencies 95–99
- representing model for active state 88
- steady-state surface composition 90–95
- temperature-programmed reaction spectroscopy 99–102
-- surface population in first-principles TPR kMC simulation 101
-- total TPR CO$_2$ yield 100
- time evolution of site occupation by O and CO 91
- transition from O-poisoned to CO-poisoned state 93
RxMC simulation method 169

s

scaling model 291
Schrödinger equation 1, 2, 4, 8, 41, 42, 45, 65
S-curve 284
selective catalytic reduction (SCR) 307
- analysis of performance 333, 334
- fast SCR and N$_2$O formation 318
- global reactions for SCR catalysts 315
- NH$_3$ (urea-SCR) modeling 315
- NO-SCR reaction 316
- NSRC + SCR-based exhaust system 308
- reaction in presence of NO$_2$ 316
- SCR vs. DOC + SCR performance 334, 335
- urea-SCR 307
- V-based SCR catalyst 317
self-diffusion 158
self-exchange coefficients 163
smooth field approximation (SFA) 154
solid oxide fuel cells (SOFC) 187, 189
- dual-column 189
- tubular anode-supported 188

solute–solvent interactions 18
solvent-excluded surface (SES) 18
solvers for stiff ODE and DAE systems 264
state-to-state dynamics 75
Stefan–Maxwell species momentum equations 159
Stefan velocity 258
steric effects of adsorbed species 130
Stewart model 167
stoichiometric coefficients 116
surface description
– by reactant and product patterns 119
surface diffusion 164
– model 76, 81
surface reaction rates, computation on molecular basis 116
– kinetic Monte Carlo simulations 116, 118–120
– MC simulations
– – to nanoparticles, extension of 120–124
– – potentials and limitations for derivation 125–128
– – reaction rates derived from 124, 125
– particle–support interaction, and spillover 125
– – capture zone of black particle 127
– – reverse MC analysis 126
surface roughness 152
surface science 22, 288, 289
– and breakdown of simplified approach 288–290
– concepts of heterogeneous catalysis in 24, 128
– data refer to ultrahigh vacuum (UHV) and 221
– diagnostics 222
– measuring techniques 221, 222
– progress 286
– simulation and scale-up to pilot operation 285
– – dual approach 285–287
– spectroscopic methods 165
syngas manufacture 294
– autothermal reforming 297–298
– – CFD calculations 298
– – oxygen-blown autothermal reforming (ATR) 297
– – Topsoe ATR process for FT synthesis, scale-up 298
– steam reforming 294–297
– – ab initio methods 294
– – blockage of sites 297
– – DFT calculations 295
– – in situ HREM 295

– – promotion of catalyst 295
– – scaling principles for activation of methane 295
– – SPARG process 297

t

Taylor expansion 6
technical catalytic reactors 114
– mean field approximation, and reaction kinetics 129–131
– models applicable, for numerical simulation 128–141
– multipurpose CFD codes, for computation 131
– scaleup and design of 143
– software package, for computation 129
– support design, and optimization of 141
– thermodynamic consistency 131–134
– – change in free enthalpy 132
– – separation of known and unknown variables 132
Temkin model 165
thermal energy 196
thermal heterogeneous kinetics 197, 198
thermodynamic treatments of molecules 16
Thiele modulus 168
– effectiveness factors 169
three-way catalysts (TWC) 304, 319, 321
time-dependent density functional theory (TDDFT) 62
tortuous pore model 153
transition-state optimizations 14
transition state theory (TST) 80–84
transport, evaluation of 242, 243
transport overpotentials 206
transport properties 156
TST. See transition state theory (TST)
turbulent flow 116
turbulent transport, in catalytic systems 243–245
turnover frequency (TOF) 95
– as a function of radial coordinate 172, 173

u

ultrahigh vacuum (UHV) 221
unity bond index-quadratic exponential potential (UBI-QEP) method 87

v

van der Waal's interactions 151
variable step size method (VSSM) 119
variational TST 83
vibrational energies 17

vibrational frequencies 14–16
vibrational spectroscopy 222

w

water formation
– DFT calculations 23
– from gaseous O_2 and H_2 24
– – HOOH formation 28
– – O_2 dissociation 25–27
– – OOH formation 27, 28
– Pt-catalyzed 22
wave function-based methods 4
– Hartree–Fock approximation 4, 5
– Post Hartree–Fock methods 5, 6

Wicke–Kallenbach cell 167
Wicke–Kallenbach stationary method 167

y

YSZ particles 187, 188

z

zeolites 150, 158, 160, 169
– entropy effects in 164
– free energy profiles 161, 162
– – categories 162
– interactions 162
– lattice flexibility 162
– screening of 170, 171